Frailty Syndrome

Frailty, a geriatric syndrome with a defined phenotype, was introduced by Linda Fried and Jeremy Walston; it is widely cited. The importance of this syndrome cannot be overstated: Frailty has an enormous relevance in clinical settings, and there are clear biological bases behind its phenotype. This book covers both biological and clinical aspects, including interactions with diseases like diabetes.

Key Features:

- Summarizes the history and recognition of the Frailty Syndrome
- Provides an update and summary of a fast-growing field
- Describes how free radical and oxidative stress contributed to current theories
- Documents the impact that nutrition and physical activity play in the emergence of frailty

OXIDATIVE STRESS AND DISEASE

Series Editors
Helmut Sies, MD
Enrique Cadenas, MD, PhD
University of Southern California School of Pharmacy
Los Angeles, California

Flavonoids and Related Compounds: Bioavailability and Function, *edited by Jeremy P. E. Spencer and Alan Crozier*

Mitochondrial Signaling in Health and Disease, *edited by Sten Orrenius, Lester Packer, and Enrique Cadenas*

Vitamin D: Oxidative Stress, Immunity, and Aging, *edited by Adrian F. Gombart*

Carotenoids and Vitamin A in Translational Medicine, *edited by Olaf Sommerburg, Werner Siems, and Klaus Kraemer*

Hormesis in Health and Disease, *edited by Suresh I. S. Rattan and Éric Le Bourg*

Liver Metabolism and Fatty Liver Disease, *edited by Oren Tirosh*

Nutrition and Epigenetics, *edited by Emily Ho and Frederick Domann*

Lipid Oxidation in Health and Disease, *edited by Corinne M. Spickett and Henry Jay Forman*

Diversity of Selenium Functions in Health and Disease, *edited by Regina Brigelius-Flohé and Helmut Sies*

Mitochondria in Liver Disease, *edited by Derick Han and Neil Kaplowitz*

Fetal and Early Postnatal Programming and Its Influence on Adult Health, *edited by Mulchand S. Patel and Jens H. Nielsen*

Biomedical Application of Nanoparticles, *edited by Bertrand Rihn*

The Biology of the First 1,000 Days, *edited by Crystal D. Karakochuk, Kyly C. Whitfield, Tim J. Green, and Klaus Kraemer*

Hydrogen Peroxide Metabolism in Health and Disease, *edited by Margreet C M Vissers, Mark Hampton, and Anthony J. Kettle*

Glutathione, *edited by Leopold Flohé*

Vitamin C: Biochemistry and Function, *edited by Margreet C M Vissers and Qi Chen*

Cancer and Vitamin C, *edited by Margreet C M Vissers and Qi Chen*

Mammalian Heme Peroxidases: Diverse Roles in Health and Disease, *edited by Clare Hawkins and William M. Nauseef*

Proteostasis and Proteolysis, *edited by Niki Chondrogianni, Elah Pick and Anna Gioran*

Frailty Syndrome: Basic and Clinical Aspects

María Carmen Gómez-Cabrera, José Viña, and Leocadio Rodríguez-Mañas

For more information about this series, please visit:
https://www.crcpress.com/Oxidative-Stress-and-Disease/book-series/CRCOXISTRDIS

Frailty Syndrome
Basic and Clinical Aspects

Edited by
María Carmen Gómez-Cabrera, José Viña,
and Leocadio Rodríguez-Mañas

CRC Press
Taylor & Francis Group
Boca Raton London New York

CRC Press is an imprint of the
Taylor & Francis Group, an **informa** business

Designed cover image: Getty Images

First edition published 2026
by CRC Press
2385 NW Executive Center Drive, Suite 320, Boca Raton FL 33431

and by CRC Press
4 Park Square, Milton Park, Abingdon, Oxon, OX14 4RN

CRC Press is an imprint of Taylor & Francis Group, LLC

ISBN: 978-1-032-50388-2 (hbk)
ISBN: 978-1-032-50389-9 (pbk)
ISBN: 978-1-003-39829-5 (ebk)

DOI: 10.1201/9781003398295

Typeset in Times
by KnowledgeWorks Global Ltd.

In Memoriam

While the final texts, tables, and figures for this book were being completed, we received the terrible news that Dr. Jeremy D. Walston had passed away. The person of Dr. Walston, of Jeremy, goes beyond what is usual for many of us who dedicate ourselves to trying to advance through small steps in knowledge to make people's lives more enjoyable. He is one of that small handful of privileged individuals who are making a leap. A leap that changes paradigms, the way of approaching and understanding problems. The process of disability in older adults is understood and explained very differently today than it was when Jeremy began his work. Work that, along with a few other colleagues, would eventually lead to the construct of frailty. Today, one cannot talk about disability without mentioning frailty. And one cannot talk about frailty without mentioning Jeremy Walston. In fact, his contribution to this book was to write the introductory chapter, recounting the history of the concept of frailty, its ups and downs, and its meaning. We'll never be able to read it because his illness prevented him from writing it, and death robbed us of our last hope of ever completing it. But his stature as a scientist, as a researcher, is magnified when seen in light of his character as a person. Generous, kind, always with a warm word on his lips and a friendly expression on his face, Jeremy made friends wherever he went. The editors of this book attest to this, and as a humble thank you, we would like to dedicate this book you hold in your hands to him.

Contents

Foreword

Leocadio Rodríguez-Mañas

The 20th century witnessed, for the first time in human history, an event that transformed societies: population aging, driven by a remarkable increase in longevity and life expectancy. This phenomenon, which began in the most prosperous societies, eventually affected virtually the entire planet. This event, which must undoubtedly be considered a success and was largely due to improved living conditions, has led to a series of enormous improvements in people's lives, but it has also raised a series of challenges in many areas of life: from models of coexistence to architecture and urban planning, from economics to technology, and, of course, it has also affected the field of health. While the history of medicine has focused since the times of Aesculapius and Galen on healing, on prolonging life, the focus has shifted in our time from the quantity of life to the quality of life, from how much one lives to how one lives, from total life expectancy to healthy life expectancy. This shift is so radical that it has led to impressive conceptual changes, including that of healthy aging, which is defined (WHO) as the maintenance of functional autonomy, regardless of the individual's disease burden. This means that research priorities must also consider a shift in focus. While we increasingly understand the reasons for longevity, we continue to ignore almost the entire process of functional decline, which is what determines quality of life and where there is enormous ground to be gained. While the probability of prolonging life expectancy is very limited, the probability of prolonging healthy life expectancy is very high, since those additional years we gained in the past century have not been accompanied by a parallel delay in the age at which we begin to functionally deteriorate. Therefore, part of that gain in years has been in years of dependency, in years of poor quality of life. In this progressive, still incipient, study of the process of functional decline, there are some relevant milestones. One of them is the one that defines the model of functional decline during the aging process, which consists of noncatastrophic decline. This mode of losing function assumes a gradual process in which, along with the loss of function, functional reserve capacity is also lost, and consequently, the susceptibility to suffer a severe loss of function in the face of increasingly insignificant environmental challenges increases. Terms such as resilience, intrinsic capacity, or frailty merely define, from different perspectives, the same process. What we do know today is that, while interventions can be made throughout the process, the most efficient approach is probably to do so at the point where disability has not yet appeared but is close, and where functional reserve remains, but is beginning to dwindle. This evolutionary stage of the path from robustness to severe disability is frailty. The concept of frailty emerged at the end of the past century and has not been free of controversy to date. From its very conceptualization to its definition, including screening and diagnostic methods and how to address it, to the mechanisms underlying it, controversy has been a constant. Fortunately, in recent years, a series of experimental evidence has emerged that has helped substantiate the concept and its consequences. And, this book aims to highlight them.

Preface

Jose Viña and Maria Carmen Gomez-Cabrera

The splendid increase in life expectancy that has occurred during the 20th century has brought, as a side effect, an increase in the number of people who are frail, leading to dependency.

Frailty was identified as a phenotype in older adults by Linda Fried and her colleagues. This concept has revolutionized the field of gerontology and geriatrics and has placed emphasis on increasing quality of life rather than increasing longevity itself. We aim to increase healthspan, not just lifespan.

In the first chapter of this book, we attempt to analyze vulnerability and associate it with the intrinsic vulnerability of older adults. This leads us directly to the second chapter, which analyzes aspects of the clinical management of frailty, highlighting the advances and challenges presented by the diagnosis and management of frailty in older adults.

It is naturally very difficult to analyze the effects of interventions in frail individuals without good biomarkers of frailty. Considerable efforts have recently been made to improve the quality of these biomarkers. These advances are summarized and interpreted in Chapter 3.

Naturally, one of the fundamental objectives of this book is to compile the most useful interventions for the prevention and treatment of frailty. Chapter 4 addresses one of the most important aspects of frailty treatment: physical exercise. However, physical exercise is not enough. It is important to improve the quality of nutrition in older adults to delay frailty. We must point out that optimizing exercise programs and nutrition in older adults has synergistic effects, and optimal results are not achieved without improving the quality of these two aspects. In Chapters 4–6, the reader will find detailed and up-to-date information on the importance of physical exercise and nutrition in delaying the onset of frailty and its treatment. A fundamental objective of all clinicians and researchers involved in the study of frailty is to prevent the transition to dependency, an extremely serious situation that leads to a very substantial decrease in quality of life and even life expectancy.

Frailty can occur in older adults who are otherwise healthy. However, the situation is seriously aggravated in many metabolic diseases, one of the most important of which is diabetes. This is why we dedicate Chapter 7 to the relationship between diabetes and frailty.

By now, the reader will understand the complexity of frailty syndrome, which has significant psychological and sociological implications. We dedicate Chapter 8 of this book to analyzing this complexity to address cognitive aspects of frailty in Chapters 9 and 10. Finally, taking into account the brain-gut axis, we analyze the role of the microbiota in the development of frailty in older adults.

The book concludes with a chapter dedicated to experimental aspects of frailty. The development of frailty models is essential for analyzing basic aspects that, for obvious reasons, cannot be studied in humans. The final chapter of this book deals specifically with analyzing basic aspects that allow us to understand the mechanisms that lead to frailty in older adults.

Editor Biographies

Maria Carmen Gómez-Cabrera is a full professor in the Department of Physiology at the Faculty of Medicine, University of Valencia. Her research focuses on the physiology of skeletal muscle and the mechanisms underlying muscle dysfunction, including the loss of strength and muscle mass associated with aging, periods of immobilization or physical inactivity, and certain pathological conditions. Maria Carmen is the author of over 140 peer-reviewed international publications. Her contributions to exercise redox biology, healthy aging, frailty, and sarcopenia have garnered over 14,000 citations, reflecting an h-index of 50.

José Viña studied Medicine at the University of Valencia with Hans Krebs in Oxford. He is Professor of Physiology in Valencia and leads the FRESHAGE group that works on healthy aging, exercise, and Alzheimer's disease. He has received two honorary doctorates from the Universities of Buenos Aires and Rennes. President of the International Society of Free Radicals (SFRR-I), Dr. Viña has been cited more than 21.000 times, resulting in an h-index of 79 (data taken from: https://exaly.com/author/2255005/jose-vina/rankings).

Leocadio Rodríguez-Mañas is Head of the Geriatrics Service of the Getafe University Hospital (Madrid) and Scientific Director of the Spanish Center for Biomedical Research on Frailty and Healthy Aging (CIBERFES) (Ministry of Science and Innovation and Ministry of Health), as well as Coordinator of the Toledo Study of Healthy Aging. His research has been mainly focused on frailty and functional decline and diabetes in older people, the mechanisms involved, its operational definition (diagnostics tools) and its management (RCTs), including the use of technology in covering the different requirements of older frail people or at risk of frailty. He has participated in the preparation of technical reports on these topics for national and international organizations.

Contributors

A. H. Abdelhafiz
Department of Geriatric Medicine
Rotherham General Hospital
Rotherham, United Kingdom

Alejandro Alvarez-Bustos
Centro de Investigación Biomédica en Red de Fragilidad y Envejecimiento Saludable CIBERFES
Instituto de Salud Carlos III
Madrid, Spain

Cristina Andrés-Lacueva
Faculty of Pharmacy and Food Sciences
University of Barcelona
Barcelona, Spain

Eleftheria Antoniadou
Technical University of Crete
Crete, Chania, Greece

Stefano Cacciatore
Department of Geriatrics, Orthopedics and Rheumatology
Università Cattolica del Sacro Cuore
Rome, Italy

Riccardo Calvani
Department of Geriatrics, Orthopedics and Rheumatology
Università Cattolica del Sacro Cuore
Rome, Italy

Brenda Amelia Casasola Espinosa
Research Assistant
Instituto Nacional de Geriatría
Mexico City, México

Antonio Cherubini
Department of Molecular and Clinical Sciences
Università Politecnica delle Marche
Ancona, Italy

Hélio José Coelho-Júnior
Department of Geriatrics, Orthopedics and Rheumatology
Università Cattolica del Sacro Cuore
Rome, Italy

Ana María Ayala Copete
Hospital Universitario San Ignacio
Pontificia Universidad Javeriana
Bogotá, Colombia

Mónica De la Fuente
Department of Physiology
Complutense University of Madrid
Madrid, Spain

Gustavo Duque
McGill University Health Centre: Montreal
Quebec, Canada

Mariam El Assar de la Fuente
Centro de Investigación Biomédica en Red de Fragilidad y Envejecimiento Saludable
Instituto de Salud Carlos III
Madrid, Spain
and
Fundación para la Investigación Biomédica del Hospital
Universitario de Getafe
Getafe, Spain

Luis Miguel F. Gutiérrez Robledo
Collaborative Centre on Integrated Care and Healthy Ageing
Instituto Nacional de Geriatría
Mexico City, México

Francesco Landi
Department of Geriatrics, Orthopedics and Rheumatology
Università Cattolica del Sacro Cuore
Rome, Italy

Emanuele Marzetti
Department of Geriatrics, Orthopedics and Rheumatology
Università Cattolica del Sacro Cuore
Rome, Italy

Eduardo Ortega
Department of Physiology
University of Extremadura University
Badajoz, Spain

M. Cristina Polidori
Ageing Clinical Research
Department of Internal Medicine
University Hospital of Cologne
Cologne, Germany

Daniel Rivas
McGill University Health Centre: Montreal
Quebec, Canada

Leocadio Rodríguez-Mañas
Servicio de Geriatría
Hospital Universitario de Getafe
Getafe, Spain

Isabel Rodríguez-Sánchez
Servicio de Geriatria
Hospital Clinico Universitario San Carlos
Madrid, Spain

Alan J. Sinclair
King's College London
London, United Kingdom

LaDora V. Thompson
Department of Physical Therapy
Boston University
Boston, MA, United States

Alba Tor-Roca
Faculty of Pharmacy and Food Sciences
University of Barcelona
Barcelona, Spain

Mireia Urpi-Sardà
Faculty of Pharmacy and Food Sciences
University of Barcelona
Barcelona, Spain

1 Unveiling Vulnerability
Frailty in the Lens of Intrinsic Capacity

Luis Miguel F. Gutiérrez Robledo and Brenda Amelia Casasola Espinosa

INTRINSIC CAPACITY: THE FOUNDATIONS OF RESILIENCE

Intrinsic capacity (IC) is a comprehensive construct encompassing physical and mental abilities in older adults, proposed by the World Health Organization (WHO) as an alternative to traditional disease-focused approaches. Beyond the model of frailty, the new concept of IC is defined as "the composite of all the physical and mental capacities of an individual" in the context of healthy aging, including five key domains: locomotion, vitality, cognitive, psychological, and sensory functions.[1] In this regard, higher IC has been associated with a lower risk of frailty onset; conversely, decline in IC has been already associated with a higher risk. Thus, we assume that, beyond frailty, a capacity-centered model, with detection and management of IC declines in older adults, is an important step toward a better global and integrative approach to older adults' health screening aiming to prevent frailty. High IC reflects a robust reserve, enabling individuals to adapt to challenges and maintain well-being. It partially underlies physical resilience (PR), and so it signifies a greater capacity to withstand stressors and recover from illness or injury. The relationship between IC and PR has been supported by structural equation modeling, suggesting that better IC may lead to improved PR and clinical outcomes.[2]

Studies have shown that IC predicts mortality and adverse health outcomes better than comorbidity measures.[3,4] Low IC scores are associated with biological markers of aging, such as inflammation, hyperglycemia, and low DHEA-S levels.[3] Longitudinal studies reveal that IC tends to decline over time, with distinct patterns of deterioration across domains.[5] IC assessment can be used as a screening tool in public health to categorize older adults into risk groups and develop targeted interventions.[6] Further research is needed to explore the underlying pathophysiology of IC and its potential for promoting healthy aging.[7]

FRAILTY: A PRECARIOUS STATE

The two dominant models of frailty, the phenotype model[8] and the cumulative deficit model,[9] were first described in 2001. Frailty portrays a vulnerable state characterized by diminished physiological reserves. Individuals with frailty exhibit a heightened susceptibility to adverse health events, even from minor stressors. This vulnerability arises from a cumulative decline across multiple physiological systems, exceeding the body's ability to compensate. Frailty has also been described as "the clinical viewpoint of a glass half empty, where it highlights functional deficits accumulated with aging and existing comorbidities, while IC represents a glass half full, accounting for the functional reserves of the patient".[10] In the last decade, research on frailty has focused on refining biomedical tools and putting them into practice, leading to a reductionist approach. The development of a new horizon in frailty research requires a holistic approach to health and disease in old age. This would be possible through considering healthy aging in terms of functional ability, in the sense of the IC in interaction with the social environment, while emphasizing positive attributes.[11]

DOI: 10.1201/9781003398295-1

In the WHO framework of reference, frailty is recognized as originating in both the social and biological spheres, and coexisting with positive attributes and resilience, and as a threshold on a continuum between health and disease. In relation to this, the consideration of the social and biographical context in which frailty arises contributes to the understanding of the impact on the subsequent trajectory of frailty beyond the stress that precipitates it. This approach, which considers the opinions and experiences of older people living with frailty, adds to the holistic tradition of geriatrics. Such a positive and integrated vision may lead to a better involvement of participants and healthcare professionals. Notably, even if frailty has not been conceived as a condition for excluding patients from interventions,[12] it is usually viewed by clinicians in a negative, potentially exclusionary way. Moreover, to provide a comprehensive global and integrative approach to older adults beyond the "perimeter of geriatric medicine", and particularly at the primary care level, IC appears to be an important step forward.

BRIDGING THE GAP

Intrinsic Capacity as a Predictor of Negative Outcomes

Together with environmental factors, IC (the composite of all the physical and mental capacities of an individual) influences healthy aging. However, whether IC predicts major clinical outcomes (e.g., functional decline or mortality) remains unclear. A recently published systematic search on this topic is the first to incorporate a quantitative synthesis of evidence on the predictive ability of IC. This study partly overcomes issues related to heterogeneity across studies and settings in the operationalization of IC and demonstrates how it is possible to explore the strength of the associations.[13] Nonetheless, more evidence is needed to confirm these findings. Besides, the authors stress that further efforts must be made to standardize the operationalization of IC. And they conclude saying that: "intrinsic capacity is inversely associated with functional decline and mortality risk in older adults. Therefore, intrinsic capacity assessment could be useful for risk stratification in older adults. Moreover, preservation of intrinsic capacity should be a key focus". At the same time, the same authors accept that longitudinal and interventional studies as well as a standardization of the way of measuring IC are needed.

Intrinsic Capacity as a Predictor of Frailty

Frailty and IC are two constructs stemming from the same need of overcoming traditional medical paradigms that negatively impact the correct way clinical and research practice should be conducted in older persons. Frailty and IC are complementary constructs that aim to overcome traditional medical paradigms and promote a more comprehensive, person-centered approach to aging, with frailty representing deficits and IC representing reserves that can be used together to provide a holistic assessment. The link between IC and frailty lies in their inverse relationship. Lower levels of IC, particularly across several domains, often precede the development of frailty. This association underscores the predictive power of IC. By measuring an individual's IC, we can identify those at increased risk of transitioning to a frail state. Early detection offers a valuable window for intervention, potentially mitigating or delaying frailty's progression.

There is a need for longitudinal monitoring to better understand how they change over time and impact disability in older adults. It seems that transitions of IC, but not frailty, can predict future disability and that baseline frailty status, locomotion impairment, and vitality impairment are associated with transitions of frailty. So, monitoring IC trajectory would be a priority in order to develop interventions to prevent frailty and disability.[14]

IC emerges as a significant predictor of frailty and adverse health outcomes in older adults. Multiple studies have demonstrated that higher IC scores are associated with a lower risk of developing frailty.[15] Specifically, the vitality and locomotor domains of IC appear to be the strongest predictors of incident frailty. IC assessment using the WHO's Integrated Care for Older People (ICOPE) Step 1 tool has shown utility in identifying frail older persons in community settings. Longitudinal studies have revealed that IC can predict various adverse outcomes, including physical function decline, falls, and mortality.[16] Furthermore, IC impairment often precedes clinical frailty manifestation, suggesting its potential for early intervention.[17] The complementary nature of IC and frailty measures allows for better risk profiling and personalized care planning for older adults.[18,19]

Recent research has also explored the concept of IC in older adults and its relationship to resilience. Studies have identified factors associated with IC, including demographic characteristics, socioeconomic factors, health conditions, and lifestyle behaviors.[20] IC has been shown to predict functional decline and is influenced by physiological reserve.[21] A clinical construct of IC has been developed and validated, demonstrating its ability to predict declines in activities of daily living.[22] Enhancing resilience in older adults involves promoting positive reframing, personal meaning, and community engagement.[23] A better understanding of the relationship between IC and resilience will inform interventions to promote healthy aging and maintain functional ability in older populations.

BEYOND PREDICTION: THE LIMITATIONS AND POTENTIAL SCOPE OF INTRINSIC CAPACITY

However, the relationship between IC and frailty is not deterministic. Not everyone with low IC will inevitably become frail. This highlights a crucial limitation of using IC solely for frailty prediction. Other factors, such as environmental exposures, allostatic load (AL), and access to healthcare, significantly influence an individual's trajectory. Moreover, IC focuses on baseline capacities, neglecting the complex and thus dynamic nature of resilience in the context of the ever-changing homeodynamic space. Individuals can demonstrate remarkable adaptability, even with declining IC, through compensatory mechanisms and lifestyle modifications (Figure 1.1).

The relationship between IC and AL is emerging as a key area of research in aging. IC has been shown to have a strong association with AL, which represents the cumulative biological wear and tear on the body. Studies have demonstrated that higher AL is associated with lower IC scores, even when adjusted for socioeconomic factors and chronic diseases.[24]

The concept of homeodynamics offers a more comprehensive understanding of human aging than traditional homeostasis. It recognizes the complex, dynamic nature of biological systems and their interactions across various organizational levels.[25] Aging is characterized by the progressive shrinkage of the homeodynamic space, leading to reduced stress tolerance and increased vulnerability to diseases. This process involves the accumulation of molecular damage and heterogeneity, impacting maintenance and repair pathways. However, aging is not a disease but a natural process that can be modulated through the enhancement of IC. Hormesis, involving mild stress-induced stimulation of protective mechanisms, offers a promising strategy for extending healthspan and promoting healthy aging. Understanding aging as a dynamic, malleable process opens avenues for interventions that enhance homeodynamic capabilities and could delay age-related pathologies.

The relationship between homeodynamics and allostasis is complex and evolving. While homeostasis aims to maintain a steady internal state, allostasis involves dynamic adaptation to environmental challenges.[26] Allostasis recognizes that there is no single ideal set of conditions and allows for flexible responses to stressors. This adaptive process involves multiple regulatory systems, including cognitive, somatic, and autonomic. The concept of AL describes the cumulative wear and tear on the body due to repeated or chronic stress.[27] AL can lead to disease over time and is influenced by individual differences in stress susceptibility. Understanding the mechanisms of allostasis and homeostasis is crucial for comprehending physiological regulation and its implications for healthy aging.

The homeodynamic space is a concept that encompasses the theoretical and practical aspects of defining and measuring biological health. This space shrinks with aging; homeodynamics contrasts with traditional homeostasis, emphasizing dynamic self-organization and complexity in living.

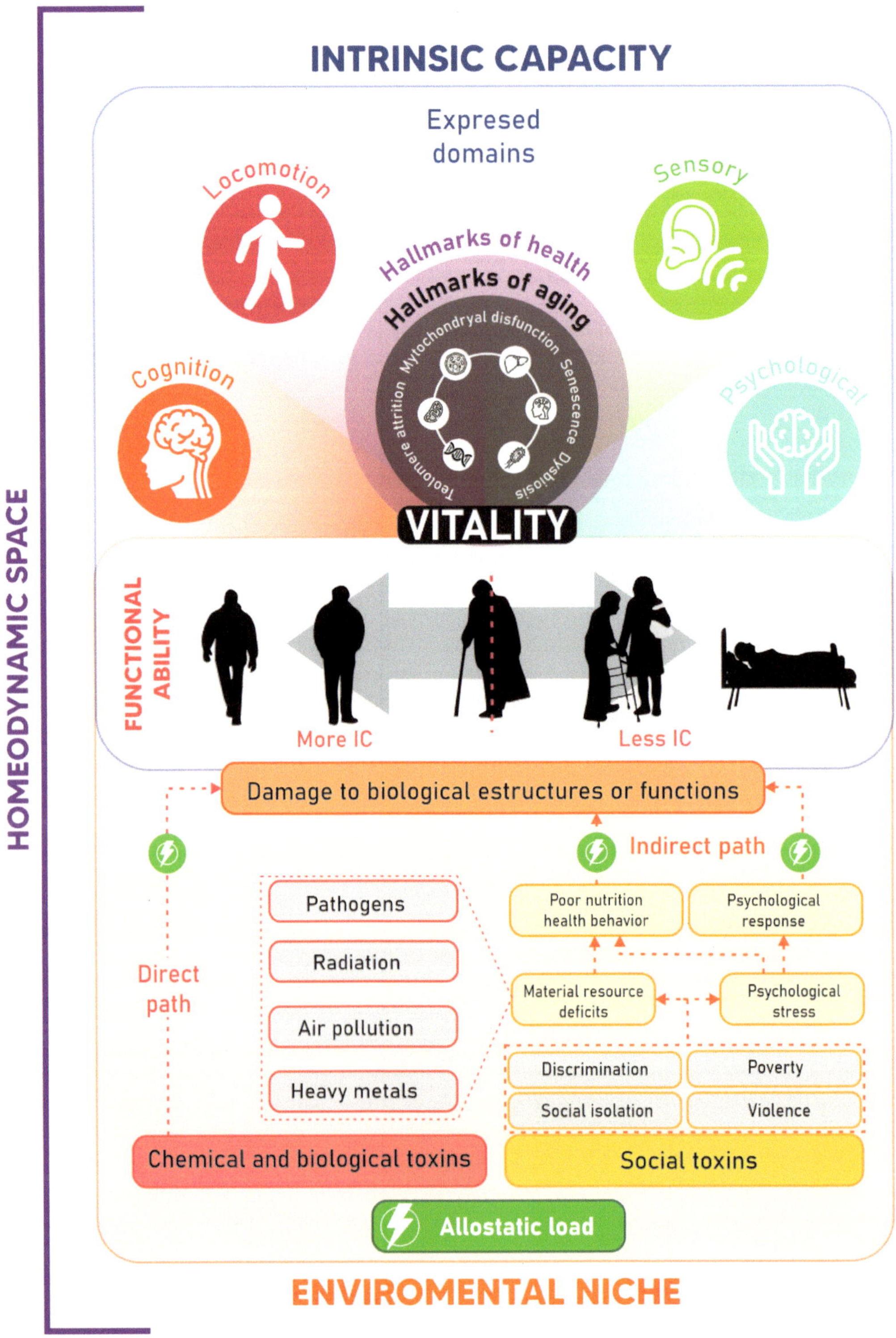

FIGURE 1.1 Scheme of the interactions between different components of the IC, its functional consequences, and the biological and social agents involved in its deterioration rate

Quantitative features include stress response profiles and redox set points, while qualitative aspects involve spatial compartmentalization, maintenance of homeostasis, and stress responses.[28] The homeodynamic space delimits the resilience of biological systems and hence determines the ability to survive and maintain health by damage control, adequate stress responses, reduction of biological noise, and constant remodeling. Its expression in the view of the framework of healthy aging that is proposed would be the definition of the dynamic frailty threshold that moves in accordance with and increasing or decreasing IC (Figure 1.1). This view allows us to move toward the following:

- **Maintaining independence and function:** preserving IC across domains allows older adults to remain active, manage daily tasks, and live independently.
- **Resilience against age-related challenges:** preserving a high IC enhances resilience against health declines, cognitive impairment, and functional limitations.
- **Promoting preventative strategies:** focusing on enhancing and maintaining IC across ages can be a proactive approach to healthy aging and potentially delay the onset of frailty.

A SYNERGISTIC APPROACH: OPTIMIZING INTERVENTIONS

The true strength lies in leveraging the complementary nature of IC and frailty. By understanding an individual's IC profile alongside their frailty status, specialists can tailor interventions with greater precision. Here is how this synergy translates to action:

- **Targeted Interventions:** Low IC in specific domains (e.g., low muscle strength) can guide targeted interventions like exercise programs to improve that particular domain, potentially reducing frailty risk.
- **Prioritizing at-risk populations:** Individuals with both low IC and pre-frail status can be prioritized for early intervention programs, potentially delaying frailty onset.
- **Monitoring Treatment Efficacy:** Tracking changes in IC alongside frailty measures can provide a more comprehensive picture of treatment effectiveness.

This can be achieved through the ICOPE framework that includes person-centered and integrated guidelines at the community-level to manage declines in IC.[29,30] ICOPE notably recommends using tools (including self-administered tools) to screen for loss in IC in primary care. This also aims to support the "self-empowerment of each individual for his/her health status". In the future, if ICOPE is developed on a large scale in older adults, cognition-specialized physicians (e.g., neurologists, geriatricians, or psychiatrists) will see patients with a cognitive complaint, but perhaps also with declines in IC in other domains (i.e., sensory, psychological, vitality, or locomotion), screened with ICOPE tools. As clinicians, it will be important to understand the relationships between these declines in IC and to propose personalized and tailored interventions in order to improve IC and functional abilities. A decline in IC will lead to targeted interventions; for example, sensory loss could be managed, which could lead to improve functional abilities. IC being independently associated with incident frailty,[31] screening and management of IC loss[32] may lead to reduce frailty occurrence, and then improve functional abilities and prevent negative events, for example, in patients with Alzheimer's disease (AD). Therapeutic and preventive approaches based on IC by using competences of the people rather than their deficits (comparable to the neurocognitive rehabilitation approach in AD) may be proposed. Interventions should not only focus on the individuals but also focus on the environmental characteristics, and actions are needed against the social determinants of frailty and IC declines.[33] The WHO defines healthy aging as "the process of developing and maintaining the functional ability that enables well-being in older age",[1] encompassing the relationships between IC and environmental characteristics. Environment comprises "all the factors in the extrinsic world that form the context of an individual's life", including factors from the micro level to the macro level (home, communities, and the broader society). Thus, contrary to the concept of frailty, the concept of functional abilities has been developed in a more global and less individual way: functional abilities are made up of the IC (at the individual level) and relevant environmental characteristics, both at the collective and at the individual level (the AL).[34] This approach may lead to specific interventions both at the individual and at the collective level:

- **Multifaceted interventions:** targeting specific domains like physical activity for vitality, cognitive training for cognition, and social activities for psychological well-being.
- **Personalized and holistic approach:** advocate for tailoring interventions to individual needs and preferences, addressing physical, mental, and social dimensions.
- **Early intervention and lifelong focus:** recognize the importance of promoting healthy habits and building capacities throughout life, not just when frailty symptoms emerge.

This approach also has implications for Healthcare and Aging Services:

- **Shifting the paradigm:** Healthcare systems have to move beyond solely managing disease toward promoting and enhancing IC as a key marker of well-being throughout life.
- **Comprehensive assessments:** We must incorporate assessments of IC alongside traditional medical evaluations to better inform care plans and interventions.
- **Interdisciplinary collaboration:** Promote collaboration between healthcare professionals, physical therapists, social workers, and community services to address diverse needs and optimize well-being across domains.

CONCLUSION: A SPECTRUM OF VULNERABILITY

Frailty, within the framework of IC, unveils a spectrum of vulnerability. While IC follow-up offers a valuable tool for identifying those at risk, it's the interplay between these concepts, frailty and IC, that empowers specialists to craft effective preventive interventions. By acknowledging both the predictive power and the limitations of IC, we can move beyond simply identifying frailty to fostering resilience and optimizing well-being in our aging population.

REFERENCES

1. World Health Organization. *World report on ageing and health.* World Health Organization, Geneva, 2015.
2. Hu F-W, Yueh F-R, Fang T-J, et al. Testing a conceptual model of physiologic reserve, intrinsic capacity, and physical resilience in hospitalized older patients: A structural equation modelling. *Gerontology* 2024; 70: 165–172.
3. Lee W-J, Peng L-N, Lin M-H, et al. Intrinsic capacity differs from functional ability in predicting 10-year mortality and biological features in healthy aging: Results from the I-Lan longitudinal aging study. *Aging* 2023; 15: 748–764.
4. Lu F, Liu S, Liu X, et al. Comparison of the predictive value of intrinsic capacity and comorbidity on adverse health outcome in community-dwelling older adults. *Geriatr Nur (Lond)* 2023; 50: 222–226.
5. Gonzalez-Bautista E, Llibre-Guerra JJ, Sosa AL, et al. Exploring the natural history of intrinsic capacity impairments: Longitudinal patterns in the 10/66 study. *Age Ageing* 2023; 52: afad137.
6. Nascimento LMD, Cruz TGCD, Silva JFDLE, et al. Use of intrinsic capacity domains as a screening tool in public health. *Int J Environ Res Public Health* 2023; 20: 4227.
7. Meng L-C, Huang S-T, Peng L-N, et al. Biological features of the outcome-based intrinsic capacity composite scores from a population-based cohort study: Pas de Deux of biological and functional aging. *Front Med* 2022; 9: 851882.
8. Fried LP, Tangen CM, Walston J, et al. Frailty in older adults: Evidence for a phenotype. *J Gerontol A Biol Sci Med Sci* 2001; 56: M146–M157.
9. Mitnitski AB, Mogilner AJ, Rockwood K. Accumulation of deficits as a proxy measure of aging. *Sci World J* 2001; 1: 323–336.

10. Fan BE, Cheong CY. Enhancing geriatric assessment for non-Hodgkin lymphoma with intrinsic capacity: Can we move beyond existing models of frailty? *J Geriatr Oncol* 2022; 13: 904–905.
11. Pickard S, Cluley V, Danely J, et al. New horizons in frailty: The contingent, the existential and the clinical. *Age Ageing* 2019; 48: 466–471.
12. Belloni G, Cesari M. Frailty and intrinsic capacity: Two distinct but related constructs. *Front Med* 2019; 6: 133.
13. Sánchez-Sánchez JL, Lu W-H, Gallardo-Gómez D, et al. Association of intrinsic capacity with functional decline and mortality in older adults: A systematic review and meta-analysis of longitudinal studies. *Lancet Healthy Longev* 2024; 5: e480–e492.
14. Jia S, Zhao W, Ge M, et al. Associations between transitions of intrinsic capacity and frailty status, and 3-year disability. *BMC Geriatr* 2023; 23: 96.
15. Tay L, Tay E-L, Mah SM, et al. Association of intrinsic capacity with frailty, physical fitness and adverse health outcomes in community-dwelling older adults. *J Frailty Aging*. Epub ahead of print 2022. DOI: 10.14283/jfa.2022.28.
16. Zhou J, Chang H, Leng M, et al. Intrinsic capacity to predict future adverse health outcomes in older adults: A scoping review. *Healthcare* 2023; 11: 450.
17. Liu S, Kang L, Liu X, et al. Trajectory and correlation of intrinsic capacity and frailty in a Beijing elderly community. *Front Med* 2021; 8: 751586.
18. Gutiérrez-Robledo LM, García-Chanes RE, González-Bautista E, et al. Validation of two intrinsic capacity scales and its relationship with frailty and other outcomes in Mexican community-dwelling older adults. *J Nutr Health Aging* 2021; 25: 33–40.
19. Chew J, Lim JP, Yew S, et al. Disentangling the relationship between frailty and intrinsic capacity in healthy community-dwelling older adults: A cluster analysis. *J Nutr Health Aging* 2021; 25: 1112–1118.
20. Wei X, Chen Y, Qin J, et al. Factors associated with the intrinsic capacity in older adults: A scoping review. *J Clin Nurs* 2024; 33: 1739–1750.
21. Chhetri JK, Xue Q-L, Ma L, et al. Intrinsic capacity as a determinant of physical resilience in older adults. *J Nutr Health Aging* 2021; 25: 1006–1011.
22. Beard JR, Jotheeswaran AT, Cesari M, et al. The structure and predictive value of intrinsic capacity in a longitudinal study of ageing. *BMJ Open* 2019; 9: e026119.
23. Woo J. Combating frailty and sarcopenia in aging populations: Switching to a more positive paradigm. *Aging Med* 2019; 2: 7–10.
24. Gutiérrez-Robledo LM, García-Chanes RE, Pérez-Zepeda MU. Allostatic load as a biological substrate to intrinsic capacity: A secondary analysis of CRELES. *J Nutr Health Aging* 2019; 23: 788–795.
25. Rattan SIS. Biological Health and Homeodynamic Space. In: Sholl J, Rattan SIS (eds) *Explaining Health Across the Sciences*. Cham: Springer International Publishing, pp. 43–51.
26. Mushiake H. [Allostasis and homeostasis: Dynamic adaptive systems from a neurophysiological perspective]. *Brain Nerve* 2023; 11: 1189–1196.
27. Beese S, Postma J, Graves JM. Allostatic load measurement: A systematic review of reviews, database inventory, and considerations for neighborhood research. *Int J Environ Res Public Health* 2022; 19: 17006.
28. López-Otín C, Kroemer G. Hallmarks of health. *Cell* 2021; 184: 1929–1939.
29. Banerjee A, Sadana R. Integrated care for older people (ICOPE): From guidelines to demonstrating feasibility. *J Frailty Aging* 2020; 1–2.
30. World Health Organization. *Integrated care for older people: Guidelines on community-level interventions to manage declines in intrinsic capacity*. World Health Organization, Geneva, 2017.
31. Yu R, Leung J, Leung G, et al. Towards healthy ageing: Using the concept of intrinsic capacity in frailty prevention. *J Nutr Health Aging* 2022; 26: 30–36.
32. Blancafort Alias S, Cuevas-Lara C, Martínez-Velilla N, et al. A multi-domain group-based intervention to promote physical activity, healthy nutrition, and psychological wellbeing in older people with losses in intrinsic capacity: AMICOPE development study. *Int J Environ Res Public Health* 2021; 18: 5979.
33. Daly T. The vital need for action against the social determinants of frailty. *Aging Med* 2022; 5: 73–73.
34. Belsky DW, Baccarelli AA. To promote healthy aging, focus on the environment. *Nat Aging* 2023; 3: 1334–1344.

2 Clinical Management of Frailty
Advances and Challenges

Isabel Rodríguez Sánchez, Alejandro Alvarez-Bustos, Mariam El Assar de la Fuente, and Leocadio Rodríguez-Mañas

INTRODUCTION

Frailty was first described in the general population living in the community. Its characteristics, components, underlying mechanisms, screening and diagnostic criteria, prognosis, and indeed prevention and treatment have been developed and tested usually in older people in non-clinical settings. However, many of the adverse outcomes are closely related with events leading to clinical settings or taking place in those settings, so it is not surprising that the prevalence of frailty is extremely high in hospital and other clinical environments, being hospitalization one of the main factors accounting for the poor prognosis of older frail people.

Increasing evidence has been raised in the last decade providing new insights into those consequences of frailty, their particular characteristics, how the presence of frailty will change the management of the patient with several other diseases and conditions, and the potential treatments of the condition in some of those clinical settings.

FRAILTY: A UNIQUE CLINICAL SYNDROME OR A SYNDROME WITH SEVERAL CLINICAL SUBTYPES

When frailty appeared more than 20 years ago as a syndrome, it was conceived as a unique syndrome that was the final consequence of several, usually converging, mechanisms and diseases and with similar outcomes once the syndrome had been developed. Since that time, there has been a fragmentation of the syndrome (physical frailty, cognitive frailty, social frailty, etc.) but also a definition of different characteristics clustered around some type of diseases or subtypes of frailty with different profiles and consequences. While the first fragmentation of the syndrome is still a matter of debate, the definition of different subtypes of frailty is becoming increasingly popular, and the best knowledge of them appears as one of the fields of research in frailty. Is it the same frailty the one experienced by patients with heart failure than the one in those with cancer? Is frailty the same one in the absence of disability that when disability starts to be present? Among those who are frail, are there any accompanying characteristics, like the presence of sarcopenia, modifying the clinical profile, the associated risk, and the management of the syndrome? The existence of these particularities opens a new perspective in the clinical management of frailty, a new clinical way in the provision of care to these patients. Once a patient is diagnosed with frailty, it is needed to implement some diagnostic procedures to assess accurately its subtype and, accordingly, to promote the best therapeutic scheme.

HOW FREQUENT IS FRAILTY IN CLINICAL SETTINGS

There are a huge number of publications, including some narrative and systematic reviews, assessing the prevalence and incidence of frailty in different countries. However, most of them are done in community-dwelling settings, while there are very few studies focused on clinical settings, especially in hospitals, although hospitalization is one of the main factors linked to a poor evolution of the frailty syndrome. In this regard, it is important to highlight that older frail people are distributed

DOI: 10.1201/9781003398295-2

in all the clinical settings and services, and because of that they should be properly detected and managed when they attend to these different clinical facilities (outpatient offices, medical and surgical departments, in-hospital services, etc.).

It is widely known that the prevalence of frailty varies depending on the tool used to assess it. For instance, 6% of oncology inpatients were classified as frail people according to the Rockwood Frailty Index, whereas the prevalence of frailty in this population was 81.6% when the G8 scale was used (FRAILCLINIC study data) (Checa-López et al., 2023). Moreover, the presence of frailty also fluctuates regarding the clinical scenario. In the above-mentioned study, the prevalence of frailty in patients hospitalized in non-geriatric settings ranged from 21.2% (elective surgery) up to 60% (Emergency department). In Primary care or Geriatric settings, prevalent frailty also differs across different settings (i.e., geriatric wards, outpatient clinics, nursing homes). For example, when assessing frailty by Fried phenotype, the FRAILTOOLS study (Oviedo-Briones et al., 2020) showed a frailty prevalence of 16.2% in primary care, 25.5% in outpatient clinics, 60.4% in nursing homes, or 65.9% in geriatric wards. Corresponding figures for FTS5 were 7.5%, 14.2%, 74.3%, and 50.6%, respectively.

These tools, again, have been usually validated in community settings, but few of them in clinical settings. Moreover, the usual way of validating these tools has been through the assessment of their prognostic value (i.e., OR, HR), which is the clinical way of establishing that value, although not the best one for establishing a diagnosis. There are very few studies assessing those parameters through the appropriate instruments (sensitivity, specificity, negative and positive predictive value or area under the curve) (Faller et al., 2019), including, of course, the studies run in clinical settings. This huge disparity of prevalence of frailty could be explained by the low concordance between all the frailty tools (less than 0.65 according to Oviedo-Briones et al., 2020), with a high heterogeneity (Aguayo et al., 2017).

HOW RELEVANT IS FRAILTY IN SOME TYPES OF PATIENTS

It has already been noted that the prevalence of frailty varies depending on the clinical setting and the assessment tool used to evaluate it. Regardless of these factors, assessing frailty in older adults is of critical importance, as it represents the strongest prognostic predictor of adverse events. Furthermore, frailty assessment plays a key role in guiding diagnostic and therapeutic strategies, anticipating disease progression, and identifying potential complications (Figure 2.1).

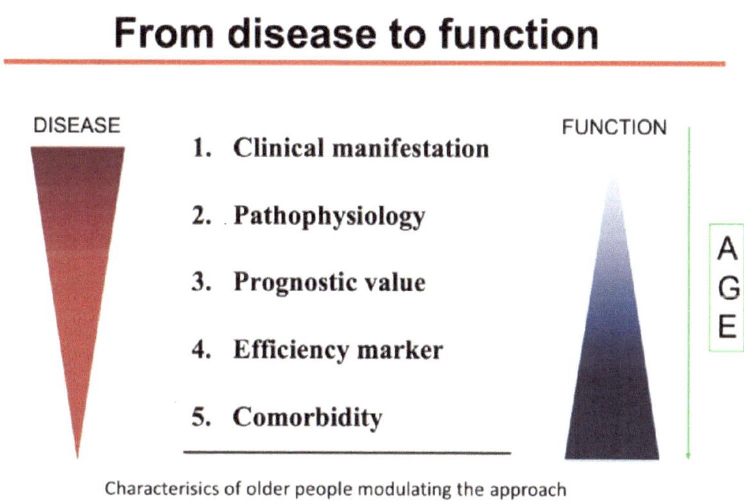

FIGURE 2.1 As people age, the role of function becomes predominant in the manifestation of the diseases, its prognostic value, and as a marker of effectiveness and efficiency, thus modulating and determining the management of the patient.

Certain clinical conditions closely intersect with frailty syndrome, including cancer, heart failure, malnutrition, or the postoperative state, with a higher prevalence of frailty in this population. However, treating the underlying disease can potentially reverse frailty, and vice versa, reversing frailty could improve the underlying disease.

Renal Failure

Frailty has been identified as a risk factor for developing acute kidney injury in hospitalized patients (Baek et al., 2016), as well as for mortality (Ali H et al., 2018) and institutionalization in care homes among patients with advanced chronic kidney disease or post-transplant complications (Roshanravan et al., 2012; McAdams-DeMarco et al., 2019). Current management guidelines for advanced chronic kidney disease (stages 4 and 5) highlight the necessity of performing a comprehensive geriatric assessment (CGA) (Farrington et al., 2016), which includes frailty assessment (nephrology-tailored geriatric assessment), a practice widely endorsed by health care professionals (Berkhout-Byrne et al., 2024).

Moreover, frailty assessment plays a crucial role in decision-making regarding kidney transplantation, the initiation of dialysis, the selection of dialysis modality, and dialysis vintage and adequacy (Berger et al., 2016; Basile et al., 2019; Harhay et al., 2020; Bulbul & Selimoglu Namoglu, 2024). Notably, Chess et al. (2024) observed that age and frailty were the only two factors influencing the choice to undergo home dialysis, with younger and less frail patients demonstrating a stronger preference for home-based treatment.

Frailty has also been associated with an increased risk of adverse events for patients on the kidney transplant waiting list, including death or removal from the list (Chen X et al., 2020). Evidence suggests that centers implementing thorough frailty assessments prior to transplantation report lower rates of adverse events and post-transplant complications (Chen et al., 2022). Therefore, in this patient population, conducting a robust frailty assessment may not only reduce certain adverse outcomes but also enhance overall quality of life.

Surgical Patients

Frailty is a key determinant in the outcomes of patients undergoing both urgent and elective surgery. Its impact on morbidity and mortality has been extensively studied in recent years. In urgent surgery, frailty is associated with a higher prevalence of mortality. According to the UK observational Emergency Laparotomy and Frailty (ELF) study (Parmar et al., 2021), frail patients undergoing emergency laparotomy had a 30-day mortality rate of 42%, compared to 16% in non-frail patients. Similarly, geriatric assessment in the emergency department has shown that frailty is associated with a 60% increase in the risk of postoperative adverse events (Covino et al., 2022). Additionally, a recent meta-analysis (Aguilar-Frasco et al., 2023) confirmed that preoperative frailty is linked to higher rates of severe complications and prolonged hospital stays.

In elective surgery, although mortality is lower, frailty remains a key predictor of postoperative complications. In patients undergoing lower extremity bypass surgery, the modified frailty index has been shown to be a significant predictor of adverse outcomes and mortality (Ali TZ et al., 2018). Likewise, in colorectal oncologic surgery, frail patients exhibited a higher prevalence of postoperative complications, with rates reaching up to 50% (Argillander et al., 2022). A recent study (Theodorakis et al., 2024) highlighted the usefulness of clinical and radiological tools in preoperative frailty assessment, improving the identification of high-risk patients.

Frailty has been shown to outperform traditional surgical risk assessment scales, such as ASA or APACHE, in predicting adverse events in this population. In cardiac surgery, for instance, the Frailty in Cardiac Surgery (FICS) Copenhagen study demonstrated that frailty was the primary determinant of 30-day mortality (Back et al., 2019).

Frailty should be considered a fundamental pillar in preoperative assessment and surgical planning, enabling better patient selection and perioperative management optimization. The implementation of systematic geriatric evaluations and prehabilitation strategies may improve surgical outcomes and reduce the incidence of complications in frail patients.

Oncogeriatric

Frailty in oncological patients is a critical component in clinical management and therapeutic decision-making, particularly in older adults. The prevalence of frailty in oncology patients varies depending on the type of cancer and the tool used to assess it, but recent studies have found that between 40 and 50% (with a wide range from 5% to 90%) of patients with cancer over 65 years old exhibit some degree of frailty (Goede, 2023). In these patients, frailty is associated with increased mortality, with the risk of death being two to three times higher in those with frail characteristics compared to non-frail individuals. Additionally, frailty predicts a range of adverse events, including postoperative complications, falls, hospitalizations, and poorer tolerance to oncological treatments. In fact, the incidence of postoperative complications in frail oncology patients can exceed 50%, compared to less than 20% in non-frail patients (Niemeläinen et al., 2021).

Frailty has become a fundamental pillar in oncological treatment decision-making. Through its assessment, therapeutic strategies can be individualized, optimizing the type of treatment offered, whether surgical, chemotherapy, radiotherapy, or palliative. These patients often show a diminished response to treatments such as chemotherapy and radiotherapy, which limits treatment options and frequently leads to interruptions (Hormigo-Sanchez et al., 2023). This is why doing a robust frailty assessment could help to choose the suitable treatment for a specific patient, trying to minimize undertreatment and overtreatment (Dumontier et al., 2021). Farcet et al. (2016) analyzed whether five frailty markers (nutrition, physical activity, energy, mobility, and strength) were related to treatment recommendations in an oncogeriatric clinic. They observed that, as the number of frailty markers increased, treatment recommendations were accordingly oriented toward modified treatment, whereas no frailty markers were associated with standard treatment. In frail patients, less invasive approaches or those better suited to their functional capacity are favored, prioritizing quality of life improvement and symptom control (Hormigo-Sanchez et al., 2023). This approach enables a more comprehensive and personalized management, improving long-term outcomes and reducing the risks of adverse events.

In terms of predicting adverse events, frailty has proven to be a more effective tool than traditional scales, such as the Charlson comorbidity index or the Karnofsky performance scale. The Clinical Frailty Scale (CFS) has emerged as a more precise tool for assessing frailty and predicting complications in these patients, showing greater sensitivity in identifying those at risk (Niemeläinen et al., 2021; Goede, 2023).

Heart Failure

Heart failure and frailty share common pathophysiological mechanisms such as chronic inflammation, mitochondrial dysfunction, and sarcopenia (Bielecka-Dabrowa et al., 2020; Albulushi, 2024). The evolving concept of the "sarcopenic heart" (Curcio et al., 2020) presents an intriguing area of investigation in contemporary cardiovascular and geriatric medicine, referring to the interplay between heart failure (HF) and sarcopenia. This condition is particularly relevant in heart failure patients, as both disorders share common pathophysiological mechanisms, including chronic inflammation, oxidative stress, mitochondrial dysfunction, and hormonal imbalances.

Both frailty and sarcopenia (understood as a biological substrate of frailty) are key factors in the progression of HF patients, influencing mortality (Aguilar Ginés, 2020; Aguilar-Iglesias et al., 2024), adverse events [higher rates of hospital readmission (Aida et al., 2020), major cardiovascular events, higher likelihood of intolerance to essential heart failure drugs (Abe et al., 2023)], and therapeutic decision-making. Recognizing sarcopenia in HF patients is essential for optimizing treatment strategies that may help mitigate disease progression (i.e., targeted interventions, including physical rehabilitation, nutritional supplementation, and anti-inflammatory therapies).

Frailty scales have proven superior to traditional risk assessment tools in predicting adverse events in HF. Aida et al. (2020) validated the Simplified Frailty Scale as a reliable predictor of mortality and hospital readmission, surpassing conventional tools such as CHA_2DS_2-VASc or EUROScore, including multidimensional aspects that best reflect the vulnerability of the elderly patient with HF.

Additionally, frailty is a key factor in selecting therapeutic strategies in cardiology. For instance, it has been shown that, in frail patients with severe aortic stenosis, TAVI is more beneficial than conventional surgery due to its lower surgical burden (Abdelgawad et al., 2019).

Specific Considerations in Some Clinical Settings/Patients

Acute Care Units

The clinical presentation of frailty in acute geriatric units can vary from non-specific symptoms (i.e., weakness, fatigue) to geriatric syndromes (i.e., falls, delirium, complications related to polypharmacy), which could mask underlying acute pathology.

Systematic frailty screening is recommended for older adults in acute care settings. Several validated tools exist for frailty identification, including the CFS, the Frailty Phenotype (FP), the Frailty Index, and the FRAIL Scale. Other validated instruments include the Tilburg Frailty Indicator, Groningen Frailty Indicator, Hospital Frailty Risk Score, and the Frailty Index-LAB. These tools vary in terms of validity, reliability, and feasibility, as previously mentioned. Taking into account the characteristics of hospitalized patients, which are not at their baseline situation, some frailty tools (such as Fried phenotype) could be not the best to assess frailty in this clinical scenario due to the two objectives measurements needed (walking speed and hand-grip), which could be altered because of the acute illness.

A multicomponent intervention is recommended by several clinical practice guidelines (Mehta et al., 2021), including exercise [showing improvements on functional, cognitive, and quality of life (Martínez-Velilla et al., 2019)], nutritional interventions with early dietary support and protein supplementation showing a reduced mortality, hospital readmissions (EFFORT trial; Baumgartner et al., 2021), or an increase in muscle strength and improvement on protein intake with whey protein supplementation (Niccoli et al., 2017) and optimizing drug management to avoid potentially inappropriate medications.

Orthogeriatric Units

Fragility fractures, particularly hip fractures, are among the most serious conditions in geriatrics, as they result from multiple geriatric syndromes, including falls, osteosarcopenia, polypharmacy, cognitive decline, and frailty. These fractures often lead to permanent disability, institutionalization, functional deterioration, and an increased one-year mortality risk (Dreinhöfer et al., 2018).

Osteoporosis, falls, and hip fractures impose a substantial burden on individuals and health care systems. Annually, over 8.9 million fragility fractures occur worldwide, with one osteoporotic fracture happening every 3 seconds. Hip fractures carry a 22.4% one-year mortality risk, with elevated mortality persisting for at least 5 years. Among survivors, 40% lose independent mobility, 60% require assistance, and 33% become fully dependent or require nursing home care within a year.

Osteosarcopenia, a newly recognized geriatric syndrome, plays a crucial role in fracture risk by integrating both bone fragility and fall susceptibility. Frailty and osteosarcopenia create a vicious cycle of falls and fractures, reinforcing each other and accelerating functional decline (Patel, 2017; Duque, 2021). This interdependence underscores the need for a comprehensive approach to fracture prevention and care. Advances in understanding muscle-bone interactions have shifted treatment strategies toward both pharmacologic and non-pharmacologic interventions.

To improve fragility fracture management, the Fragility Fracture Network (FFN) proposes a four-pillar approach: prevention, acute care, discharge planning, and long-term follow-up (Dreinhöfer et al., 2018). This structured framework aims to break the cycle of frailty, osteosarcopenia, and fractures, ultimately enhancing patient outcomes.

Optimal hip fracture management necessitates a multidisciplinary approach, with a narrow collaboration between orthopedic surgeons and geriatricians. This model of orthogeriatrics has demonstrated benefits in reducing hospital stays, improving functionality, and minimizing complications and costs. The role of the geriatrician is crucial in every single phase of the hospitalization: preoperative care to prevent post-surgical adverse events; postoperative phase enhancing early mobilization, monitoring anemia and preventing delirium; and rehabilitation and follow-up ensuring an appropriate pain management, physical therapy, osteoporosis treatment, and social support for a successful recovery (Sabharwal & Wilson, 2015; Falaschi & Marsh, 2021).

Prehabilitation

Prehabilitation in surgical or oncology patients is defined as the implementation of perioperative interventions to enhance the patient's physiological reserve and reduce postoperative complications. It is a multidisciplinary strategy aimed at optimizing physical, nutritional, and functional status before surgery or oncological treatment, with the objective of improving postoperative outcomes and quality of life, demonstrating a reduction in morbidity and mortality (Ramírez-Martín et al., 2022). Additionally, studies such as Andersson et al. (2023) demonstrate that prehabilitation reduces complications and improves physical function following surgery in older patients with gastrointestinal cancer.

In this population, CGA is essential for identifying areas requiring optimization, such as frailty, malnutrition, and polypharmacy (Rodríguez Couso, 2020), predicting postoperative outcomes (Argillander et al., 2022). The following interventions are recommended:

a. Exercise: Enhances functional capacity and reduces postoperative complications. Physical prehabilitation programmes have demonstrated benefits in patients undergoing neoadjuvant treatment for colorectal cancer (Alejo et al., 2019).
b. Nutrition: Perioperative immunonutrition decreases complications and improves functional recovery in frail patients with colorectal cancer (Achilli et al., 2020).
c. Pharmacological optimization: Reviewing and adjusting treatments according to CGA to minimize adverse effects and improve therapeutic tolerance.
d. Management of geriatric syndrome, addressing frailty, falls syndrome, or cognitive impairment as part of the preoperative approach (Bai et al., 2024).

In conclusion, prehabilitation in oncology is a key strategy to enhance treatment tolerance and postoperative recovery, particularly in frail patients.

FRAILTY ASSESSMENT

The detection of frailty is one of the greatest challenges in geriatric medicine. Dozens of tools have tried to capture this geriatric syndrome, evaluating both subjective (such as fatigue, self-reported weight loss, or physical activity levels) and objective domains (such as the assessment of functional variables). In this section, we summarize the different scales proposed, as well as other functional variables that have been related to frailty.

COMPREHENSIVE GERIATRIC ASSESSMENT (CGA)

The efficacy of CGA-based interventions has been evaluated in both community-dwelling older adults and hospitalized individuals. Li et al. found that the likelihood of improvement or worsening in frailty status was not significantly different between the CGA group and the control group at 6-month follow-up (Li et al., 2010). Moreover, Checa-López and colleagues (Checa-López et al., 2023) proved that an intervention targeting hospitalized frail older adults (>75 years), admitted to various medical and surgical wards, consisting of a CGA leading to a coordinated approach to provide recommendations and to implement them about polypharmacy, delirium, falls, nutrition and physical exercise plus a discharge plan were effective in reducing mortality and incident disability at discharge and at 3 months later.

FRAILTY TOOLS

Assessment of frailty in different care settings, such as clinical care and research, focusing on different patient characteristics, has allowed different concepts of frailty with a wide spectrum of frailty assessment tools with poor agreement between them and different sensitivity and specificity depending on the setting and event studied (Oviedo-Briones et al., 2022).

The Frailty Phenotype, proposed by Linda P Fried and colleagues (Fried et al., 2001a), stands as one of the most widely embraced approaches for evaluating frailty (Buta et al., 2016). Their framework portrays frailty as a biological syndrome closely tied to aging, marked by diminished functional reserves. Despite the presence of organic decline predisposing individuals to adverse events (Fried et al., 2001a), the potential for reversing this state remains plausible (Gill et al., 2006; Kojima et al., 2019; Álvarez-Bustos et al., 2022). This methodological approach delineates frailty through five key indicators: unintentional weight loss, self-reported exhaustion, weakness in grip strength, slow gait speed, and low physical activity (Fried et al., 2001a). Individuals meeting three or more criteria are classified as frail, while those meeting one or two are deemed prefrail, and those meeting none are categorized as robust (Fried et al., 2001a). Although Fried et al. detailed the cutoff points for the different items to consider whether the criterion was met or not, it is advisable to adapt these cutoff points based on various factors, such as geographic region or ethnicity, among others (Abizanda Soler et al., 2011; Alonso Bouzón et al., 2017; Siriwardhana et al., 2018; Hanlon et al., 2020).

This scale allows a quick and simple classification of individuals that is easily achievable in routine health care practice. However, it does not allow the evaluation of small changes in the individual that are associated with a decrease in the risk of adverse events or the identification of potential therapeutic targets to guide health care professionals (Cesari et al., 2014; García-García et al., 2014, 2020). With the aim of resolving these aspects, other scales based on this concept have been designed to solve some of these handicaps.

Three of them were proposed by García-García and colleagues (García-García et al., 2014, 2020). These authors generated a scale based on the evaluation of 12 items, including seven domains (energy balance-nutrition, physical activity, nervous system, vascular system, strength, endurance, and gait speed) (García-García et al., 2014). For the construction of this tool, the so-called Frailty Trait Scale (FTS), data from a population-based cohort of older Spanish adults from the Toledo region were used, and it was created with the aim of capturing small changes, both improvement and worsening, that would allow not only the assessment of frailty but also the monitoring of the individual on the basis of frailty. To this end, the scale allows a simple dichotomous classification (distinguishing between frail and non-frail individuals) and a continuous version ranging from 0 to 100 (García-García et al., 2014). However, given that the assessment of certain items proposed in this scale, such as albumin levels or ankle-arm index, may not be easily achievable in routine clinical practice, the authors generated two reduced versions of 3 (FTS3) and 5 (FTS5) items, respectively (García-García et al., 2020). Like the FP, FTS5 has been shown to capture changes in frailty, not only through transitions between states (Álvarez-Bustos et al., 2022) but also by drawing trajectories that have been associated with the incidence of adverse events (Álvarez-Bustos et al., 2022). It is noteworthy that this scale has been evaluated in other cohorts and different clinical settings (Oviedo-Briones et al., 2022).

Another remarkable frailty model was conceived by Kenneth Rockwood and Arnold Mitnitski (Rockwood et al., 2005; Rockwood & Mitnitski, 2007). While the Fried phenotype characterizes frailty as a syndrome reflecting declining functional reserves from a biological perspective (Fried et al., 2001a), Rockwood and Mitnitski conceived frailty as the cumulative outcome of deficits across several systems (Rockwood et al., 2005; Rockwood & Mitnitski, 2007). The number of items proposed to be evaluated has oscillated from 10 (Nishijima et al., 2021) to 83 (Hogan et al., 2012). One of these systems would include the locomotor system, so this scale would include disability as an aspect to be assessed within frailty. This factor would be a determining factor in distinguishing between the previous concept, originally proposed by Fried, and this one, since according to the previous model, although frailty and disability could agree in the same individual, ideally frailty would be an entity that would appear prior to disability and would allow its identification and management to prevent its progression from leading to disability (Fried et al., 2004; Rockwood et al., 2005; Rockwood & Mitnitski, 2007).

However, as previously mentioned, the lack of concordance among the various existing tools for assessing frailty, as well as the significant heterogeneity between them, raises the following question: Can the same tool be used across all different socio-health care settings? Should the chosen tool be adapted according to the level of care (hospital, outpatient, community) or the specific medical-surgical department where the patient is being treated? As highlighted earlier, based on data from the FRAILTOOLS (Oviedo-Briones et al., 2020) and FRAILCLINIC (Checa-López et al., 2023) studies, not all diagnostic tools adequately detect frailty depending on the level of care.

Additionally, it is important to emphasize the feasibility of each tool in different care settings. For instance, the tools with the highest feasibility are the FRAIL scale (ranging from 95.1% in Emergency Departments to 100% in urgent surgery), the CFS (ranging from 96.7% in urgent surgery to 100% in elective surgery), and the Tilburg Frailty Indicator (ranging from 88% in Emergency Departments to 98.9% in urgent surgery) (Checa-López et al., 2023).

This underscores the necessity of using different diagnostic tools depending on the care level, although some tools may be applicable across multiple settings (see results from FRAILCLINIC and FRAILTOOLS). This variability poses a significant challenge not only for assessing frailty but also for monitoring the patient's frailty status over time. For this reason, efforts have been made to standardize the use of these tools (albeit with mixed results) to enable a standardized patient follow-up across care levels.

PHYSICAL PERFORMANCE-BASED MEASUREMENTS IN DETECTING FRAILTY

Physical function decline starts in the third decade of life and becomes more evident at older ages (Dodds et al., 2016; Cruz-Jentoft et al., 2019; Alcazar et al., 2021; Prince et al., 2021). Physical performance, defined as "an objectively measured whole body function related to mobility" (Kell et al., 2001), involves not just muscle function but also neurological, cardiovascular, and psychosocial aspects (Hart & Zernicke, 2020). Affordable and simple tools designed to mimic real-life scenarios have been used to objectively assess performance and their association with frailty.

CARDIORESPIRATORY FITNESS

Maximal oxygen uptake (VO2 max) is widely regarded as the premier indicator of cardiorespiratory fitness across various clinical contexts due to its strong correlation with adverse events (Mancini et al., 1991; Blair et al., 1996; Kodama et al., 2009; Steell et al., 2019). Indeed, VO2 max was found to be associated with frailty status in individuals over 65 years old and could serve as a predictor of mortality in those considered as robust (Fung et al., 2020). However, assessing maximal fitness in sedentary older adults at risk of mobility disability is often impractical (Church et al., 2008). Alternative tests like the 400-m test or the 6-minute walking test have shown a strong correlation with VO2 max measurement, offering greater feasibility and reproducibility (Jones et al., 2017). In fact, not only the time taken to perform the test, but also the self-perceived exertion when performing the gait test, appears to be an important predictor of physical disability. Older adults who perceived the effort of performing a 400-m test as "hard" presented almost threefold higher risk of becoming disabled in two and a half years than those who felt the effort to be light (Cenko et al., 2021). Additionally, several equations have been proposed to estimate peak VO2 using the 6-minute walking test (Ross et al., 2010; Deka et al., 2021) or the 400-m walking test (Garcia et al., 2024).

MUSCLE STRENGTH

The strength of various muscle groups has been assessed in relation to numerous clinical outcomes, such as diminished quality of life (Sayer et al., 2006), incident disability (Rantanen et al., 1999; al Snih et al., 2004; Davies et al., 2022), and even mortality (Sayer & Kirkwood, 2015; Celis-Morales et al., 2018; Lee, 2020; Davies et al., 2022).

The isometric handgrip strength has been the most commonly measured (Fried et al., 2001b; García-García et al., 2014, 2020), due to its simplicity and reproducibility (Mijnarends et al., 2013). Various cutoff points for identifying frailty have been proposed, taking into account the characteristics of the population under evaluation (Lourenço et al., 2015; Alonso Bouzón et al., 2017; Ramírez-Vélez et al., 2019; Fernandes et al., 2022), as well as gender (Roberts et al., 2011; Dodds et al., 2016).

In addition to the handgrip strength test, which is the most reliable and commonly used isometric strength test, several similar tests have been proposed to identify frailty. In this regard, Buckinx et al. (2016) demonstrated that hip, knee, shoulder, or elbow strength was significantly distinguishable between robust, prefrail, or frail individuals. This distinction occurred regardless of whether the musculature assessed was flexor or extensor (Buckinx et al., 2016). For their part, Van Roie and colleagues evaluated the association between strength and knee extensor speed of movement in 123 institutionalized women aged 80 years on average (Van Roie et al., 2011). However, the knee extension strength of 18 kg in men and 16 kg in women, cutoff points proposed as a definition of dynapenia by Iwamura and Kanauchi, had a sensitivity of 17% for identifying community-dwelling older adults with frailty (Iwamura & Kanauchi, 2017).

CHAIR STAND TEST

The ability to rise from a chair is perhaps the most closely related clinical biomarker of disability (Braun et al., 2022) and serves as a good prognostic factor for adverse events such as falls (Buatois et al., 2010; Applebaum et al., 2017). This simple test, requiring only a chair and a stopwatch, is widely used.

One version of this test consists of sitting down and standing up five times as fast as possible. Taking more than ≥10.8 or ≥12.8 second to perform this test has been associated with an increased risk of worsening the older adults' physical function (Buatois et al., 2010; Kim & Won, 2022). The time to complete this test increases with the progression of frailty (Kang et al., 2009b; Batista et al., 2012; Meng et al., 2019; L. B. Tay et al., 2019).

Muscle power, defined as the ability to perform a movement as fast as possible, has also been associated with the frailty status. Individuals with low muscle power were 1.82 times and 4.33 times more likely to be prefrail and frail, respectively (Burbank et al., 2023). Furthermore, at 4 years of follow-up, robust individuals with low power had a decreased risk (OR 0.86) of maintaining their robustness and an increased risk of transitioning to prefrail (OR 1.24) and frail (OR 1.70). On the other hand, prefrail subjects with low muscle power presented a decreased risk of improving to robustness (OR 0.71) and a 62% increased risk of worsening to frail.

Another version of this test involves performing the maximum number of repetitions (sitting, standing, and sitting again) within 30 seconds (Jones et al., 1999). This test can effectively differentiate between subjects with different frailty levels (Theou et al., 2011; Furtado et al., 2017; L. B. Tay et al., 2019). Moreover, completing more than five repetitions in 30 seconds has been associated with a lower risk of worsening frailty status (Rodríguez-Laso et al., 2022), while individuals who transitioned to a robust status performed significantly more repetitions compared to those who remained prefrail (Rodríguez-Laso et al., 2023). In addition, phases of standing up and sitting down, as well as velocity peaks, would be different between frailty categories (Millor et al., 2013).

BALANCE

The performance in different balance tests, including the Tinetti (Buckinx et al., 2016), the Berg Balance Scale (Vieira et al., 2018), or the BEST test (Marques et al., 2019), has shown to be better among robust individuals compared to those who are not robust. Furthermore, according to the progressive Romberg balance test, a higher percentage of robust individuals (84–86.6%) were able to maintain the most demanding position compared to prefrail or frail individuals (22–66.7%), while no differences were observed in the two easier positions (J. Tay et al., 2019).

Likewise, the balance evaluated using a force platform according to the state of frailty has been evaluated in different studies with disparate results. While Marques et al. (2019) did not observe greater oscillations in this population according to this technique, in the study by Kang and collaborators, frail individuals presented greater oscillations than those considered as robust and prefrail (Kang et al., 2009a).

GAIT SPEED

Short distances (from 2.4 to 10 m) and long distances (400 m to a mile; or even by the maximum distance walked in 6 minutes) walking tests have been proposed for assessing this functional parameter. In fact, gait speed as a single domain has been proposed as a screening measure of frailty (Woo et al., 2018; Sanchez-Sanchez et al., 2022), and it may present a similar predictive ability of the entire Short Physical Performance Battery (SPPB) in predicting disability both in mobility and activities of daily living (Guralnik et al., 2000).

However, the cutoff points have to be adapted to the different populations (Woo et al., 2018; Sanchez-Sanchez et al., 2022) in addition to gender and height (Sanchez-Sanchez et al., 2022).

On the other hand, there are different modalities in the measurement of this test. While some authors evaluate walking speed by performing the test at the usual speed, others evaluate it by enhancing individuals to complete the distance to be evaluated at the maximum possible speed. In both approaches, prefrail and frail individuals present a poorer performance (Theou et al., 2011; Buckinx et al., 2016; L. B. Tay et al., 2019; Navarrete-Villanueva et al., 2021; Sanchez-Sanchez et al., 2022).

SHORT PHYSICAL PERFORMANCE BATTERY

The SPPB is a battery of tests that comprises three assessments designed for the objective evaluation of lower extremity functioning (Guralnik et al., 1994). This test evaluates balance, gait speed, and the time to perform five rises from a chair as quickly as possible (Guralnik et al., 1994). However, given that the correct performance of these tests requires a proper global functioning of the organism, this test has been suggested as an easily accessible "vital sign" for evaluating the overall health of older adults (Studenski et al., 2003).

Different cutoff points have been proposed in this scale to try to identify frail individuals. In this regard, scores ≤7 for men and ≤4 for women has demonstrated excellent sensitivity (0.88 and 0.93, in men and women, respectively) and specificity (0.84 to 0.88) in recognizing community-dwelling older adults that met frailty criteria according to the FP (Sanchez-Sanchez et al., 2022). On the other hand, Jung and colleagues (Jung et al., 2021) found a lower specificity (0.55) with a cutoff point of ≤9. Additionally, these authors also suggested alternative cutoff points (≤6 and ≤7) for the Frailty Index (Jung et al., 2021). Ramírez-Vélez and colleagues, utilizing data from 4125 individuals aged ≥60, obtained areas under the curve of 0.74 for cutoff points of 8 and 7 for men and women, respectively. In this study, both the FP and FRAIL scales were used (Ramírez-Vélez et al., 2021).

TIMED UP AND GO TEST

The Timed Up and Go test (TUG) is a single assessment that measures the time spent for a participant to complete several tasks, including standing up from a chair, walking 3 m, turning around a cone or another object, returning to the chair, and sitting back down as fast as possible (Mathias et al., n.d.; Podsiadlo & Richardson, 1991). In a systematic review, Clegg and colleagues (Clegg et al., 2015) showed that spending more than 10 seconds on this test has high sensitivity for identifying frailty (0.93), although it has modest specificity (0.62).

CLINICAL PHENOTYPES OF FRAILTY

Frailty is understood as the expression of the interaction between aging and deterioration in different systems. Thus, some authors have proposed different clinical subtypes of frailty based on the different biological substrates that might predominantly account for the frailty state in that individual. This concept is interesting as different clinical subtypes of frailty could have different approaches and different risks of adverse events. In this regard, researchers from the Toledo Study of Healthy Ageing demonstrated that the presence of sarcopenia (understood as the presence of low muscle mass and function) increased the risk of disability in community-dwelling older adults (Davies et al., 2022). However, sarcopenia did not increase the risk associated with frailty in terms of mortality (Álvarez-Bustos et al., 2022; Davies et al., 2022). These findings suggest that sarcopenia influences the relationship between frailty and its key outcomes—mortality and disability—in distinct ways. Considering the varying impact of frailty on disability based on the presence of sarcopenia, it is plausible that frailty comprises distinct clinical subtypes—specifically, sarcopenic, and non-sarcopenic. Differentiating frailty subtypes based on sarcopenia status has clinical and diagnostic implications. In the absence of sarcopenia, other etiologies of frailty must be studied. Assessing sarcopenia in frail patients could improve prognosis and treatment, particularly in tailoring exercise interventions and decreasing rates of non-responders. This could be because only those frail subjects with sarcopenia will improve their functional status with strength exercises, while non-sarcopenic frail subjects may not improve due to an inadequate exercise program.

Regarding the presence of other pathologies, some authors have evaluated whether the presence of cancer (Pamoukdjian et al., 2020) or HIV (McMillan et al., 2020) could modify the outcome of individuals with frailty.

Likewise, other authors have explored this same idea based on different phenotypes of physical frailty. This is the case of Tsung Huang and colleagues' manuscript, in which they found that those individuals with mobility-subtype frailty presented a higher risk of functional decline and progression in multimorbidity (Huang et al., 2020).

INTERVENTIONS FOR TREATING FRAILTY

Several interventions have been proposed for treating frailty. Due to its complex nature, it seems that the inclusion of multiple domains is more effective than those that only include one. In this sense, lifestyle interventions, especially those that are composed by nutritional and physical exercise, seems to be the most effective one (Negm et al., 2019; Travers et al., 2019; Oliveira et al., 2020). Nevertheless, the evidence is still scarce, especially by comparing different types of interventions (Negm et al., 2019).

NUTRITIONAL INTERVENTION

Few studies have evaluated the isolated effect of nutritional interventions on frailty status (de Moraes et al., 2021). Park et al. (2018) conducted a randomized clinical trial (RCT) in prefrail and frail undernourished individuals comparing the intake of 0.8, 1.2, and 1.5 g of protein per kilogram of body weight per day. At the end of 12 weeks of intervention, all three groups decreased their mean frailty score, but no differences between the three groups were found in terms of frailty.

Badrasawi et al. (2016) evaluated the efficacy of L-carnitine on frailty status. For this purpose, they performed a RCT including 58 older adults with prefrailty. After 10 weeks of follow-up, the group receiving L-carnitine supplementation decreased their frailty score as assessed by the Frailty Index. In addition, a greater number of subjects ended up being robust according to Fried's criteria compared to the control group (3 vs. 1), although this difference was not statistically significant. (Badrasawi et al., 2016).

On the other hand, Jadczak et al. (2021) evaluated the feasibility and efficacy of a multicomponent exercise program with the addition of a nutritional supplement in individuals who met at least one criterion of the FRAIL scale. Both groups performed the multicomponent exercise program 5 days per week: three multicomponent sessions (one of them supervised and two at home) and two walking sessions per week. In addition to this intervention, a nutritional supplement of whey or rice should be ingested twice a day. After the 6 months of intervention, the number of individuals with more than one criterion was reduced in both groups. In fact, 61% and 50% of the individuals who remained in the study became robust.

PHYSICAL EXERCISE INTERVENTION FOR TREATING FRAILTY

Physical exercise is defined as the physical activity performed with a controlled and planned frequency and whose aim is the maintenance or improvement of any component of the physical fitness (Caspersen et al., 1985).

Among all the physical exercise components, it seems that exercise programs that include strength training are the most efficient in improving physical function in older adults with frailty (Lopez et al., 2018). Chen et al. (2020) evaluated the effect of a physical activity program performed in individuals between 65 and 85 years of age who met at least one frailty criterion. While the control group did not receive any special intervention, the intervention group was treated with a program of eight exercises performed between 10 and 15 repetitions in each of two sets. This intervention was performed three times a week for 8 weeks. Four weeks after starting the intervention, 6% (control) vs 52% (intervention) of the subjects did not meet any frailty criteria, and these percentages increased to 9% (control) and 81% (intervention) at 8 weeks.

Coelho-Júnior and Uchida evaluated the effectiveness of two different training programs in improving frailty status in prefrail and frail adults aged 60 or over (Coelho-Júnior & Uchida, 2021). One training group performed a methodology in which the concentric phase was performed at maximum speed (high speed resistance training, HSRT), with an eccentric phase at medium speed (2.5 seconds), while the other group performed both the concentric and eccentric phases at medium speeds (2.5 seconds) (low speed resistance training, LSRT). In this randomized controlled trial, physical exercise total volume was equalized between groups, considering the series (HSRT: 8 vs LSRT: 4), repetitions (HSRT: 3–5 vs LSRT: 8–10), and load (70–75% of the repetition maximum). Both interventions significantly reduced both the prevalence and mean number of frailty criteria met, and this occurred in both prefrail and frail individuals. Among the prefrail, 55% (LSRT) vs 18% (HSRT) turned robust after the intervention. In those who were frail at the beginning, 63% (LSRT) and 45% (HSRT) and 13% (LSRT) and 9% (LSRT) improved their frailty status to the prefrail and robust status, respectively. Among participants who were prefrail at baseline, exhaustion was the most frequently reversed criterion. In those classified as frail at the time of recruitment, the most commonly reversed criteria were physical inactivity, exhaustion, and unintentional weight loss.

MULTICOMPONENT PHYSICAL ACTIVITY OR EXERCISE INTERVENTIONS

Due to the fact that several physical components are impaired in older adults with frailty and pre-frailty, the inclusion of multiple physical exercise components has been proposed in the literature. Multicomponent exercise programs have been shown to improve physical function in frail older adults, even in the presence of other conditions such as sarcopenia (Bernabei et al., 2022) or cognitive impairment (Casas-Herrero et al., 2022). Cadore and colleagues proved that a 12-week program of balance, gait, and muscle power training significantly increased muscle strength, power, and mass in institutionalized nonagenarians with frailty (Cadore et al., 2014). Nevertheless, authors did not evaluate changes in frailty status. On the other hand, Cesari and colleagues demonstrated that a multicomponent physical activity program that includes aerobic, strength, flexibility, and balance training significantly reduced the prevalence of frailty and the mean number of frailty criteria met by the intervention group at 6 and 12 months of follow-up. In this intervention, significant differences were observed within subgroups defined by baseline frailty and race. In this sense, participants under the age of 80, those of Black race, individuals classified as frail at baseline, and those with more than three comorbidities showed a significant reduction in their mean frailty scores. Moreover, in this sample, the only single criterion that was significantly reduced between randomized groups in terms of prevalence along the follow-up was the low physical activity (Cesari et al., 2015). A personalized and almost-individualized 12-week multicomponent physical activity program that includes mobility, flexibility, endurance, and resistance training reduced the prevalence of and the number of frailty in frail older residents in a long-term care facility (Ferreira et al., 2018).

De Vries and coworkers also demonstrated that a customized physiotherapy intervention based on the needs of adults over 70 years of age with mobility problems and/or low physical activity significantly reduced frailty to a greater extent than a conventional physiotherapy intervention at 6 months of follow-up (de Vries et al., 2016). In addition, Gineta-Garriga also found that a 12-week multicomponent functional circuit intervention led to improvements in both normal and fast gait speed, as well as in balance, among individuals over the age of 80. In this study, frailty was defined by slowness, unable to stand up 5 times from a chair, or if they presented the fatigue criterion of the FP (Gineta-Garriga et al., 2014).

Tarazona-Santabalbina and colleagues conducted a supervised multicomponent exercise program involving older adults who were frail according to the FP. This exercise program includes endurance (with an intensity between 40% and 60% of maximum heart rate), strength (from 25% to 75% of the repetition maximum), proprioceptive (10–15 minutes), and flexibility exercises performed during 65 minutes, 5 days per week for 24 weeks. After the intervention, the multicomponent exercise group reduced their mean frailty score according to the FP and the Edmonton Frail Scale. Moreover, while no participant in the control group reversed frailty after the 6-month period, 31% of the intervention group did (Tarazona-Santabalbina et al., 2016).

Sadjapong and colleagues evaluated the effect of a multicomponent exercise program in reducing frailty status in community-dwelling older adults. They randomized 64 frail older adults according to Fried's criteria. This intervention was supervised by a trainer for 12 weeks and then non-supervised other 12 weeks. The supervised intervention consisted of a 12-week program that included aerobic, resistance, and balance training, with a gradual increase in intensity, duration, and sets, with a frequency of 3 days per week. Moreover, the home-based exercise program was prescribed with a total duration of 12 weeks and a frequency of 3 days per week. Authors found that the multicomponent group significantly decreased frailty scores after 12 and 24 months when compared with the basal assessment and with the usual care group. Nevertheless, no significant differences were found when comparing between the 12- and 24-week evaluations (Sadjapong et al., 2020). A similar intervention was developed by Sanchis et al. (2021), including prefrail and frail older adults after acute myocardial infarction. After the 12 weeks of supervised intervention, the intervention group was encouraged to continue the intervention at home. Both the multicomponent exercise group and the control group significantly reduced their frailty score. However, no differences were found between the two groups. On the other hand, the intervention group improved significantly in the domains of fatigue and gait speed.

Another multicomponent exercise intervention was also developed in prefrail women by Bray and colleagues (2020). These authors evaluated the effect of a 12-week exercise program, distributed in three blocks of 4 weeks each. Sessions started with low-intensity aerobic exercise and ended with stretching work. In the middle of the session, four muscular strength exercises (squat, deadlift, bench, and leg press) were performed at high intensity for two to three sets of 4 to 12 repetitions per set, depending on the block. Both strength and balance exercises increased in intensity as the program progressed. The authors found a time main effect when evaluating frailty according to FP and CFS, but did not find a significant group effect.

In a non-randomized study, Losa-Reyna (Losa-Reyna et al., 2019) performed concurrent training in prefrail and frail volunteers with a mean age of 84 years. The authors conducted two assessment sessions, two familiarization sessions, 12 training sessions, and two final evaluation sessions. Power training was performed at a frequency of 2 days per week and consisted of two exercises to improve lower limb power. The leg press was performed three to four sets of 8 to 15 repetitions with an intensity between 30% and 60% of F0, while the plantar flexion exercise was performed three sets of 4 to 12 repetitions in a bipodal or monopodal manner, using body weight as resistance (leg press and plantar flexion). As for aerobic exercise, after 2 weeks of conditioning, 6 to 10 intervals were performed with 10–30 seconds at 90% of maximum walking speed followed by 50–100 seconds at 50% of usual walking speed. At the end of the intervention, the mean frailty criteria score decreased from 3.1 to 1.5. This same group conducted a second non-randomized clinical trial in a population over 75 years of age with criteria of prefrailty or frailty according to Fried's phenotype (Baltasar-Fernandez et al., 2023). In it, while one of the groups performed the training described above, the second group performed the strength work in clusters with 10 seconds of rest every two repetitions. At the end of the intervention, both groups decreased their frailty status, as well as the mean score on both Fried's FP and the FTS5.

MULTIFACTORIAL INTERVENTION

Kim and colleagues (Kim et al., 2015) conducted a four-arm randomized controlled trial exploring the effects of an exercise and nutritional supplementation program in frail older adults. The resistance band exercise training consisted of strength, balance, and gait training 2 days per week for 3 months. On the other hand, nutritional supplementation consisted of the ingestion of six pills of milk fat globule membrane composed of 21% protein, 44% fat, and 27% carbohydrate. At the end of the intervention, exhaustion and low physical activity were reversed in all groups. In those groups in which the exercise intervention was included, the weight loss criterion was also reversed. Moreover, slowness was significantly reversed at a greater rate in participants allocated to the exercise + supplement group. Nevertheless, no significant improvements were found in any group for the weakness criterion.

The effect of an exercise program based on Nordic walking, nutritional supplementation, or the combination of both was evaluated in prefrail people over 60 years of age in India (Chatterjee et al., 2018). At 12 weeks of intervention, all groups found significant differences in frailty status, resulting in 9% robust (Nordic walking), 18% robust (nutritional supplementation), and 27% robust Nordic walking and nutritional supplementation) of the participants. In addition, the group that included the combination of both interventions was the only one in which none of the individuals ended up being frail (Chatterjee et al., 2018).

Serra-Prat et al. (2017) randomized 172 older adults with prefrail criteria and performed a multicomponent intervention that englobes nutritional assessment and a multicomponent physical activity program. Exercise intervention consisted of 30–45 minutes per day of aerobic training performed at least 4 days per week, and a home-based combination of 15 mixed exercises (strength, balance, and coordination) performed 10–15 repetitions. At follow-up, 4.9% vs 15.3% of the intervention and control groups were frail, respectively. In this sense, authors also reported that of those who were considered as good adherers, none evolved to frailty. Moreover, 21.3% and 15.3% of the intervention and control groups improved their frailty status to robustness. Once age, gender, and comorbidities were considered, a significant association between the intervention and frailty status was found. Among the specific domains of frailty, only the outdoor walking reached a statistical significance between groups.

Nagai et al. (2018) also evaluated the effect of two physical exercise programs on frailty status. Both groups performed a supervised intervention twice a week for 24 weeks with an intensity between 50% and 80% of their RM. Additionally, one group received information on the benefits of physical activity, nutrition, and adequate protein intake (PAR). In addition, physical activity levels and the number of steps were assessed by accelerometry, and participants in this group were encouraged to reduce their sedentary lifestyle. Although at the end of the intervention the frailty status did not differ between training groups, a significant reduction in the frailty score was found when analyzing the time × group interaction in the PAR group. Regarding the different domains, the prevalence of weakness and involuntary weight loss was significantly reduced in this latter group. On the other hand, the prevalence of individuals reporting self-reported fatigue increased significantly in the group that only performed strength exercises.

The effect of a nutritional and strength exercise intervention in the prefrail and frail was also evaluated in older adults in Austria (Luger et al., 2016). The nutritional intervention sought to ensure sufficient intake of fluids, energy, and animal and vegetable protein. Muscle strength training consisted of performing two sets of six exercises for 15 repetitions until muscle exhaustion. Nutritional and strength interventions were supervised by non-professional volunteers twice a week. Moreover, participants were encouraged to perform one additional day per week of strength work. The control group, on the other hand, also received visits from these volunteers, but without nutrition or exercise recommendations. Although both groups significantly reduced their frailty scores as assessed by the SHARE-FI (Romero-Ortuno et al., 2010) at 12 weeks, no significant differences were found between groups.

Ng and colleagues (2015) compared five different interventions in participants who met one or more of the FP criteria. The intervention consisted of moderate physical exercise performed 2 days a week for 24 weeks, 12 of which were supervised. This intervention consisted of a balance workout with three levels of difficulty and a strength program performed at moderate intensity, performing 8 to 15 repetitions in a single set. Another intervention included the intake of a nutritional supplement that included iron, folate, vitamins B, D, and calcium for 24 weeks. Additionally, another intervention included 2 hours of cognitive training to recall information, improve attention, or reasoning, among others, performed for 12 weeks, followed by 12 weeks of reinforcement training performed for 2 hours every 2 weeks. Added to these three groups, one of the groups underwent all three interventions, while another group received no intervention in addition to standard care. At 12 months of follow-up, all interventions significantly reduced the mean frailty score, but the control group did not. Furthermore, relative to the control group, the probability of improvement of reducing frailty ranged from 2.89 for the cognition intervention to five times higher odds for the group that received a combined intervention.

The rate of improvement in frailty status at 3 months was also significantly higher in prefrail and frail people over 65 years of age in Taiwan (Chan et al., 2012). In this study, the authors evaluated the effect of an exercise program performed 3 days a week for 3 months by performing 10 to 15 repetitions of major muscle groups using rubber bands and water bottles of up to 1 L as resistance, as well as a balance program performed for 10 minutes. In addition, nutritional recommendations and advice were also given to the participants. Compared to the control group, individuals in this group significantly improved their frailty to a greater extent at 3 months, but not at 6 or 12 months. This study also showed that six sessions of psychotherapy did not show significant benefits compared to those who did not receive such intervention in terms of frailty.

Cameron et al. (2013) performed a standardized multifactorial intervention according to the criteria the patient met: nutritional intervention (if weight loss criteria was present), ten sessions of home physiotherapy and exercises to improve mobility, increase physical activity and prevent falls (weakness, slow walking speed, or low physical activity), or if they were considered for evaluation by a psychologist or psychiatrist (fatigue). In addition, if the patient was socially isolated, participation in group activities was encouraged and telephone support was provided by volunteers. Moreover, participants were provided with or recommended a CGA. Individuals in the intervention group significantly reduced both frailty status and the mean number of criteria at 12-month follow-up compared to the control group.

Hsieh et al. (2019) conducted a four-arm (control, exercise, nutrition, and a combination of exercise plus nutrition), single-blind, randomized controlled trial. They randomized 319 subjects over 65 years of age with prefrailty or frailty criteria and assessed frailty at 1 month, 3 months, and 6 months. The physical exercise was a multicomponent home-based exercise that followed the recommendations of the American College of Sports Medicine with a frequency of 3 to 7 days per week performed between 5 and 60 minutes, adapted to the patient's capabilities. For its part, the nutritional intervention sought to maintain a desirable body weight by ensuring caloric requirements based on age, sex, height, weight, and physical activity levels, in addition to daily milk and nut supplementation. At 6 months of intervention, the results of the post hoc analysis revealed that the exercise, nutrition, and the combination of the two groups significantly decreased the mean frailty score.

In another RCT conducted in institutionalized prefrail and frail women, two physical exercise protocols were compared with a control group (Furtado et al., 2020). Both exercise programs were performed 3 days a week for 28 weeks. One exercise program consisted of a functional circuit, with nine multimodal exercises performed in a chair for two to three sets of 10 to 20 repetitions. The other exercise protocol included ten strengthening exercises using one's own body weight or elastic bands performed for two to three sets of 10 to 15 repetitions. In both training protocols, intensity was monitored objectively (% of maximum heart rate) and subjectively. At the end of the intervention, both training protocols significantly decreased the mean score according to Fried's criteria. Regarding the different domains, both interventions improved the criterion of low physical activity, while the multimodal exercise program furthermore significantly reduced low gait speed.

Chan et al. compared the effect of two interventions in adults over 65 who met at least one frailty criterion. One consisted of a 1-hour lecture on mainly geriatric syndromes and healthy habits, while during the second an adapted exercise protocol was applied. In addition, the subjects in this group received an educational booklet and an exercise CD. On the other hand, the comparator group received the same educational course, to which were added 48 exercise sessions and six sessions of psychotherapy focused on improving both mental and physical aspects. Of the individuals studied, 35%, 40%, and 39% improved in their frailty status at 3, 6, and 12 months after the start of the study; however, no significant differences were found between the two except for grip strength at 12 months.

More recently, a RCT in older frail people admitted to a hospital has shown that the intervention led by a geriatric team and based on a CGA, including recommendations about polypharmacy, delirium, falls, nutrition and physical exercise plus a discharge plan, showed benefit in terms of lower mortality and incident disability at 3 months after discharge (Checa-López et al., 2024).

TECHNOLOGICAL INTERVENTIONS

Technological interventions have also been shown to improve frailty status in older adults. For example, Tan et al. evaluated the impact of a technological intervention on frailty status in prefrail individuals according to the CFS. This intervention was conducted without a comparison group and consisted of three modules such as polypharmacy management, nutritional recommendations, and physical exercise based on Tai chi and cycling. The authors reported significant improvements

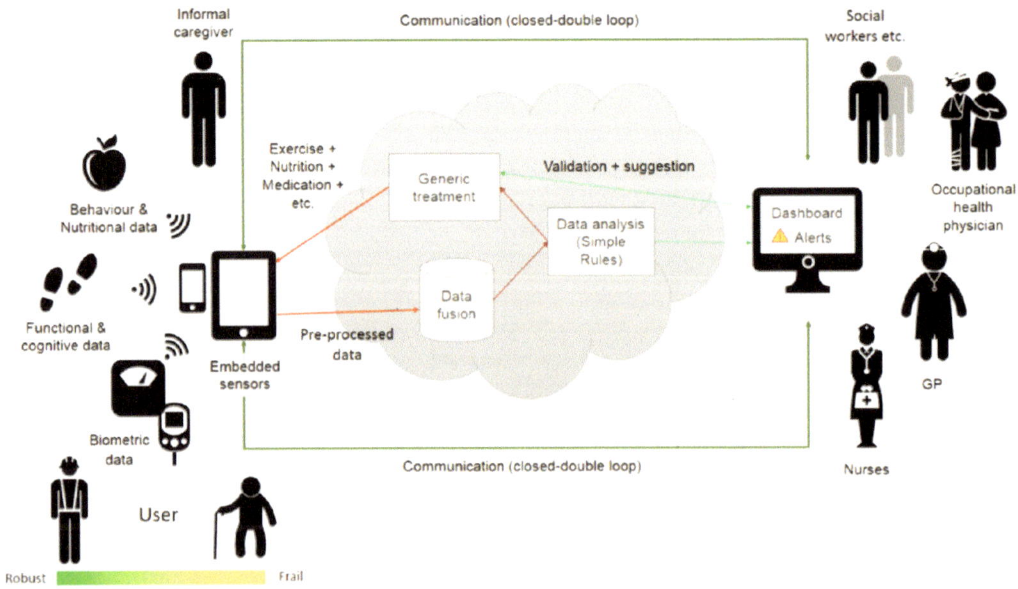

FIGURE 2.2 General deployment, structure, and components of a technological platform to detect, manage, and monitor older people with frailty.

in mean frailty score according to the modified Fried criteria, but not in frailty status according to this scale or mean CFS score (Tan et al., 2022). Similarly, Pérez-Rodríguez et al. (2020) conducted a remote intervention and monitoring focused on disability prevention in individuals with prefrailty and frailty (Figure 2.2).

They found that those to whom the technology was applied were eight times more likely to improve by 2.5 points according to the FTS5 (Pérez-Rodríguez et al., 2020). Recently, they have also shown a 74% and 92% risk decrease of frailty status deterioration by FTS5 score and Fried phenotype, respectively, using a technological home-monitoring ecosystem (FACET). Not to mention the improvement on both FTS5 and Fried criteria at 3 months and FTS5 at 6 months of follow-up (Valdés-Aragonés et al., 2024). Additionally, with their POSITIVE project consisting of managing frailty through a tailored home-monitoring intervention with a technological platform, they demonstrated a significant improvement in frailty status and physical performance by SPPB after 6 months of follow-up. These results were similar in a validation study in usual clinical practice called INTEGRACAM, with a high rate of usability (Rodríguez-Mañas et al., 2024).

Liao et al. (2019) compared the effect of Kinect-based exergaming with a combined exercise program in older adults from daycare centers over 65 years of age who met at least one frailty criterion. Both protocols were performed three times a week for 12 weeks. The combined program included an aerobic component (performed at an intensity between 50% and 75% of maximum heart rate or a perceived exertion between 13 and 14), strength (using elastic bands for three sets of 10 to 15 repetitions), and balance (including both dynamic and static exercises performed for 20 minutes). The exergame program, on the other hand, included Tai-chi, resistance, aerobic, and balance exercises. At the end of the intervention, 44% (exergame) and 38% (combined exercise) of the individuals had improved their frailty status. In addition, both groups had decreased the number of frailty criteria met. Regarding the different domains, the exergame group significantly reduced the prevalence of all items met except intentional weight loss, while the combined exercise group significantly reduced the number of individuals meeting the criteria of weakness, slowness, and low physical activity.

REFERENCES

Abdelgawad, A. M. E., Hussein, M. A., Naeim, H., Abuelatta, R., & Alghamdy, S. (2019). A comparative study of TAVR versus SAVR in moderate and high-risk surgical patients: Hospital outcome and midterm results. Heart Surgery Forum, 22(5), E331–E339.

Abe, T., Jujo, K., Maeda, D. et al. (2023). The interaction between physical frailty and prognostic impact of heart failure medication in elderly patients. ESC Heart Failure, 10(3), 1698–1705. https://doi.org/10.1002/ehf2.14114

Abizanda Soler, P., López-Torres Hidalgo, J., Romero Rizos, L., López Jiménez, M., Sánchez Jurado, P. M., Atienzar Núñez, P., Esquinas Requena, J. L., García Nogueras, I., Hernández Zegarra, P., Bardales Mas, Y., Campos Rosa, R., Martínez Peñalver, M., de la Osa Nieto, E., Carión González, M., Ruiz Gómez, Á, Aguilar Cantos, C., Mañueco Delicado, P., & Oliver Carbonell, J. L. (2011). Fragilidad y dependencia en Albacete (estudio FRADEA): razonamiento, diseño y metodología. Revista Espanola de Geriatria y Gerontologia, 46(2), 81–88. https://doi.org/10.1016/j.regg.2010.10.004

Achilli, P., Mazzola, M., Bertoglio, C. L. et al. (2020). Preoperative immunonutrition in frail patients with colorectal cancer: An intervention to improve postoperative outcomes. International Journal of Colorectal Disease, 35(1), 19–27.

Aguayo, G. A., Donneau, A. F., Vaillant, M. T., Schritz, A., Franco, O. H., Stranges, S., Malisoux, L., Guillaume, M., & Witte, D. R. (2017). Agreement between 35 published frailty scores in the general population. American Journal of Epidemiology, 186(4), 420–434. https://doi.org/10.1093/aje/kwx061

Aguilar-Frasco, J. L., Moctezuma-Velázquez, P., Rodríguez-Quintero, J. H. et al. (2023). Preoperative frailty assessment in older patients with colorectal cancer: Use of clinical and radiological tool. Langenbecks Arch Surg, 408(1), 19.

Aguilar Ginés, S. A. (2020). Prognosis in heart failure: Importance of physical frailty at the time of admission. Emergencias, 32(2), 147–148.

Aguilar-Iglesias, L., Merino-Merino, A., Sanchez-Corral, E., Garcia-Sanchez, M. J., Santos-Sanchez, I., Dominguez-Calvo, J., Saez-Maleta, R., & Perez-Rivera, J. A. (2024). Prognostic effect of systematic geriatric assessment on patients with acute heart failure. ESC Heart Failure, 11(2), 1194–1204. https://doi.org/10.1002/ehf2.14692

Aida, K., Kamiya, K., Hamazaki, N. et al. (2020). Usefulness of the simplified frailty scale in predicting risk of readmission or mortality in elderly patients hospitalized with cardiovascular disease. International Heart Journal, 61(3), 571–578.

Albulushi, A. (2024). Exploring the pathophysiologic interplay between multidimensional frailty and heart failure prognosis. Canadian Journal of Cardiology, 40(9), 1755.

Alcazar, J., Alegre, L. M., Van Roie, E., Magalhães, J. P., Nielsen, B. R., González-Gross, M., Júdice, P. B., Casajús, J. A., Delecluse, C., Sardinha, L. B., Suetta, C., & Ara, I. (2021). Relative sit-to-stand power: Aging trajectories, functionally relevant cut-off points, and normative data in a large European cohort. Journal of Cachexia, Sarcopenia and Muscle, 12(4), 921. https://doi.org/10.1002/JCSM.12737

Alejo, L. B., Pagola-Aldazabal, I., Fiuza-Luces, C. et al. (2019). Exercise prehabilitation program for patients under neoadjuvant treatment for rectal cancer: A pilot study. Journal of Cancer Research and Therapeutics, 15(1), 20–25.

Ali, H., Abdelaziz, T., Abdelaal, F., & Baharani, J. (2018). Assessment of prevalence and clinical outcome of frailty in an elderly pre dialysis cohort using simple tools. Saudi Journal of Kidney Diseases and Transplantation, 29(1), 63–70.

Ali, T. Z., Lehman, E. B., & Aziz, F. (2018). Modified Index can be used to predict adverse outcomes and mortality after lower extremity bypass surgery. Annals of Vascular Surgery, 46, 168–177.

Alonso Bouzón, C., Rodríguez-Mañas, L., Carnicero, J., García-García, A., Turín, F. J., Rodríguez-Mañas, J. G., & Esteban, L. (2017). The standardization of frailty phenotype criteria improves its predictive ability: The Toledo Study for Healthy Aging. Journal of the American Medical Directors Association, 18(5), 402–408. https://doi.org/10.1016/j.jamda.2016.11.003

al Snih, S., Markides, K. S., Ottenbacher, K. J., & Raji, M. A. (2004). Hand grip strength and incident ADL disability in elderly Mexican Americans over a seven-year period. Aging Clinical and Experimental Research, 16(6), 481–486. https://doi.org/10.1007/BF03327406

Álvarez-Bustos, A., Carnicero-Carreño, J. A., Davies, B., Garcia-Garcia, F. J., Rodríguez-Artalejo, F., Rodríguez-Mañas, L., & Alonso-Bouzón, C. (2022). Role of sarcopenia in the frailty transitions in older adults: A population-based cohort study. Journal of Cachexia, Sarcopenia and Muscle, 13(5), 2352–2360. https://doi.org/10.1002/jcsm.13055

Álvarez-Bustos, A., Carnicero-Carreño, J. A., Sanchez-Sanchez, J. L., Garcia-Garcia, F. J., Alonso-Bouzón, C., & Rodríguez-Mañas, L. (2022). Associations between frailty trajectories and frailty status and adverse outcomes in community-dwelling older adults. Journal of Cachexia, Sarcopenia and Muscle, 13(1), 230–239. https://doi.org/10.1002/jcsm.12888

Andersson, M., Egenvall, M., Danielsson, J. et al. (2023). CANOPTIPHYS study protocol: Optimising PHYSical function before CANcer surgery: Effects of pre-operative optimisation on complications and physical function after gastrointestinal cancer surgery in older people at risk – a multicentre, randomised, parallel-group study. Trials, 24(1), 41.

Applebaum, E., Breton, D., Feng, Z. W., Ta, A. T., Walsh, K., Chassé, K., & Robbins, S. M. (2017). Modified 30-second Sit to Stand test predicts falls in a cohort of institutionalized older veterans. PloS One, 12(5). https://doi.org/10.1371/JOURNAL.PONE.0176946

Argillander, T. E., Schafer, S., van Westreenen, H. L. et al. (2022). The predictive value of preoperative frailty screening for postoperative outcomes in older patients undergoing surgery for non-metastatic colorectal cancer. Journal of Geriatric Oncology, 13(6), 888–891.

Back, C., Hornum, M., Olsen, P. S., & Moller, C. H. (2019). 30-day mortality in frail patients undergoing cardiac surgery: The results of the frailty in cardiac surgery (FICS) Copenhagen study. Scandinavian Cardiovascular Journal, 53(6), 348–354.

Badrasawi, M., Shahar, S., Zahara, A. M., Nor Fadilah, R., & Singh, D. K. A. (2016). Efficacy of L-carnitine supplementation on frailty status and its biomarkers, nutritional status, and physical and cognitive function among prefrail older adults: A double-blind, randomized, placebo-controlled clinical trial. Clinical Interventions in Aging, 1675. https://doi.org/10.2147/CIA.S113287

Baek H.S., Lee, S.W., Kim, S-W., Ahn S.Y., Yu, M-Y., Kim K-I. et al. (2016). Frailty as a predictor of acute kidney injury in hospitalized elderly patients: A single center, retrospective cohort study. PLoS One, 11(6), e0156444.

Bai, Z., Solomon, M., Shahab, R., Koh, C., & Steffens, D. (2024). The good, the bad, and the frail – The role of prehabilitation in frail patients undergoing cancer surgery. ANZ Journal of Surgery, 94(6), 993–995.

Baltasar-Fernandez, I., Alcazar, J., Martín-Braojos, S., Ara, I., Alegre, L. M., García-García, F. J., Alfaro-Acha, A., & Losa-Reyna, J. (2023). Power-oriented resistance training combined with high-intensity interval training in pre-frail and frail older people: Comparison between traditional and cluster training set configurations on the force–velocity relationship, physical function and frailty. European Journal of Applied Physiology. https://doi.org/10.1007/s00421-023-05298-x

Basile, C., Casino, F. G., & Acuella, F. (2019). Incremental hemodialysis, a valuable option for the frail elderly patient. Journal of Nephrology, 32(5), 741–750.

Batista, F. S., Gomes, G. A. O., Neri, A. L., Guariento, M. E., Cintra, F. A., Sousa, M. L. R., & D'Elboux, M. J. (2012). Relationship between lower-limb muscle strength and frailty among elderly people. Sao Paulo Medical Journal, 130(2), 102–108. https://doi.org/10.1590/S1516-31802012000200006

Baumgartner, A., Pachnis, D., & Parra, L. et al. (2021). The impact of nutritional support on malnourished inpatients with aging-related vulnerability. Nutrition, 89, 111279.

Berger, J. R., Jaikaransingh, V., & Hedayati, S. S. (2016). End-stage kidney disease in the elderly: Approach to dialysis initiation, choosing modality, and predicting outcomes. Advances in Chronic Kidney Disease, 23(1), 36–43.

Berkhout-Byrne, N. C., Voorend, C. G. N., Meuleman, Y. et al. (2024). Nephrology-tailored geriatric assessment as decision-making tool in kidney failure. Journal of Renal Care, 50(2), 112–127.

Bernabei, R., Landi, F., Calvani, R., Cesari, M., Del Signore, S., Anker, S. D., Bejuit, R., Bordes, P., Cherubini, A., Cruz-Jentoft, A. J., Di Bari, M., Friede, T., Gorostiaga Ayestarán, C., Goyeau, H., Jónsson, P. V., Kashiwa, M., Lattanzio, F., Maggio, M., Mariotti, L., & Marzetti, E. (2022). Multicomponent intervention to prevent mobility disability in frail older adults: Randomised controlled trial (SPRINTT project). BMJ, 377, e068788. https://doi.org/10.1136/bmj-2021-068788

Bielecka-Dabrowa, A., Ebner, N., Rodrigues dos Santos, M., Ishida, J., Hasenfuss, G., & von Haehling, D. (2020). Cachexia, muscle wasting, and frailty in cardiovascular disease. European Journal Heart Failure, 22(12), 2314–2326.

Blair, S. N., Kampert, J. B., Kohl, H. W., Barlow, C. E., Macera, C. A., Paffenbarger, R. S., & Gibbons, L. W. (1996). Influences of cardiorespiratory fitness and other precursors on cardiovascular disease and all-cause mortality in men and women. JAMA, 276(3), 205–210. https://doi.org/10.1001/JAMA.1996.03540030039029

Braun, T., Thiel, C., Peter, R. S., Bahns, C., Büchele, G., Rapp, K., Becker, C., & Grüneberg, C. (2022). Association of clinical outcome assessments of mobility capacity and incident disability in community-dwelling older adults – A systematic review and meta-analysis. Ageing Research Reviews, 81. https://doi.org/10.1016/J.ARR.2022.101704

Bray, N. W., Jones, G. J., Rush, K. L., Jones, C. A., & Jakobi, J. M. (2020). Multi-component exercise with high-intensity, free-weight, functional resistance training in pre-frail females: A quasi-experimental, pilot study. The Journal of Frailty & Aging, 9(2), 111–117. https://doi.org/10.14283/JFA.2020.13

Buatois, S., Perret-Guillaume, C., Gueguen, R., Miget, P., Vançon, G., Perrin, P., & Benetos, A. (2010). A simple clinical scale to stratify risk of recurrent falls in community-dwelling adults aged 65 years and older. Physical Therapy, 90(4), 550–560. https://doi.org/10.2522/PTJ.20090158

Buckinx, F., Reginster, J. Y., Petermans, J., Croisier, J. L., Beaudart, C., Brunois, T., & Bruyère, O. (2016). Relationship between frailty, physical performance and quality of life among nursing home residents: The SENIOR cohort. Aging Clinical and Experimental Research, 28(6), 1149–1157. https://doi.org/10.1007/S40520-016-0616-4

Bulbul, E., & Selimoglu Namoglu, S. (2024). The frailty in older hemodialysis patients and association with depression, functional status, dialysis adequacy. Therapeutic Apheresis and Dialysis, 28(5), 727–734.

Burbank, C. M., Branscum, A., Bovbjerg, M. L., Hooker, K., & Smit, E. (2023). Muscle power predicts frailty status over four years: A retrospective cohort study of the National Health and Aging Trends Study. Journal of Frailty, Sarcopenia and Falls, 8(1), 1–8. https://doi.org/10.22540/JFSF-08-001

Buta, B. J., Walston, J. D., Godino, J. G., Park, M., Kalyani, R. R., Xue, Q. L., Bandeen-Roche, K., & Varadhan, R. (2016). Frailty assessment instruments: Systematic characterization of the uses and contexts of highly-cited instruments. Ageing Research Reviews, 26, 53–61. https://doi.org/10.1016/j.arr.2015.12.003

Cadore, E. L., Casas-Herrero, A., Zambom-Ferraresi, F., Idoate, F., Millor, N., Gómez, M., Rodriguez-Mañas, L., & Izquierdo, M. (2014). Multicomponent exercises including muscle power training enhance muscle mass, power output, and functional outcomes in institutionalized frail nonagenarians. Age (Dordr), 36(2), 773–785. https://doi.org/10.1007/s11357-013-9586-z

Cameron, I. D., Fairhall, N., Langron, C., Lockwood, K., Monaghan, N., Aggar, C., Sherrington, C., Lord, S. R., & Kurrle, S. E. (2013). A multifactorial interdisciplinary intervention reduces frailty in older people: Randomized trial. BMC Medicine, 11, 65. https://doi.org/10.1186/1741-7015-11-65

Casas-Herrero, Á, Sáez de Asteasu, M. L., Antón-Rodrigo, I., Sánchez-Sánchez, J. L., Montero-Odasso, M., Marín-Epelde, I., Ramón-Espinoza, F., Zambom-Ferraresi, F., Petidier-Torregrosa, R., Elexpuru-Estomba, J., Álvarez-Bustos, A., Galbete, A., Martínez-Velilla, N., & Izquierdo, M. (2022). Effects of Vivifrail multicomponent intervention on functional capacity: A multicentre, randomized controlled trial. Journal of Cachexia, Sarcopenia and Muscle, 13(2), 884–893. https://doi.org/10.1002/jcsm.12925

Caspersen, C. J., Powell, K. E., & Christenson, G. M. (1985). Physical activity, exercise, and physical fitness: Definitions and distinctions for health-related research. Public Health Reports (Washington, DC: 1974), 100(2), 126–131. http://www.ncbi.nlm.nih.gov/pubmed/3920711

Celis-Morales, C. A., Welsh, P., Lyall, D. M., Steell, L., Petermann, F., Anderson, J., Iliodromiti, S., Sillars, A., Graham, N., MacKay, D. F., Pell, J. P., Gill, J. M. R., Sattar, N., & Gray, S. R. (2018). Associations of grip strength with cardiovascular, respiratory, and cancer outcomes and all cause mortality: Prospective cohort study of half a million UK Biobank participants. BMJ, 361. https://doi.org/10.1136/BMJ.K1651

Cenko, E., Chen, H., Gill, T. M., Glynn, N. W., Henderson, R. M., King, A. C., Pahor, M., Qiu, P., Rego, A., Reid, K. F., Tudor-Locke, C., Valiani, V., You, L., & Manini, T. M. (2021). Ratings of perceived exertion during walking: Predicting major mobility disability and effect of structured physical activity in mobility-limited older adults. The Journals of Gerontology. Series A, Biological Sciences and Medical Sciences, 76(10), E264–E271. https://doi.org/10.1093/GERONA/GLAB036

Cesari, M., Gambassi, G., Abellan van Kan, G., & Vellas, B. (2014). The frailty phenotype and the frailty index: Different instruments for different purposes. Age and Ageing, 43(1), 10–12. https://doi.org/10.1093/ageing/aft160

Cesari, M., Vellas, B., Hsu, F. C., Newman, A. B., Doss, H., King, A. C., Manini, T. M., Church, T., Gill, T. M., Miller, M. E., Pahor, M. & LIFE Study Group. (2014). A physical activity intervention to treat the frailty syndrome in older persons- results from the LIFE-P study. The Journals of Gerontology. Series A, Biological Sciences and Medical Sciences, 70(2), 216–222. https://doi.org/10.1093/gerona/glu099

Chan, D. C. D., Tsou, K. K., Yang, R. S., Tsauo, J. Y., Chen, C. Y., Hsiung, C. A., & Kuo, K. N. (2012). A pilot randomized controlled trial to improve geriatric frailty. BMC Geriatrics, 12, 58. https://doi.org/10.1186/1471-2318-12-58

Chatterjee, P., Kumar, P., Kandel, R., Madan, R., Tyagi, M., Kumar, D. A., Khan, M. A., Desai, G., Chaudhary, P., Gupta, S., Grover, K., & Dey, A. B. (2018). Nordic walking training and nutritional supplementation in pre-frail older Indians: An open-labelled experimental pre-test and post-test pilot study to develop intervention model. BMC Geriatrics, 18(1), 212. https://doi.org/10.1186/s12877-018-0890-4

Checa-López, M., Costa-Grille, A., Álvarez-Bustos, A., Carnicero-Carreño, J. A., Sinclair, A., Scuteri, A., Landi, F., Solano-Jaurrieta, J. J., Bellary, S., & Rodríguez-Mañas, L. (2024). Effectiveness of a randomized intervention by a geriatric team in frail hospital inpatients in non-geriatric settings: FRAILCLINIC project. Journal of Cachexia, Sarcopenia and Muscle, 15(1), 361–369. https://doi.org/10.1002/JCSM.13374

Checa-López, M., Rodriguez-Laso, A., Carnicero, J. A. et al. (2023). Differential utility of various frailty diagnostic tools in non-geriatric hospital departments of several countries: A longitudinal study. European Journal of Clinical Investigation, 53(7), e13979.

Chen, R., Wu, Q., Wang, D., Li, Z., Liu, H., Liu, G., Cui, Y., & Song, L. (2020). Effects of elastic band exercise on the frailty states in pre-frail elderly people. Physiotherapy Theory and Practice, 36(9), 1000–1008. https://doi.org/10.1080/09593985.2018.1548673

Chen, X., Liu, Y., Thompson, V. et al. (2022). Transplant centers that assess frailty as part of clinical practice have better outcomes. BMC Geriatric, 22(1), 82.

Chen, X. S., Myers, J., Han, J. et al. (2020). Physical performance testing in kidney transplant candidates at the top of the waitlist. American Journal of Kidney Diseases, 76(6), 815–825.

Chess, J., Roberts, G., McLaughlin, L., Williams, G., & Noyes, J. (2024). What are the factors that determine treatment choices in patients with kidney failure: A retrospective cohort study using data linkage of routinely collected data in Wales. BMJ Open, 14(2), e082386.

Church, T. S., Gill, T. M., Newman, A. B., Blair, S. N., Earnest, C. P., & Pahor, M. (2008). Maximal fitness testing in sedentary elderly at substantial risk of disability: LIFE-P study experience. Journal of Aging and Physical Activity, 16(4), 408–415. https://doi.org/10.1123/JAPA.16.4.408

Clegg, A., Rogers, L., & Young, J. (2015). Diagnostic test accuracy of simple instruments for identifying frailty in community-dwelling older people: A systematic review. Age and Ageing, 44(1), 148–152. https://doi.org/10.1093/AGEING/AFU157

Coelho-Júnior, H. J., & Uchida, M. C. (2021). Effects of low-speed and high-speed resistance training programs on frailty status, physical performance, cognitive function, and blood pressure in prefrail and frail older adults. Frontiers in Medicine, 8, 702436. https://doi.org/10.3389/fmed.2021.702436

Covino, M., Salini, S., & Russo, A. et al. (2022). Frailty assessment in the emergency department for patients ≥80 years undergoing urgent major surgical procedures. JAMDA, 23(4), 581–588.

Cruz-Jentoft, A. J., Bahat, G., Bauer, J., Boirie, Y., Bruyère, O., Cederholm, T., Cooper, C., Landi, F., Rolland, Y., Sayer, A. A., Schneider, S. M., Sieber, C. C., Topinkova, E., Vandewoude, M., Visser, M., Zamboni, M., Writing group for the European Working Group on Sarcopenia in Older People 2 (EWGSOP2), & the E G. for E. (2019). Sarcopenia: Revised European consensus on definition and diagnosis. Age and Ageing, 48(1), 16–31. https://doi.org/10.1093/ageing/afy169

Curcio, F., Testa, G., & Liguori, I. et al. (2020). Sarcopenia and heart failure. Nutrients, 12(1), 211.

Davies, B., Walter, S., Rodríguez-Laso, A., Carnicero Carreño, J. A., García-García, F. J., Álvarez-Bustos, A., & Rodríguez-Mañas, L. (2022). Differential association of frailty and sarcopenia with mortality and disability: Insight supporting clinical subtypes of frailty. Journal of the American Medical Directors Association, 23(10), 1712–1716.e3. https://doi.org/10.1016/j.jamda.2022.03.013

de Moraes, M. B., Avgerinou, C., Fukushima, F. B., & Vidal, E. I. O. (2021). Nutritional interventions for the management of frailty in older adults: Systematic review and meta-analysis of randomized clinical trials. Nutrition Reviews, 79(8), 889–913. https://doi.org/10.1093/nutrit/nuaa101

de Vries, N. M., Staal, J. B., van der Wees, P. J., Adang, E. M., Akkermans, R., Olde Rikkert, M. G., & Nijhuis-van der Sanden, M.W. (2016). Patient-centred physical therapy is (cost-) effective in increasing physical activity and reducing frailty in older adults with mobility problems: a randomized controlled trial with 6 months follow-up. Journal of Cachexia, Sarcopenia and Muscle, 7(4), 422–435. https://doi.org/10.1002/jcsm.12091

Deka, P., Pozehl, B. J., Pathak, D., Williams, M., Norman, J. F., Alonso, W. W., & Jaarsma, T. (2021). Predicting maximal oxygen uptake from the 6 min walk test in patients with heart failure. ESC Heart Failure, 8(1), 47–54. https://doi.org/10.1002/EHF2.13167

Dodds, R. M., Syddall, H. E., Cooper, R., Kuh, D., Cooper, C., & Sayer, A. A. (2016). Global variation in grip strength: A systematic review and meta-analysis of normative data. Age and Ageing, 45(2), 209–216. https://doi.org/10.1093/AGEING/AFV192

Dreinhöfer, K. E., Mitchell, P. J., Bégué, T. et al. (2018). A global call to action to improve the care of people with fragility fractures. Injury, 49(8), 1393–1397.

Dumontier, C., Por Loh, K., Soto-Perez-de-Celis, E., & Dale, W. (2021). Decision making in older adults with cancer. Journal of Clinical Oncology, 39(19), 2164–2175.

Duque, G. (2021). Osteosarcopenia: A geriatric giant of the XXI century. Journal of Nutrition, Health & Aging, 25(6), 716–719.

Falaschi, P., & Marsh, D., editors. (2021). Orthogeriatrics: The Management of Older Patients with Fragility Fractures. Springer International Publishing. https://doi.org/10.1007/978-3-030-48126-1.

Faller, J. W., do Nascimento Pereira, D., Souza, de, Nampo, S., de Souza Orlandi, F. K., & Matumoto, F. (2019). Instruments for the detection of frailty syndrome in older adults: A systematic review. PLoS One, 14(4), e0216166.

Farcet, A., de Decker, L., Pauly, V. et al. (2016). Frailty markers and treatment decisions in patients seen in oncogeriatric clinics: Results from the ASRO pilot study. PLoS One, 11(2), e0149732.

Farrington, K., Covic, A., Aucella, F. et al. (2016). Clinical practice guideline on management of older patients with chronic kidney disease stage 3b or higher (eGFR <45 mL/min/1.73m2). Nephrol Dialysis Transplant, 31, ii1–ii66.

Fernandes, S., Rodrigues da Silva, E., New York, B., Macedo, P., Gonçalves, R., Camara, S., Larco, R., & Maciel, A. (2022). Cutoff points for grip strength in screening for sarcopenia in community-dwelling older-adults: A systematic review. The Journal of Nutrition, Health & Aging, 26(5), 452–460. https://doi.org/10.1007/S12603-022-1788-6

Ferreira, C. B., Teixeira, P. D. S., Alves Dos Santos, G., Dantas Maya, A. T., Americano Do Brasil, P., Souza, V. C., Córdova, C., Ferreira, A. P., Lima, R. M., & Nóbrega, O. D. T. (2018). Effects of a 12-week exercise training program on physical function in institutionalized frail elderly. Journal of Aging Research, 2018. https://doi.org/10.1155/2018/7218102

Fried, L. P., Ferrucci, L., Darer, J., Williamson, J. D., & Anderson, G. (2004). Untangling the concepts of disability, frailty, and comorbidity: Implications for improved targeting and care. The Journals of Gerontology Series A: Biological Sciences and Medical Sciences, 59(3), M255–M263. https://doi.org/10.1093/gerona/59.3.m255

Fried, L. P., Tangen, C. M., Walston, J., Newman, A. B., Hirsch, C., Gottdiener, J., Seeman, T., Tracy, R., Kop, W. J., Burke, G., & McBurnie, M. A. (2001a). Frailty in older adults: Evidence for a phenotype. The Journals of Gerontology Series A: Biological Sciences and Medical Sciences, 56(3), M146–M157. https://doi.org/10.1093/gerona/56.3.m146

Fung, E., Ting Lui, L., Gustafsson, F., Yau, F. C. F., Leung, J. C. S., Wiklund, P., Järvelin, M. R., MacDonald, P. S., & Woo, J. (2020). Predicting 10-year mortality in older adults using VO_2max, oxygen uptake efficiency slope and frailty class. European Journal of Preventive Cardiology, 28(10), 1148–1151. https://doi.org/10.1177/2047487320914435

Furtado, G., Patrício, M., Loureiro, M., Teixeira, A. M., & Ferreira, J. P. (2017). Physical fitness and frailty syndrome in institutionalized older women. Perceptual and Motor Skills, 124(4), 754–776. https://doi.org/10.1177/0031512517709915

Furtado, G. E., Carvalho, H. M., Loureiro, M., Patrício, M., Uba-Chupel, M., Colado, J. C., Hogervorst, E., Ferreira, J. P., & Teixeira, A. M. (2020). Chair-based exercise programs in institutionalized older women: Salivary steroid hormones, disabilities and frailty changes. Experimental Gerontology, 130, 110790. https://doi.org/10.1016/j.exger.2019.110790

Garcia, R. E., Cawthon, P. M., Nicklas, B. J., Goodpaster, B. H., Coen, P. M., Forman, D. E., Cummings, S. R., Newman, A. B., & Glynn, N. W. (2024). Usual-paced 400 m long distance corridor walk estimates cardiorespiratory fitness among older adults: The Study of Muscle, Mobility and Aging. Journal of the American Geriatrics Society, 72(3), 858–865. https://doi.org/10.1111/JGS.18713

García-García, F. J., Carcaillon, L., Fernandez-Tresguerres, J., Alfaro, A., Larrion, J. L., Castillo, C., & Rodriguez-Mañas, L. (2014). A new operational definition of frailty: The frailty trait scale. Journal of the American Medical Directors Association, 15(5). https://doi.org/10.1016/j.jamda.2014.01.004. https://doi.org/10.1016/j.jamda.2014.01.004

García-García, F. J., Carnicero, J. A., Losa-Reyna, J., Alfaro-Acha, A., Castillo-Gallego, C., Rosado-Artalejo, C., Gutiérrrez-Ávila, G., & Rodriguez-Mañas, L. (2020). Frailty trait scale-short form: A frailty instrument for clinical practice. Journal of the American Medical Directors Association, 21(9), 1260–1266.e2. https://doi.org/10.1016/j.jamda.2019.12.008

Gill, T. M., Gahbauer, E. A., Allore, H. G., & Han, L. (2006). Transitions between frailty states among community-living older persons. Archives of Internal Medicine, 166(4), 418. https://doi.org/10.1001/archinte.166.4.418

Gineta-Garriga, M., Roqué-Fíguls, M., Coll-Planas, L., Sitjà-Rabert, M., & Salvadora, A. (2014). Physical exercise interventions for improving performance-based measures of physical function in community-dwelling, frail older adults: a systematic review and meta-analysis. Archives of Physical Medicine and Rehabilitation, 95(4), 753–769.e.3. https://doi.org/10.1016/j.apmr.2013.11.007

Goede, V. (2023). Frailty and cancer: Current perspectives on assessment and monitoring. Clinical Interventions in Aging, 18, 505–521.

Guralnik, J. M., Ferrucci, L., Pieper, C. F., Leveille, S. G., Markides, K. S., Ostir, G., Studenski, S., Berkman, L. F., & Wallace, R. B. (2000). Lower extremity function and subsequent disability: Consistency across studies, predictive models, and value of gait speed alone compared with the short physical performance battery. The Journals of Gerontology. Series A, Biological Sciences and Medical Sciences, 55(4). https://doi.org/10.1093/GERONA/55.4.M221

Guralnik, J. M., Simonsick, E. M., Ferrucci, L., Glynn, R. J., Berkman, L. F., Blazer, D. G., Scherr, P. A., & Wallace, R. B. (1994). A short physical performance battery assessing lower extremity function: Association with self-reported disability and prediction of mortality and nursing home admission. Journal of Gerontology, 49(2), M85–M94. https://doi.org/10.1093/geronj/49.2.M85

Hanlon, P., Fauré, I., Corcoran, N., Butterly, E., Lewsey, J., McAllister, D., & Mair, F. S. (2020). Frailty measurement, prevalence, incidence, and clinical implications in people with diabetes: A systematic review and study-level meta-analysis. The Lancet Healthy Longevity, 1(3), e106–e116. https://doi.org/10.1016/s2666-7568(20)30014-3

Harhay, M. N., Rao, M. K., & Woodside, K. J. et al. (2020). An overview of frailty in kidney transplantation: Measurement, management and future considerations. Nephrology Dialysis Transplantation, 35, 1099–1112.

Hart, D. A., & Zernicke, R. F. (2020). Optimal human functioning requires exercise across the lifespan: Mobility in a 1g environment is intrinsic to the integrity of multiple biological systems. Frontiers in Physiology, 11. https://doi.org/10.3389/fphys.2020.00156

Hogan, D. B., Freiheit, E. A., Strain, L. A., Patten, S. B., Schmaltz, H. N., Rolfson, D., & Maxwell, C. J. (2012). Comparing frailty measures in their ability to predict adverse outcome among older residents of assisted living. BMC Geriatrics, 12, 56. https://doi.org/10.1186/1471-2318-12-56

Hormigo-Sanchez, A. I., Lopez-Garcia, A., Mahillo-Fernandez, I. et al. (2023). Frailty assessment to individualize treatment in older patients with lymphoma. European Geriatric Medicine, 14(6), 1393–1402.

Hsieh, T. J., Su, S. C., Chen, C. W., Kang, Y. W., Hu, M. H., Hsu, L. L., Wu, S. Y., Chen, L., Chang, H. Y., Chuang, S. Y., Pan, W. H., & Hsu, C. C. (2019). Individualized home-based exercise and nutrition interventions improve frailty in older adults: A randomized controlled trial. The International Journal of Behavioral Nutrition and Physical Activity, 16(1). https://doi.org/10.1186/S12966-019-0855-9

Huang, S., Tange, C., Otsuka, R., Nishita, Y., Peng, L., Hsiao, F., Tomida, M., Shimokata, H., Arai, H., & Chen, L. (2020). Subtypes of physical frailty and their long-term outcomes: A longitudinal cohort study. Journal of Cachexia, Sarcopenia and Muscle. https://doi.org/10.1002/jcsm.12577

Iwamura, M., & Kanauchi, M. (2017). A cross-sectional study of the association between dynapenia and higher-level functional capacity in daily living in community-dwelling older adults in Japan. BMC Geriatrics, 17(1), 1–6. https://doi.org/10.1186/S12877-016-0400-5

Jadczak, A. D., Visvanathan, R., Barnard, R., & Luscombe-Marsh, N. (2021). A randomized controlled pilot exercise and protein effectiveness supplementation study (EXPRESS) on reducing frailty risk in community-dwelling older people. Journal of Nutrition in Gerontology and Geriatrics, 40(1), 26–45. https://doi.org/10.1080/21551197.2021.1886222

Jones, C. J., Rikli, R. E., & Beam, W. C. (1999). A 30-s chair-stand test as a measure of lower body strength in community-residing older adults. Research Quarterly for Exercise and Sport, 70(2), 113–119. https://doi.org/10.1080/02701367.1999.10608028

Jones, S., Tillin, T., Williams, S., Coady, E., Chaturvedi, N., & Hughes, A. D. (2017). Assessment of exercise capacity and oxygen consumption using a 6 min stepper test in older adults. Frontiers in Physiology, 8(June). https://doi.org/10.3389/FPHYS.2017.00408

Jung, H. W., Baek, J. Y., Jang, I. Y., Guralnik, J. M., Rockwood, K., Lee, E., & Kim, D. H. (2021). Short physical performance battery as a crosswalk between frailty phenotype and deficit accumulation frailty index. The Journal of Gerontology. Series A, Biological Sciences and Medical Sciences, 76(12), 2249–2255. https://doi.org/10.1093/GERONA/GLAB087

Kang, H. G., Costa, M. D., Priplata, A. A., Starobinets, O. V., Goldberger, A. L., Peng, C. K., Kiely, D. K., Cupples, L. A., & Lipsitz, L. A. (2009a). Frailty and the degradation of complex balance dynamics during a dual-task protocol. The Journals of Gerontology. Series A, Biological Sciences and Medical Sciences, 64(12), 1304–1311. https://doi.org/10.1093/GERONA/GLP113

Kell, R. T., Bell, G., & Quinney, A. (2001). Musculoskeletal fitness, health outcomes and quality of life. Sports Medicine, 31(12), 863–873. https://doi.org/10.2165/00007256-200131120-00003

Kim, H., Suzuki, T., Kim, M., Kojima, N., Ota, N., Shimotoyodome, A., Hase, T., Hosoi, E., & Yoshida, H. (2015). Effects of exercise and milk fat globule membrane (MFGM) supplementation on body composition, physical function, and hematological parameters in community-dwelling frail Japanese women: A randomized double blind, placebo-controlled, follow-up trial. PLoS ONE, 10(2). https://doi.org/10.1371/JOURNAL.PONE.0116256

Kim, M., & Won, C. W. (2022). Cut points of chair stand test for poor physical function and its association with adverse health outcomes in community-dwelling older adults: A cross-sectional and longitudinal study. Journal of the American Medical Directors Association, 23(8), 1375–1382.e3. https://doi.org/10.1016/J.JAMDA.2021.11.007

Kodama, S., Saito, K., Tanaka, S., Maki, M., Yachi, Y., Asumi, M., Sugawara, A., Totsuka, K., Shimano, H., Ohashi, Y., Yamada, N., & Sone, H. (2009). Cardiorespiratory fitness as a quantitative predictor of all-cause mortality and cardiovascular events in healthy men and women: A meta-analysis. JAMA, 301(19), 2024–2035. https://doi.org/10.1001/JAMA.2009.681

Kojima, G., Taniguchi, Y., Iliffe, S., Jivraj, S., & Walters, K. (2019). Transitions between frailty states among community-dwelling older people: A systematic review and meta-analysis. In Ageing Research Reviews (Vol. 50, pp. 81–88). Elsevier Ireland Ltd. https://doi.org/10.1016/j.arr.2019.01.010

Lee, J. (2020). Associations between handgrip strength and disease-specific mortality including cancer, cardiovascular, and respiratory diseases in older adults: A meta-analysis. Journal of Aging and Physical Activity, 28(2), 320–331. https://doi.org/10.1123/JAPA.2018-0348

Li, C. M., Chen, C. Y., Li, C. Y., Wang, W. D., & Wu, S. C. (2010). The effectiveness of a comprehensive geriatric assessment intervention program for frailty in community-dwelling older people: A randomized, controlled trial. Archives of Gerontology and Geriatrics, 50, S39–S42. https://doi.org/10.1016/S0167-4943(10)70011-X

Liao, Y. Y., Chen, I. H., & Wang, R. Y. (2019). Effects of Kinect-based exergaming on frailty status and physical performance in prefrail and frail elderly: A randomized controlled trial. Scientific Reports, 9(1). https://doi.org/10.1038/S41598-019-45767-Y

Lopez, P., Pinto, R. S., Radaelli, R., Rech, A., Grazioli, R., Izquierdo, M., & Cadore, E. L. (2018). Benefits of resistance training in physically frail elderly: A systematic review. Aging Clinical and Experimental Research, 30(8), 889–899. https://doi.org/10.1007/S40520-017-0863-Z

Losa-Reyna, J., Baltasar-Fernandez, I., Alcazar, J., Navarro-Cruz, R., Garcia-Garcia, F. J., Alegre, L. M., & Alfaro-Acha, A. (2019). Effect of a short multicomponent exercise intervention focused on muscle power in frail and pre frail elderly: A pilot trial. Experimental Gerontology, 115, 114–121. https://doi.org/10.1016/J.EXGER.2018.11.022

Lourenço, R. A., Pérez-zepeda, M., Gutiérrez-robledo, L., García-garcía, F. J., & Rodríguez mañas, L. (2015). Performance of the European Working Group on Sarcopenia in Older People algorithm in screening older adults for muscle mass assessment. Age and Ageing, 44(2), 334–338. https://doi.org/10.1093/AGEING/AFU192

Luger, E., Dorner, T. E., Haider, S., Kapan, A., Lackinger, C., & Schindler, K. (2016). Effects of a home-based and volunteer-administered physical training, nutritional, and social support program on malnutrition and frailty in older persons: A randomized controlled trial. Journal of the American Medical Directors Association, 17(7). https://doi.org/10.1016/J.JAMDA.2016.04.018. https://doi.org/10.1016/J.JAMDA.2016.04.018

Mancini, D. M., Eisen, H., Kussmaul, W., Mull, R., Edmonds, L. H., & Wilson, J. R. (1991). Value of peak exercise oxygen consumption for optimal timing of cardiac transplantation in ambulatory patients with heart failure. Circulation, 83(3), 778–786. https://doi.org/10.1161/01.CIR.83.3.778

Marques, L. T., Rodrigues, N. C., Angeluni, E. O., dos Santos Pessanha, F., da Cruz Alves, P. A., Freire, N. M., Ferriolli, R. C., & de Abreu, E. (2019). Balance evaluation of prefrail and frail community-dwelling older adults. Journal of Geriatric Physical Therapy, 42(3), 176–182. https://doi.org/10.1519/JPT.0000000000000147

Martínez-Velilla, N., Casa-Herrero, A., Zambom-Ferraresi, F. et al. (2019). Effect of exercise intervention on functional decline in very elderly patients during acute hospitalization: A randomized clinical trial. JAMA Internal Medicine, 179(1), 28–36.

Mathias, S., Nayak, U. S., & Isaacs, B. (n.d.). Balance in elderly patients: The "get-up and go" test. Retrieved February 9, 2023, from https://pubmed.ncbi.nlm.nih.gov/3487300/

McAdams-DeMarco, M. A., Chu, N. M., & Segev, D. L. (2019). Frailty and long-term post-kidney transplant outcomes. Current Transplantation Reports, 6, 45–51.

McMillan, J. M., Gill, M. J., Power, C., Fujiwara, E., Hogan, D. B., & Rubin, L. H. (2020). Comorbidities in older persons with controlled HIV infection: Correlations with frailty index subtypes. AIDS Patient Care and STDs, 34(7), 284–294. https://doi.org/10.1089/apc.2020.0051

Mehta, P., Lemon, G., Hight, L. et al. (2021). A systematic review of clinical practice guidelines for identification and management of frailty. Journal of Nutrition, Health and Aging, 25(3), 382–391.

Meng, L., Shi, H., Shi, J., Yu, P.-L., & Xi, H. (2019). Differences in clinical characteristics, muscle mass, and physical performance among different frailty levels in Chinese older men. Chinese Medical Journal, 132(3), 352–355. https://doi.org/10.1097/CM9.0000000000000035

Mijnarends, D. M., Meijers, J., Halfens, M., ter Borg, R. J. G., Luiking, S., Verlaan, Y. C., Schoberer, S., Cruz Jentoft, D., van Loon, A. J., & Schols, J. C. (2013). Validity and reliability of tools to measure muscle mass, strength, and physical performance in community-dwelling older people: A systematic review. Journal of the American Medical Directors Association, 14(3), 170–178. https://doi.org/10.1016/j.jamda.2012.10.009

Millor, N., Lecumberri, P., Gómez, M., Martínez-Ramírez, A., & Izquierdo, M. (2013). An evaluation of the 30-s chair stand test in older adults: Frailty detection based on kinematic parameters from a single inertial unit. Journal of Neuroengineering and Rehabilitation, 10(1). https://doi.org/10.1186/1743-0003-10-86

Nagai, K., Miyamato, T., Okamae, A., Tamaki, A., Fujioka, H., Wada, Y., Uchiyama, Y., Shinmura, K., & Domen, K. (2018). Physical activity combined with resistance training reduces symptoms of frailty in older adults: A randomized controlled trial. Archives of Gerontology and Geriatrics, 76, 41–47. https://doi.org/10.1016/j.archger.2018.02.005

Navarrete-Villanueva, D., Gómez-Cabello, A., Marín-Puyalto, J., Moreno, L. A., Vicente-Rodríguez, G., & Casajús, J. A. (2021). Frailty and physical fitness in elderly people: A systematic review and meta-analysis. Sports Medicine, 51(1), 143–160). https://doi.org/10.1007/s40279-020-01361-1

Negm, A. M., Kennedy, C. C., Thabane, L., Veroniki, A. A., Adachi, J. D., Richardson, J., Cameron, I. D., Giangregorio, A., Petropoulou, M., Alsaad, S. M., Alzahrani, J., Maaz, M., Ahmed, M. M., Kim, E., Tehfe, H., Dima, R., Sabanayagam, K., Hewston, P., Abu Alrob, H., & Papaioannou, A. (2019). Management of frailty: A systematic review and network meta-analysis of randomized controlled trials. Journal of the American Medical Directors Association, 20(10), 1190–1198. https://doi.org/10.1016/J.JAMDA.2019.08.009

Ng, T. P., Feng, L., Nyunt, M. S. Z., Feng, L., Niti, M., Tan, B. Y., Chan, G., Khoo, S. A., Chan, S. M., Yap, P., & Yap, K. B. (2015). Nutritional, physical, cognitive, and combination interventions and frailty reversal among older adults: A randomized controlled trial. The American Journal of Medicine, 128(11), 1225–1236.e1. https://doi.org/10.1016/J.AMJMED.2015.06.017

Niccoli, S., Kolobov, A., Bon, T. et al. (2017). Whey protein supplementation improves rehabilitation outcomes in hospitalized geriatric patients: A double blinded, randomized controlled trail. Journal of Nutrition in Gerontology and Geriatrics, 36(4), 149–165.

Niemeläinen, S., Huhtala, H., & Andersen, J. et al. (2021). The clinical frailty scale is a useful tool for predicting postoperative complications following elective colon cancer surgery at the age of 80 years and above: A prospective, multicentre observational study. Colorectal Disease, 23(7), 1824–1836.

Nishijima, T. F., Shimokawa, M., Esaki, T., Morita, M., Toh, Y., & Muss, H. B. (2021). A 10-item frailty index based on a comprehensive geriatric assessment (FI-CGA-10) in older adults with cancer: Development and construct validation. The Oncologist, 26(10), e1751–e1760. https://doi.org/10.1002/ONCO.13894

Oliveira, J. S., Pinheiro, M. B., Fairhall, N., Walsh, S., Franks, T. C., Kwok, W., Bauman, A., & Sherrington, C. (2020). Evidence on physical activity and the prevention of frailty and sarcopenia among older people: A systematic review to inform the world health organization physical activity guidelines. Journal of Physical Activity and Health, 17(12), 1247–1258. https://doi.org/10.1123/jpah.2020-0323

Oviedo-Briones, M., Laso, Á. R., Carnicero, J. A., Cesari, M., Grodzicki, T., Gryglewska, B., Sinclair, A., Landi, F., Vellas, B., Checa-López, M., & Rodriguez-Mañas, L. (2020). A comparison of frailty assessment instruments in different clinical and social care settings: The Frailtools project. Journal of the American Medical Directors Association, 22(3). https://doi.org/10.1016/j.jamda.2020.09.024. https://doi.org/10.1016/j.jamda.2020.09.024

Oviedo-Briones, M., Rodríguez-Laso, Á, Carnicero, J. A., Gryglewska, B., Sinclair, A. J., Landi, F., Vellas, B., Rodríguez Artalejo, F., Checa-López, M., & Rodriguez-Mañas, L. (2022). The ability of eight frailty instruments to identify adverse outcomes across different settings: The FRAILTOOLS project. Journal of Cachexia, Sarcopenia and Muscle, 13(3), 1487–1501. https://doi.org/10.1002/jcsm.12990

Pamoukdjian, F., Laurent, M., Martinez-Tapia, C., Rolland, Y., Paillaud, E., & Canoui-Poitrine, F. (2020). Frailty parameters, morbidity and mortality in older adults with cancer: A structural equation modelling approach based on the Fried phenotype. Journal of Clinical Medicine, 9(6), 1826. https://doi.org/10.3390/jcm9061826

Park, Y., Choi, J. E., & Hwang, H. S. (2018). Protein supplementation improves muscle mass and physical performance in undernourished prefrail and frail elderly subjects: A randomized, double-blind, placebo-controlled trial. The American Journal of Clinical Nutrition, 108(5), 1026–1033. https://doi.org/10.1093/AJCN/NQY214

Parmar, K., Law, J., Carter, B. et al. (2021). Frailty in older patients undergoing emergency laparotomy: Results from the UK observational emergency laparotomy and frailty (ELF) study. Annals of Surgery, 273, 709–718.

Patel, H. P., Clift, E., Lewis, L., & Cooper, C. (2017). Epidemiology of sarcopenia and frailty. In: Dionyssiotis Y, editor. Frailty and Sarcopenia—Onset, Development and Clinical Challenges. InTech. https://doi.org/10.5772/intechopen.69771

Pérez-Rodríguez, R., Guevara-Guevara, T., Moreno-Sánchez, P. A., Villalba-Mora, E., Valdés-Aragonés, M., Oviedo-Briones, M., Carnicero, J. A., & Rodríguez-Mañas, L. (2020). Monitoring and intervention technologies to manage diabetic older persons: The CAPACITY case—A pilot study. Frontiers in Endocrinology, 11, 300. https://doi.org/10.3389/FENDO.2020.00300

Podsiadlo, D., & Richardson, S. (1991). The timed "Up & Go": A test of basic functional mobility for frail elderly persons. Journal of the American Geriatrics Society, 39(2), 142–148. https://doi.org/10.1111/j.1532-5415.1991.tb01616.x

Prince, M. J., Acosta, D., Guerra, M., Huang, Y., Jacob, K. S., Jimenez-Velazquez, I. Z., Jotheeswaran, A. T., Llibre Rodriguez, J. J., Salas, A., Sosa, A. L., Acosta, I., Mayston, R., Liu, Z., Llibre-Guerra, J. J., Prina, A. M., & Valhuerdi, A. (2021). Intrinsic capacity and its associations with incident dependence and mortality in 10/66 Dementia Research Group studies in Latin America, India, and China: A population-based cohort study. PLOS Medicine, 18(9), e1003097. https://doi.org/10.1371/JOURNAL.PMED.1003097

Ramírez-Martín, R., Pérez-Rodríguez, P., & Menéndez-Colino, R. et al. (2022). Prehabilitation and perioperative geriatric care in patients aged over 80 years with colorectal cancer: Results of a cross-speciality geriatrics program. Journal of Geriatric Oncology, 13(6), 813–820.

Ramírez-Vélez, R., Correa-Bautista, J. E., García-Hermoso, A., Cano, C. A., & Izquierdo, M. (2019). Reference values for handgrip strength and their association with intrinsic capacity domains among older adults. Journal of Cachexia, Sarcopenia and Muscle, 10(2), 278–286. https://doi.org/10.1002/JCSM.12373

Ramírez-Vélez, R., López Sáez De Asteasu, M., Morley, J. E., Cano-Gutierrez, C. A., & Izquierdo, M. (2021). Performance of the short physical performance battery in identifying the frailty phenotype and predicting geriatric syndromes in community-dwelling elderly. The Journal of Nutrition, Health & Aging, 25(2), 209–217. https://doi.org/10.1007/S12603-020-1484-3

Rantanen, T., Guralnik, J. M., Foley, D., Masaki, K., Leveille, S., Curb, J. D., & White, L. (1999). Midlife hand grip strength as a predictor of old age disability. JAMA, 281(6), 558–560. https://doi.org/10.1001/JAMA.281.6.558

Roberts, H. C., Denison, H. J., Martin, H. J., Patel, H. P., Syddall, H., Cooper, C., & Sayer, A. A. (2011). A review of the measurement of grip strength in clinical and epidemiological studies: Towards a standardised approach. Age and Ageing, 40(4), 423–429. https://doi.org/10.1093/ageing/afr051

Rockwood, K., & Mitnitski, A. (2007). Frailty in relation to the accumulation of deficits. Journals of Gerontology – Series A Biological Sciences and Medical Sciences, 62(7), 722–727. https://doi.org/10.1093/gerona/62.7.722

Rockwood, K., Song, X., MacKnight, C., Bergman, H., Hogan, D. B., McDowell, I., & Mitnitski, A. (2005). A global clinical measure of fitness and frailty in elderly people. CMAJ, 173(5), 489–495. https://doi.org/10.1503/cmaj.050051

Rodríguez-Couso, M. (2020). Geriatric interventions in the older with cancer based on comprehensive geriatric assessment. Optimization areas: what the geriatrician can contribute to the multidisciplinar team. Nutrición Hospitalaria, 34, 38–47.

Rodríguez-Laso, Á, García-García, F. J., & Rodríguez-Mañas, L. (2022). Transitions between frailty states and its predictors in a cohort of community-dwelling Spaniards. Journal of the American Medical Directors Association, 23(3), 524.e1–524.e11. https://doi.org/10.1016/J.JAMDA.2021.07.008

Rodríguez-Laso, Á, García-García, F. J., & Rodríguez-Mañas, L. (2023). Predictors of maintained transitions between robustness and prefrailty in community-dwelling older Spaniards. Journal of the American Medical Directors Association, 24(1), 57–64. https://doi.org/10.1016/J.JAMDA.2022.10.010

Rodríguez-Mañas, L., Costa-Grille, A. M., Molina-Hermosilla, I. et al. (2024). Effectiveness of a technological platform to manage frailty: results from the POSITIVE and INTEGRACAM studies. Congress of the Gerontological Society of America.

Romero-Ortuno, R., Walsh, C. D., Lawlor, B. A., & Kenny, R. A. (2010). A frailty instrument for primary care: Findings from the survey of health, ageing and retirement in Europe (SHARE). BMC Geriatrics, 10. https://doi.org/10.1186/1471-2318-10-57

Roshanravan, B., Khatri, M., Robinson-Cohen, C., Levin, G., Patel, K. V., de Boer, I. H. et al. (2012). A prospective study of frailty in nephrology-referred patients with CKD. American Journal of Kidney Diseases, 60, 912–21.

Ross, R. M., Murthy, J. N., Wollak, I. D., & Jackson, A. S. (2010). The six minute walk test accurately estimates mean peak oxygen uptake. BMC Pulmonary Medicine, 10(1), 1–9. https://doi.org/10.1186/1471-2466-10-31

Sabharwal, S., & Wilson, H. (2015). Orthogeriatrics in the management of frail older patients with a fragility fracture. Osteoporosis International, 26(10), 2387–2399.

Sadjapong, U., Yodkeeree, S., Sungkarat, S., & Siviroj, P. (2020). Multicomponent exercise program reduces frailty and inflammatory biomarkers and improves physical performance in community-dwelling older adults: A randomized controlled trial. International Journal of Environmental Research and Public Health, 17(11), 3760. https://doi.org/10.3390/ijerph17113760

Sanchez-Sanchez, J. L., Carnicero-Carreño, J. A., Garcia-Garcia, F. J., Álvarez-Bustos, A., Rodríguez-Sánchez, B., & Rodríguez-Mañas, L. (2022). Physical performance measures in frailty screening: Diagnostic and prognostic accuracy in the Toledo Study of Healthy Ageing. Maturitas, 165, 18–25. https://doi.org/10.1016/J.MATURITAS.2022.07.004

Sanchis, J., Sastre, C., Ruescas, A., Ruiz, V., Valero, E., Bonanad, C., García-Blas, S., Fernández-Cisnal, A., González, J., Miñana, G., & Núñez, J. (2021). Randomized comparison of exercise intervention versus usual care in older adult patients with frailty after acute myocardial infarction. The American Journal of Medicine, 134(3), 383–390.e2. https://doi.org/10.1016/j.amjmed.2020.09.019

Sayer, A. A., & Kirkwood, T. B. L. (2015). Grip strength and mortality: A biomarker of ageing? Lancet (London, England), 386(9990), 226–227. https://doi.org/10.1016/S0140-6736(14)62349-7

Sayer, A. A., Syddall, H. E., Martin, H. J., Dennison, E. M., Roberts, H. C., & Cooper, C. (2006). Is grip strength associated with health-related quality of life? Findings from the Hertfordshire Cohort Study. Age and Ageing, 35(4), 409–415. https://doi.org/10.1093/AGEING/AFL024

Serra-Prat, M., Sist, X., Domenich, R., Jurado, L., Saiz, A., Roces, A., Palomera, E., Tarradelles, M., & Papiol, M. (2017). Effectiveness of an intervention to prevent frailty in pre-frail community-dwelling older people consulting in primary care: A randomised controlled trial. Age and Ageing. https://doi.org/10.1093/ageing/afw242

Siriwardhana, D. D., Hardoon, S., Rait, G., Weerasinghe, M. C., & Walters, K. R. (2018). Prevalence of frailty and prefrailty among community-dwelling older adults in low-income and middle-income countries: A systematic review and meta-analysis. BMJ Open, 8(3). https://doi.org/10.1136/bmjopen-2017-018195

Steell, L., Ho, F. K., Sillars, A., Petermann-Rocha, F., Li, H., Lyall, D. M., Iliodromiti, S., Welsh, P., Anderson, J., MacKay, D. F., Pell, J. P., Sattar, N., Gill, J. M. R., Gray, S. R., & Celis-Morales, C. A. (2019). Dose-response associations of cardiorespiratory fitness with all-cause mortality and incidence and mortality of cancer and cardiovascular and respiratory diseases: The UK Biobank cohort study. British Journal of Sports Medicine, 53(21), 1371–1378. https://doi.org/10.1136/BJSPORTS-2018-099093

Studenski, S., Perera, S., Wallace, D., Chandler, J. M., Duncan, P. W., Rooney, E., Fox, M., & Guralnik, J. M. (2003). Physical performance measures in the clinical setting. Journal of the American Geriatrics Society, 51(3), 314–322. https://doi.org/10.1046/j.1532-5415.2003.51104.x

Tan, R. S., Goh, E. F., Wang, D., Chan, R. C. L., Zeng, Z., Yeo, A., Pek, K., Kua, J., Wong, W. C., Shen, Z., & Lim, W. S. (2022). Effectiveness and usability of the system for assessment and intervention of frailty for community-dwelling pre-frail older adults: A pilot study. Frontiers in Medicine, 9. https://doi.org/10.3389/fmed.2022.955785

Tarazona-Santabalbina, F. J., Gómez-Cabrera, M. C., Pérez-Ros, P., Martínez-Arnau, F. M., Cabo, H., Tsaparas, K., Salvador-Pascual, A., Rodriguez-Mañas, L., & Viña, J. (2016). A multicomponent exercise intervention that reverses frailty and improves cognition, emotion, and social networking in the community-dwelling frail elderly: A randomized clinical trial. Journal of the American Medical Directors Association, 17(5), 426–433. https://doi.org/10.1016/j.jamda.2016.01.019

Tay, J., Goss, A. M., Locher, J. L., Ard, J. D., & Gower, B. A. (2019). Physical function and strength in relation to inflammation in older adults with obesity and increased cardiometabolic risk. The Journal of Nutrition, Health & Aging, 23(10), 949–957. https://doi.org/10.1007/S12603-019-1260-4

Tay, L. B., Chua, M. P., Tay, E. L., Chan, H. N., Mah, S. M., Latib, A., Wong, C. Q., & Ng, Y. S. (2019). Multidomain geriatric screen and physical fitness assessment identify prefrailty/frailty and potentially modifiable risk factors in community-dwelling older adults. Annals of the Academy of Medicine, Singapore, 48(6), 171–180.

Theodorakis, N., Nikolaou, M., Hitas, C. et al. (2024). Comprehensive peri-operative risk assessment and management of geriatric patients. Diagnostics (Basel), 14(19), 2153.

Theou, O., Jones, G. R., Jakobi, J. M., Mitnitski, A., & Vandervoort, A. A. (2011). A comparison of the relationship of 14 performance-based measures with frailty in older women. Applied Physiology, Nutrition, and Metabolism, 36(6), 928–938. https://doi.org/10.1139/h11-116

Travers, J., Romero-Ortuno, R., Bailey, J., & Cooney, M. T. (2019). Delaying and reversing frailty: A systematic review of primary care interventions. The British Journal of General Practice: The Journal of the Royal College of General Practitioners, 69(678), E61–E69. https://doi.org/10.3399/BJGP18X700241

Valdés-Aragonés, M., Pérez-Rodríguez, R., Carnicero, J. A. et al. (2024). Effect of monitoring frailty through a mobile application and a sensor kit to prevent functional decline in frail and prefrail older people: The FACET pilot randomized control trial. Journal of Medical Internet Research, 2024; 26,:e58312. https://doi.org/10.2196/58312

Van Roie, E., Verschueren, S. M., Boonen, S., Bogaerts, A., Kennis, E., Coudyzer, W., & Delecluse, C. (2011). Force-velocity characteristics of the knee extensors: An indication of the risk for physical frailty in elderly women. Archives of Physical Medicine and Rehabilitation, 92(11), 1827–1832. https://doi.org/10.1016/j.apmr.2011.05.022

Vieira, E. R., da Silva, R. A., Severi, M. T., Barbosa, A. C., Amick, B. C., Zevallos, J. C., Martinez, I. L., & Chaves, P. H. M. (2018). Balance and gait of frail, pre-frail, and robust older hispanics. Geriatrics (Basel, Switzerland), 3(3). https://doi.org/10.3390/GERIATRICS3030042

Woo, J., Yu, R., & Leung, J. (2018). Predictive ability of individual items of the cardiovascular health study (CHS) scale compared with the summative score. Journal of the American Medical Directors Association, 19(5), 444–449. https://doi.org/10.1016/J.JAMDA.2017.11.006

3 Biomarkers of Frailty

Mariam El Assar de la Fuente, Isabel Rodríguez-Sánchez, Alejandro Álvarez-Bustos, and Leocadio Rodríguez-Mañas

INTRODUCTION

Frailty is a biological syndrome that has provided an excellent opportunity to elucidate numerous phenomena underlying the common functional decline observed in aging individuals. One of its most important contributions is to identify people at risk of developing serious and life-threatening adverse events such as death, disability, hospitalization, and institutionalization, highlighting its potential spontaneous reversibility and offering an opportunity to stop and potentially reverse the process of the so-called non-catastrophic disability. Additionally, several well-designed large-scale clinical trials conducted in recent years have demonstrated the potential reversibility of this condition through therapeutic interventions primarily focused on physical exercise programs, nutritional interventions, and readjusting goal settings in common diseases such as diabetes and hypertension (Bernabei et al., 2022; Rodriguez-Mañas et al., 2019; Trombetti et al., 2018), albeit with some controversial findings, highlighting the need to identify frail people in order to avoid its adverse outcomes.

A common challenge in frailty research revolves around the methods for detecting and monitoring the condition over time. This challenge stems primarily from the scarcity of tools capable of detecting the syndrome with adequate sensitivity, specificity, and predictive value. Furthermore, given that the concept of frailty was developed and validated primarily in community-dwelling settings, numerous instruments have been created to detect and assess frailty, often without validating their prognostic and diagnostic accuracy (Faller et al., 2019). Furthermore, the concordance between these instruments is very poor, as well as disparities in their ability to predict various adverse events associated with frailty. This suggests the existence of multiple subtypes of frailty, each potentially captured differently by different instruments, with varying consequences for patients (Checa-Lopez et al., 2023; Oviedo-Briones et al., 2021; Oviedo-Briones et al., 2022; Vergara et al., 2019).

Finally, the prevalence, characteristics, phenotypes, and outcomes of frail older adults in healthcare settings differ from those in the community settings, underscoring the significance of the aforementioned issues (Oviedo-Briones et al., 2021; Oviedo-Briones et al., 2022). Indeed, a common challenge in integrating frailty into clinical practice is the uncertainty surrounding the suitability of existing tools for detection and ongoing monitoring in these settings, where diagnostic accuracy is imperative (Rodriguez-Mañas et al., 2017; Rodriguez-Mañas and Fried, 2015).

Taken as a whole, the quest for biomarkers to enhance the accurate diagnosis and monitoring of frailty, a highly dynamic condition, has become a pressing concern to be addressed. Consequently, it should be integrated into the research priorities within the field (Rodríguez-Mañas and Rodriguez-Sánchez, 2021). Despite several recent efforts, our understanding of frailty biomarkers, their performance, and clinical utility remains unsatisfactory. As a result, they are not commonly used and this topic is still in the research area (LeBrasseur et al., 2021; Rodriguez-Mañas et al., 2020).

For the scope of this review, we categorized biomarkers into three groups: laboratory, imaging, and clinical biomarkers. However, future research should explore the effectiveness of combining these biomarkers for enhanced diagnosis and monitoring of frailty.

 DOI: 10.1201/9781003398295-3

LABORATORY BIOMARKERS

Inflammatory Biomarkers

Studies show that cytokine network dysregulation leading to chronic low-grade inflammation plays a crucial role in aging and age-related conditions like frailty (Table 3.1), acting as a central mediator (Xu et al., 2022; L. Zhang et al., 2023). Pro-inflammatory cytokines may directly impact frailty by promoting protein degradation or indirectly by disrupting critical metabolic pathways (Soysal et al., 2016).

Several cross-sectional studies consistently demonstrated a correlation between frailty status and elevated levels of various circulating inflammatory markers (Tran Van Hoi et al., 2023) including IL-6, C-reactive protein (CRP) (Langmann et al., 2017; Saum et al., 2015), TNF-α (Kane and Sinclair, 2019), TNF-α soluble receptors (Langmann et al., 2017), CXC chemokine ligand CXCL-10 (Liu et al., 2022), CXCL-13 (Landino et al., 2021), IL-8 (Hsu et al., 2019), the vascular inflammatory intracellular soluble adhesion molecule (sICAM1) (Lee et al., 2016), and homocysteine (Álvarez-Sánchez et al., 2020) among others. Additionally, Zhang and colleagues recently reported an association between two novel systemic inflammatory biomarkers (Systemic Immune-Inflammation Index and Systemic Inflammatory Response Index) and frailty in the population aged 40 to 80 years (Zhang et al., 2024). These findings suggest that inflammatory biomarkers hold promise as valuable indicators for screening and monitoring of frailty, as previously established (Al Saedi et al., 2019). Moreover, another recent study highlighted distinct biomarker patterns related to inflammation and muscle catabolism in frail, multimorbid older adults, emphasizing that analyzing these patterns provides more valuable insights into frailty than examining individual biomarkers (Kochlik et al., 2023). Marzetti et al. found a unique inflammatory signature in older subjects with physical frailty and sarcopenia, marked by higher levels of CRP, P-selectin, and interferon-induced protein 10, while non-frail controls exhibited higher levels of myeloperoxidase, IL-8, monocyte chemoattractant protein 1 (MCP-1), and platelet-derived growth factor (PDGF) BB (Marzetti et al., 2019).

However, few longitudinal studies have established causality or have made them useful for other purposes like the monitoring of frailty. Nevertheless, some findings have been reported in longitudinal studies. For example, the association between IL-6 and IL-8 with frailty did not persist in a longitudinal study with a 5-year follow-up period (Hsu et al., 2019). In another study of healthy older people, IL-6 concentrations increased while grip strength and gait speed declined over 9 years (Sanders et al., 2014). A recent study showed a longitudinal association of higher CRP and fibrinogen with worsening frailty determined by the frailty index but not by the frailty phenotype (Welstead et al., 2020). Samson and colleagues reported that persistent high levels of IL-6 pathway markers, such as CRP and soluble IL-6 receptor, were related to increased frailty and reduced physical strength in aging individuals over a period of 20 years. In women, a strong association between increasing sCD14 levels and frailty was observed, suggesting a potential involvement of monocyte overactivation (Samson et al., 2022). Furthermore, results from the InCHIANTI Study revealed that two proteins, cyclin-dependent kinase 5 (CDK5/CDK5R1) and interleukin 1 alpha (IL1A), were associated with worsening of frailty status over time in older subjects free of frailty at baseline (Landino et al., 2021).

Table 3.1 summarizes the main above-mentioned biomarkers related to human frailty.

Oxidative Stress Biomarkers

Oxidative stress, a key factor in age-related mechanisms that lead to functional decline (El Assar et al., 2022), results from an imbalance between reactive oxygen species (ROS) production and antioxidant capacity. This imbalance leads to cellular damage, including oxidation of proteins, lipids, and DNA (Mankhong et al., 2020). In muscles, ROS are primarily generated by the mitochondrial electron transport chain, as well as by enzymes like nicotinamide adenine dinucleotide phosphate

TABLE 3.1
Circulating Inflammatory Biomarkers Related to Frailty

Inflammatory Biomarkers	Study Design	Sample Information	Frailty Assessment	Main Findings	Reference
SII, SIRI	Cross-sectional	6690 non-frail and 4544 frail	FI	• The odds ratio and 95% confidence interval for the correlation between frailty and the natural logarithm (ln) transformed lnSII and lnSIRI were: 1.38 (1.24–1.54) and 1.69 (1.53–1.88), respectively	Zhang et al, 2024
IL-6, CXCL-10, CRP	Cross-sectional	31 robust, 116 prefrail, and 118 frail	FRAIL scale	• Higher IL-6, CXCL-10, and CRP serum levels in the frail than in the robust and prefrail groups	Liu et al., 2022
sIL-6R, CRP, sCD14	Longitudinal	144 individuals aged 65–75 years at the study endpoint were followed for more than 20 years	FI	• Chronically elevated levels of IL-6 pathway markers (CRP and sIL-6R) were related to more frailty and reduced physical strength • A strong association between sCD14 concentrations and frailty was detected in women	Samson et al., 2022
CXCL-13, IL1A	Cross-sectional and longitudinal	752 men and women older than 65 years of age from the InCHIANTI study	FP	• CXCL-13 was associated with frailty status • CDK5/CDK5R1 and IL1A were related to worsening of frailty status over time in volunteers free of frailty at baseline	Landino et al., 2021
Hcy, CRP	Cross-sectional	612 robust, 500 prefrail, and 99 frail	FP	• Hcy was independently associated with frailty • hsCRP was independently associated with both prefrailty and frailty	Álvarez-Sánchez et al., 2020
Fibrinogen, CRP	Cross-sectional and longitudinal	Participants of the Lothian Birth Cohort 1936, aged 70 at baseline and tested 4 times over 12 years	FP, FI	• A significant positive cross-sectional association was observed between fibrinogen and wave 1 FI score For FP: wave 1 prefrail and frail participants had higher CRP and fibrinogen than non-frail participants • Over a 12-year follow-up, higher wave 1 CRP and fibrinogen were significantly associated with increased FI change, whereas no significant association was detected for worsening frailty over follow-up	Welstead et al., 2020

IL-6, IL-8	Cross-sectional and longitudinal	394 robust, 426 prefrail, and 81 frail	FP, FI	• A significant association for IL-6 and IL-8 was cross-sectionally detected for frail versus robust, evaluated by the FP • IL-6 remained significantly associated in the cross-sectional age-adjusted and multivariable-adjusted model • In longitudinal unadjusted analysis, IL-8 predicted incident frailty at 8-year follow-up • IL-8 was not longitudinally associated with incident frailty after multivariable adjustment. Similar results were seen using the FI	Hsu et al., 2019
P-selectin, CRP, IFN γ-induced protein 10, IL-8, MCP-1, MIP 1-α, PDGF-BB	Crcss-sectional	100 community-dwellers aged 70 years and older with PF&S, 100 non-sarcopenic, non-frail controls	Operational definition adopted in the (SPRINTT) project	• The inflammatory profile of people with PF&S was defined by higher levels of P-selectin, CRP, and IFNγ-induced protein 10. • Non PF&S participants presented higher levels of MPO, IL-8, MCP-1, MIP1-α, PDGF-BB • Gender-specific inflammatory fingerprints of PF&S were also identified	Marzetti et al., 2019
HsCRP, TNF-α, TNF-αR1 and 2, IL-6, sIL6R, IL-10	Cross-sectional	178 women ≥ age 65 with osteoporosis in long-term care. 11 robust, 53 prefrail, and 114 frail	Modified version of FP	• Frail participants had significantly higher levels of HsCRP, TNF-αR1, TNF-αR2, IL-6, and IL-6-sR than those non-frail, and higher levels of HsCRP and IL-6 compared to prefrail participants	Langmann et al., 2017
IL-6, sICAM-1	Cross-sectional	330 robust, 531 prefrail, and 85 frail	FP	• The log ICAM-1 and log IL-6 mean and median values increased stepwise across the frailty categories • A positive association was observed between IL-6, sICAM-1, and frailty	Lee et al., 2016
CRP	Cross-sectional	845 robust, 1463 prefrail, and 210 frail	FP	• Higher levels of CRP were detected in the frail group vs the robust and prefrail groups • A positive association between CRP and frailty	Saum et al., 2015

CRP: C-reactive protein; CXCL-10: CXC chemokine ligand 10; FI: Frailty Index; FP: Frailty Phenotype; IFN γ: interferon γ; hsCRP: high-sensitivity C-reactive protein; Hcy: homocysteine; IL: interleukin; MCP-1: monocyte chemoattractant protein 1; MIP1-α: macrophage inflammatory protein 1 alpha; MPO: myeloperoxidase; sICAM1: soluble intracellular soluble adhesion molecule; SIL6R: soluble IL-6 receptor; SNP: single nucleotide polymorphisms; SPRINTT: Sarcopenia and Physical fRailty IN older people: multi-componenT Treatment strategies; PDGF-BB: platelet-derived growth factor-BB; PF&S: physical frailty and sarcopenia; SII: Systemic Immune-Inflammation Index; SIRI: Inflammatory Response Index; TNF-α: tumor necrosis factor alpha; TNF-αR: tumor necrosis factor alpha receptor.

(NADPH) oxidase and xanthine oxidase (H. Zhang et al., 2023). Elevated ROS triggers protein degradation by activating the ubiquitin-protease pathway, leading to skeletal muscle atrophy (Jang et al., 2020). The influence of oxidative stress extends beyond muscle quantity, also affecting quality through several mechanisms (Baumann et al., 2016).

Given its significant impact on frailty development, the “free radical theory of frailty” has been proposed (Viña, 2019). Indeed, oxidative stress is considered one of the age-related mechanisms that has shown the most significant differential expression between frail and robust subjects reinforcing its value as a biomarker for frailty (Álvarez-Satta et al., 2020). In line with this, older individuals showed a positive correlation between frailty and diacron reactive oxygen metabolites (d-ROM), which measure total oxidant capacity (Saum et al., 2015). The Framingham Offspring Study revealed an association between increased levels of 8-isoprostane, a prostaglandin-like compound formed as a biologically active metabolite through the non-enzymatic peroxidation of fatty acids, and frailty (Liu et al., 2016).

Another potential biomarker for oxidative stress in frailty is 8-hydroxy-2′-deoxyguanosine (8-OHdG), which indicates oxidative DNA damage. Notably, in patients with Alzheimer’s disease, frailty was linked to increased urine levels of 8-OHdG, along with higher plasma levels of d-ROM metabolites, 8-isoprostane, and IL-6 (Namioka et al., 2017). In addition to the aforementioned biomarkers, a systematic review by Soysal et al. revealed that frail individuals showed heightened levels of malondialdehyde (MDA), oxidized glutathione/glutathione, 4-hydroxy-2,3-nonenal, and protein carbonylation (Soysal et al., 2017). Furthermore, other studies have demonstrated that levels of MDA (indicative of lipid peroxidation damage) and protein carbonylation (indicative of protein oxidative damage) were associated with frailty rather than age in older participants of the Toledo Study of Healthy Aging (Inglés et al., 2014; Kochlik et al., 2019).

Lipoprotein-associated phospholipase A2 (Lp-PLA2) plays a crucial role in hydrolyzing oxidized phospholipids and is primarily associated with low-density lipoprotein (LDL). It promotes vascular inflammation and atherosclerotic plaque formation through oxidative modification of the vascular wall (Sepúlveda et al., 2022). In frail older individuals, elevated levels of Lp-PLA2, isoprostanes, and IL-6 were observed, and these markers were associated with reduced gait speed (Liu et al., 2016).

It is important to highlight that the antioxidant response plays a crucial role in ROS-induced signaling. In this sense, the nuclear factor erythroid 2-related factor 2 (Nrf2) stands out as a key regulator of cellular response to stress by activating the transcription of antioxidant genes (Kitaoka et al., 2019). Its downregulation appears to be implicated in sarcopenia and cardiovascular disease (El Assar et al., 2022). In a study involving Spanish older individuals, the frailty phenotype was correlated with reduced mRNA expression of Nrf2, along with some of its downstream targets with antioxidant activities, including thioredoxin reductase-1, superoxide dismutase-2 (SOD-2), and heme oxygenase-2 (HO-2) (El Assar et al., 2017).

In a recent study involving 139 community-dwelling older adults, a significantly lower native (NT) and total thiol (TT) levels, a proxy of redox control status, in the frail group were found. Additionally, frailty status, defined by Fried criteria, was correlated with different thiol/disulfide homeostasis parameters (Hafızoğlu et al., 2024). Saum et al also observed reduced concentration of TT in frail individuals (Saum et al., 2015), but they did not find differences in the biological antioxidant potential (BAP) capacity (Saum et al., 2015), while Namioka and colleagues found lower levels of BAP and endogenous plasma antioxidants in frail older Japanese individuals (Namioka et al., 2017).

In addition, older participants exhibited a significant correlation between frailty and decreased levels of circulating fat-soluble vitamins, including vitamin E (Ble et al., 2006; Henning et al., 2023; Kochlik et al., 2019). Furthermore, lower intake of vitamins B6, C, E and folates was linked to an increased risk of frailty (Balboa-Castillo et al., 2018). However, contrasting findings emerged in other studies, with some reporting no association between vitamin E and frailty, along

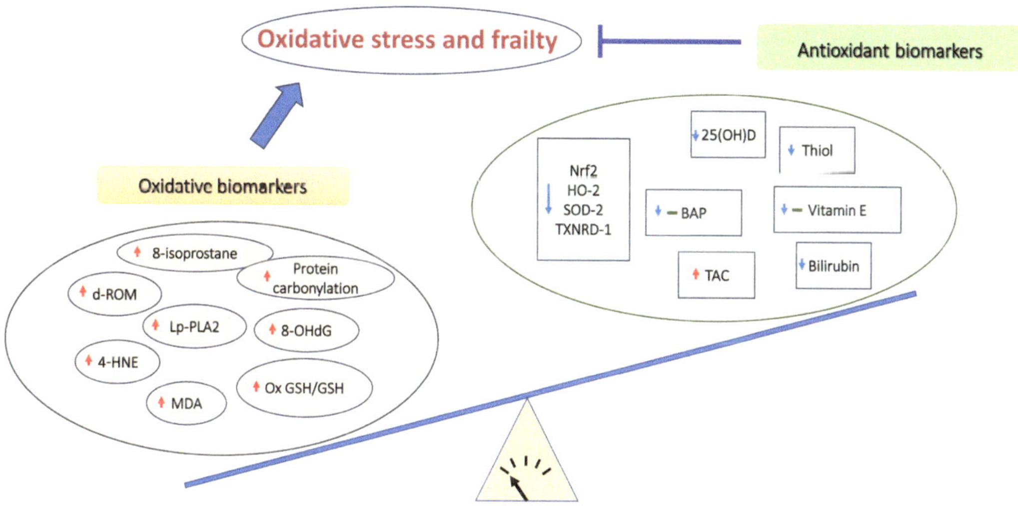

FIGURE 3.1 Key oxidative stress biomarkers associated with human frailty

(↑) Indicates increased levels/positive correlation with frailty; (↓) indicates decreased levels/negative association with frailty; (–) indicates no association with frailty.

BAP: biological antioxidant potential; d-ROM: Diacron reactive oxygen metabolites; GSH: glutathione; HO-2: heme oxygenase 2; 4-HNE: 4-hydroxy-2,3-nonenal; 8-OHdG: 8-hydroxy-2′-deoxyguanosine; 25(OH)D: 25-hydroxyvitamin D; Lp-PLA2: lipoprotein-associated phospholipase A2; MDA: malondialdehyde; Nrf2: nuclear factor erythroid-2 related factor 2; SOD-2: mitochondrial superoxide scavenger; and TXNRD1: thioredoxin reductase.

with inconsistent results for other antioxidants like vitamin A and carotenoids (Álvarez-Satta et al., 2020).

Numerous human studies consistently linked vitamin D deficiency with frailty, decreased muscle function, and increased risk of sarcopenia (Bollen et al., 2022; Gomez-Cabrero et al., 2021). Both cross-sectional (Alvarez-Ríos et al., 2015) and prospective cohort studies have shown a link between low levels of serum 25-hydroxyvitamin D (25(OH)D) and the risk or the incidence of frailty (Buta et al., 2017; Ju et al., 2018). However, the effectiveness of vitamin D supplementation in reducing frailty risk among older adults remains uncertain. A recent study reported that high dose vitamin D supplementation failed to prevent frailty in adults aged 70 and older (Cai et al., 2022).

Although there seems to be a distinct correlation between frailty, increased oxidative stress, and reduced levels of antioxidant markers (Figure 3.1), the precise direction of these associations remains unclear due to the cross-sectional design of existing studies. To achieve a more comprehensive understanding, additional prospective studies are warranted to validate these relationships and determine their predictive significance.

Mitochondrial Dysfunction Biomarkers

Growing evidence suggests that damaged and dysfunctional mitochondria are pivotal in sarcopenia development (Turkel et al., 2023), a key biological substrate of frailty. Mitochondrial dysfunction accelerates ROS accumulation leading to cellular energy depletion, particularly in skeletal muscle. This process can foster the emergence of a complex sarcopenic phenotype which contributes to functional mobility loss and frailty in older adults (Turkel et al., 2023).

In aging muscle, a correlation exists between mitochondrial changes, the downregulation of peroxisome proliferator-activated receptor-gamma coactivator 1-alpha (PGC-1α), impaired oxidative phosphorylation (OXPHOS), and changes in mitochondrial morphology (Fukunaga et al., 2021). These alterations are linked to sarcopenia, reduced physical performance, and chronic fatigue (El Assar et al., 2022).

PGC-1α is the master regulator of mitochondrial integrity, function, and biogenesis. Reduced PGC-1α expression has been observed in the muscles of aged animals and older individuals (Marzetti et al., 2013). PGC-1α regulates protein metabolism, primarily by mitigating muscle protein degradation through the ubiquitin-proteasome system (UPS) by inhibiting the activities of nuclear factor κB (NF-κB) and FoxO3 (Marzetti et al., 2013). PGC-1α seems to play a significant role in exercise physiology; in fact, the activation of its pathway through voluntary wheel running for two months or lifelong aerobic exercise has been shown to alleviate sarcopenia in aged animals (Ko and Ko, 2021; Liang et al., 2021). Given the involvement of impaired PGC-1α signaling in sarcopenia and age-related skeletal muscle changes, it is plausible to assert that PGC-1α is a potential factor in the development of the frailty phenotype (El Assar et al., 2020).

On the other hand, alterations in mitophagy, whether excessive or insufficient, contribute to muscle atrophy and decreased muscle strength in aged skeletal muscle (El Assar et al., 2022). Studies have shown a decline in various mitophagy regulators in the muscles of aging mice and humans, particularly in the context of sarcopenia (Yeo et al., 2019). Interestingly, autophagy/mitophagy gene expression markers, Beclin1, autophagy-related-7, BCL2/adenovirus E1B 19 kDa interacting protein 3 (BNIP3), dynamin-related protein 1, and Parkinson protein 2 (PARKIN) were found to be lower in inactive frail older women, correlating with decreased walking speed (Drummond et al., 2014), underscoring the crucial role of mitophagy in muscle weakness. Furthermore, research has shown that deleting parkin leads to reduced muscle mass and impaired physical performance in older mice, whereas overexpressing parkin improves skeletal muscle function (Liu et al., 2021) suggesting it as an attractive candidate for targeting mitochondrial dysfunction. Therefore, therapeutic approaches that enhance mitophagy could be crucial in delaying age-related muscle mass decline and improving healthspan.

Mitochondrial dysfunction triggers a retrograde signaling pathway from mitochondria to the nucleus, releasing myomitokine like growth differentiation factor-15 (GDF15), which orchestrates an adaptive response to mitochondrial stress (Romanello, 2020). Notably, GDF15 is a potential biomarker of mitochondrial dysfunction, the senescence-associated secretory phenotype (SASP), and inflammation and may also represent a promising frailty biomarker (Gonçalves et al., 2022). Higher plasma levels of GDF15 were detected in frail older participants but not their robust counterparts (Arauna et al., 2020), and elevated GDF15 values were observed in people with disease-related frailty as compared to those without frailty or with age-related frailty (Angioni et al., 2022). Importantly, circulating GDF15 was independently and negatively associated with relative 30 seconds sit-to-stand muscle power, suggesting its potential utility as a sensitive frailty biomarker in older people (Alcazar et al., 2021). Furthermore, Oba and colleagues have recently reported that higher serum GDF15 concentrations predicted frailty onset in older adults with cardiometabolic disease. Pointing at this biomarker as a potentially effective marker of frailty risk in interventions aimed at preventing frailty (Oba et al., 2024).

Table 3.2 summarizes the main above-mentioned biomarkers related to human frailty.

It is important to note that inflammation and mitochondrial dysfunction are intricately linked in a complex, bidirectional relationship, resulting in increased mitochondrial oxidative stress and perpetuating a vicious inflammatory cycle (López-Armada et al., 2013). In fact, mitochondrial dysfunction and systemic inflammation are considered principal factors in the development of sarcopenia. One proposed molecular link between these mechanisms involves extracellular vesicles trafficking as suggested by Picca et al. (2020).

While various biomarkers associated with mitochondria-dependent pathways show promise as indicators of frailty, further research is imperative to fully elucidate their role in functional decline.

TABLE 3.2
Description of Mitochondrial Dysfunction Biomarkers of Frailty

Biomarker	Analysis Type	Study Design	Sample Information	Frailty Assessment	Main Findings	Reference
Beclin1, autophagy-related-7, BNIP3, dynamin-related protein 1, PARKIN	Gene expression	Cross-sectional	Fourteen older female volunteers (>65 years)	Physical and performance measures for frailty identification	• The autophagy/mitophagy gene expression markers, Beclin1, autophagy-related-7, BNIP3, dynamin-related protein 1, and PARKIN, were lower in inactive frail older women vs active • Autophagy/mitophagy markers were positively correlated with the 6-minute walk and leg lean mass	Drummond et al., 2014
GDF15	Protein expression Protein expression Protein expression	Longitudinal (median 38–39 months) Cross-sectional Cross-sectional	275 outpatients (mean age: 77 ± 6 years) with cardiometabolic diseases 1199 subjects without frailty, 82 subjects with age-related frailty, and 53 with disease-related frailty 27 robust and 28 frail older than 64 years	mCHS KCL FP FP	• The highest GDF15 tertile group had a significantly higher incidence of frailty than the lowest GDF15 tertile group • Higher GDF15 values in subjects with disease-related frailty compared to subjects without frailty • Higher values in the disease-related frailty group vs the age-related frailty individuals • No difference between subjects without frailty and those with age-related frailty • Higher levels in frail vs robust subjects	Oba et al., 2024 Angioni et al., 2022 Arauna et al., 2020

BNIP3: BCL2/adenovirus E1B 19 kDa interacting protein 3; FP: Frailty Phenotype; GDF15: growth differentiation factor-15; KCL: Kihon Checklist; mCHS: modified version of the Cardiovascular Health Study; PARKIN: Parkinson protein 2.

Metabolic Biomarkers

Frailty syndrome is characterized by dysregulation of many metabolic pathways, primarily governed by nutrient sensing mechanisms (Perazza et al., 2022). Thus, there is a need to identify specific biomarkers linking these metabolic pathways to frailty and functional decline.

Among metabolic alterations related to frailty condition, insulin resistance seems to play a key role in the development of functional impairment (Angulo et al., 2016). Studies have indicated a cross-sectional association between frailty and insulin resistance, as assessed by the homeostasis model assessment method (HOMA-IR), in middle-aged and older community dwellers (Hwang et al., 2015). Moreover, recent research has unveiled a prospective relationship between HOMA-IR and the decline in physical function over 5 years follow-up, as assessed by both the incidence of frailty and the worsening of the Frailty Trait Scale (FTS) score (Rodríguez-Mañas et al., 2022), with fat mass playing a substantial role in this association (El Assar et al., 2023). Interestingly, the detrimental effects of insulin resistance on skeletal muscle may account for the decline in physical function associated with this condition (Cleasby et al., 2016). Moreover, age-related changes such as increased inflammation, oxidative stress, mitochondrial dysfunction, and intramyocellular lipid accumulation significantly impair skeletal muscle insulin signaling (Shou et al., 2020) which may promote functional decline related to aging.

With advancing age, the insulin/insulin growth factor (IGF-1) signaling pathway is impaired primarily due to insulin resistance and decreased levels of IGF-1. This leads to reduced protein synthesis and muscle growth mediated by the phosphatidylinositol-3-kinase (PI3K)/AKT/mammalian target of rapamycin (mTOR) pathway (Angulo et al., 2016). In addition, IGF-1 plays a pivotal role in regulating the synthesis of numerous transcription factors that influence gene expression associated with inflammation and cellular autophagy, both crucial mechanisms linked to frailty (Clegg and Hassan-Smith, 2018).

Previous research has established a relationship between IGF-1 and frailty. Reduced serum levels of IGF-1 were independently correlated with frailty in older adults (Doi et al., 2018). Additionally, frail individuals with low relative appendicular skeletal muscle mass (RASM) had lower levels of IGF-1 compared to their frail counterparts with normal RASM (Chew et al., 2019). An observational study reported that lower IGF-1 and higher IGF binding protein IGFBP1 levels predicted new-onset frailty among elderly men (Yeap et al., 2013).

While IGF-1 functions as a circulating hormone with anabolic effects on muscle, myostatin, a muscle derived protein belonging to the transforming growth factor β (TGF-β) superfamily, acts as a potent negative regulator of muscle mass and has been proposed as a biomarker of frailty (Chew et al., 2019). In line with this, higher levels of myostatin normalized for total body lean mass were associated with frailty and low muscle mass only in men (Chew et al., 2019). Another study explored the association between myostatin and frailty in older persons with chronic heart failure reporting higher levels of myostatin in frail participants when compared to those who were not frail (Wang et al., 2023). However, the association between myostatin and frailty remains inconclusive. For instance, findings from a randomized controlled trial (RCT) showed higher serum myostatin concentrations in non-frail individuals compared to frail long-term nursing home residents. Furthermore, subjects who underwent physical exercise intervention exhibited improved physical fitness, which was positively associated with increased myostatin levels in men (Arrieta et al., 2019).

In addition to the above-mentioned, other metabolic alterations have been proposed to significantly impact physical function in older adults. Sirtuins, a conserved family of nicotinamide adenine dinucleotide (NAD)-dependent protein deacetylases, play a key role in orchestrating adaptations to age-related metabolic changes (Angulo et al., 2016). While several authors have reported the involvement of sirtuins in frailty, the associations observed have been contradictory (Anwar et al., 2023). Kumar et al. observed that serum levels of SIRT1 and SIRT3 decrease with frailty (Kumar et al., 2014). Conversely, Ma et al. reported higher levels of SIRT1 in frail older individuals compared to robust participants and SIRT1 levels were associated with decreased physical function

(Ma et al., 2019). Additionally, it has been suggested that SIRT1 may be more closely linked to nutritional status and body composition than to aging and frailty (Razi et al., 2017). Consistent with these findings, recent evidence demonstrated a positive correlation between SIRT1 gene expression and improvements in nutritional status, irrespective of frailty status, in a Spanish cohort of older participants (El Assar et al., 2018). The potential association between sirtuins and frailty could be related to their important role in skeletal muscle health (Tonkin et al., 2012). In fact, sirtuins play a crucial role in preserving insulin sensitivity in skeletal muscle (Lantier et al., 2015), which seems to be related to their positive impact on mitochondrial function (Zhang et al., 2015). Taken together, this evidence suggests a potential role of sirtuin in the pathogenesis of frailty and underscores the importance of exploring potential related biomarkers or novel targets for interventions.

On the other hand, Klotho protein, an integral component of endocrine fibroblast growth factor (FGF) receptor complexes, is considered an aging suppressor protein that exhibits diverse functionalities encompassing energy metabolism, antioxidative and anti-inflammatory properties, among others (Kuro-O, 2018).

In recent years, there has been growing interest in exploring the involvement of Klotho in the onset of frailty. Nevertheless, the findings are inconsistent and inconclusive. For instance, while some studies found no cross-sectional association between alpha Klotho levels and frailty in older Turkish adults (Polat et al., 2020), others suggested that higher serum Klotho level was related to lower odds of physical frailty among middle-aged and older adults suggesting that Klotho might be a potential biomarker of frailty (Guan et al., 2023). Additionally, higher plasmatic Klotho concentrations were related to lower odds of incident frailty after 6 years of follow-up in older Italian individuals in the InCHIANTI Study (Shardell et al., 2019). A recent systematic review of clinical studies examining the relationship between Klotho, physical performance, and frailty in middle-aged and older adults reached the conclusion that Klotho exhibits a positive correlation with muscle strength and a negative correlation with frailty (Veronesi et al., 2021). A resume of the above-mentioned biomarkers is displayed in Table 3.3.

Endocrine Biomarkers

Insulin and its signaling pathway, along with insulin resistance, have been implicated in frailty as mentioned earlier, suggesting disruptions in endocrine regulation may contribute to this condition. Anabolic hormones are crucial for muscle growth and repair. Interestingly, the previously mentioned signaling of IGF and the production of androgens are considered crucial, as deficiencies in these hormonal systems have been linked to negative aging patterns and frailty (Clegg and Hassan-Smith, 2018). In fact, androgens notably impact skeletal muscle through several mechanisms (O'Connell and Wu, 2014). For example, the decrease in testosterone levels is closely associated with muscle mass and strength decline by impairing protein synthesis necessary for muscle function and the intramuscular IGF-1 system (Afilalo, 2014). Inadequate testosterone levels caused a decrease in mitochondrial transcription and biosynthesis in skeletal muscle cells of mice, leading to diminished energy production and impaired muscle function (Pronsato et al., 2020). However, the use of testosterone has not clearly proved to be effective. While it shows benefit in people with clinical deficit of testosterone, it does not seem to have any effect in those with normal (indeed those with low levels but in the normal range) levels. A recent meta-analysis about the effects of testosterone on musculoskeletal health in older men in well-controlled randomized clinical trials did not reach conclusive findings (Buratto et al., 2023).

Numerous studies have explored the link between frailty and testosterone levels, yielding divergent findings. While certain studies have reported a higher prevalence of frailty in individuals with lower testosterone levels, including total testosterone (TT), free testosterone (FT), and higher sex hormone-binding globulin (SHBG) (Carcaillon et al., 2012a; Mohr et al., 2007; Wu et al., 2010), a subset of these studies failed to definitively establish a link between testosterone and frailty (Mohr et al., 2007).

TABLE 3.3
Main Metabolic Biomarkers Related to Human Frailty

Metabolic Biomarkers	Study Design	Sample Information	Frailty Assessment	Main Findings	Reference
HOMA-IR	Longitudinal Longitudinal Cross-sectional	914 (mean age: 74.18 ± 5.45 years; 319 subjects experienced 2.5-point reduction in the FTS-5 score) 991 (mean age: 74.3±5.6 years) 1839 participants (mean age: 63.9 ± 9.3 years; 970 robust, 744 prefrail and 125 frail)	FTS-5 FP, FTS FP	• 1-logarithmic unit increment in HOMA-IR increased the risk of functional decline (FTS-5) and this association was meditated by body fat mass especially in the trunk region • HOMA-IR was positively associated with incident frailty (Fried) and with functional decline (FTS) • HOMA-IR was significantly associated with frailty	El Assar et al., 2023 Rodríguez-Mañas et al., 2022 Hwang et al., 2015
IGF-1 **IGF-1, IGFBP1, IGFBP3**	Cross-sectional Cross-sectional Cross-sectional and longitudinal study	200 subjects aged 50 years and older residing in the community 4133 older adults (mean age, 71.8 ± 5.4 years) 3447 community-dwelling men aged 70–89 years	FP FP FRAIL scale	• Frail/low RASM subjects presented lower IGF-1 levels when compared to frail/normal RASM subjects • In adjusted multinomial logistic regression models with the reference group being robust/normal RASM, IGF-1 was correlated with frail/low RASM status in the whole cohort • When stratified by gender, in women only, serum IGF-1 was related to frail/low RASM status • The lowest quartile group of IGF-1 had significant odds of prefrailty (adjusted model: OR 1.38, 95% CI 1.13–1.68, p = .002) and frailty (adjusted model: OR 1.54, 95% CI 1.02–2.32, p = .039) compared with the highest quartile group • Frail men had lower IGFBP3 versus not frail while IGFBP1 levels were comparable between the two groups • In multivariate analyses, higher IGFBP1 was related to increased prevalence of frailty • Lower baseline IGF-I and higher IGFBP1 predicted new-onset frailty	Chew et al., 2019 Doi et al., 2018 Yeap et al., 2013

Myostatin	Cross-sectional Cross-sectional Cross-sectional	75 older adults with CHF (29 frail) 200 subjects aged 50 years and older residing in the community 112 participants from long-term nursing homes	FP FP FP, CFS, and Tilburg frailty indicator	• Myostatin levels were notably elevated in frail participants compared to those who were not frail • In multivariable analysis, myostatin concentrations were independently related to frailty in older adults with CHF • Frail/low RASM subjects presented higher mean normalized myostatin levels when compared to frail/normal RASM subjects • In adjusted multinomial logistic regression models with the reference group being robust/normal RASM, myostatin was associated with frail/low RASM status in the whole cohort • When stratified by gender, in men only, myostatin was associated with frail/low RASM status • Higher concentrations of serum myostatin were observed in non-frail than in frail participants • Implemented physical exercise intervention significantly increased myostatin concentration in men but not in women	Wang et al., 2023 Chew et al., 2019 Arrieta et al., 2019
SIRT1	Cross-sectional Cross-sectional	130 inpatients (aged ≥60 years),16 robust, 74 prefrail, and 40 frail 350 participants (mean age: 76.5 ± 6.9; 273 non-frail and 77 frail)	FP FP	• Higher levels of SIRT1 in frail vs robust subjects • Serum SIRT1 concentrations were negatively correlated with gait speed, even after adjustment for potential confounders • An inverse association between SIRT1 gene expression and malnutrition risk independently of frailty status	Ma et al., 2019 El Assar et al., 2018

(Continued)

TABLE 3.3 (*Continued*)
Main Metabolic Biomarkers Related to Human Frailty

Metabolic Biomarkers	Study Design	Sample Information	Frailty Assessment	Main Findings	Reference
SIRT1, SIRT3 **Serum-induced SIRT1** **SIRT1 SNPs**	Cross-sectional Cross-sectional	200 patients (≥60 years of age), 119 non-frail and 81 frail 1318 participants (77.0 ± 5.5 years; 636 robust, 548 prefrail, and 105 frail)	FP FP	• Lower levels of both SIRT1 and SIRT3 in frail vs non-frail subjects • In multivariable regression analysis, lower sirtuins levels were significantly associated with frailty after adjusting by potential confounders • No difference in the serum-induced SIRT1 expression levels between the robust, prefrail, and frail groups • Serum-induced SIRT1 participants within the lowest quintile were less likely to be frail • SIRT1 SNPs were not associated with frailty	Kumar et al., 2014 Razi et al., 2017
Klotho	Cross-sectional Cross-sectional Longitudinal	7107 adults aged 45 years or older 89 patients ≥ 65 years old (44 non-frail and 45 frail) 774 aged ≥ 65 years	FP, FI FRAIL scale FP	• Lower prevalence of frailty (FP, FI) was observed in participants with a higher serum Klotho level when compared to those with a lower level • After adjustment for all covariates, the higher serum Klotho level was associated with a 26% (FP) and 17% (FI) lower odds of frailty vs robustness • No statistically significant difference in alpha Klotho concentration was observed between frail and control groups • Natural logarithm of Klotho was associated with lower odds of frailty versus robustness after adjustment for covariates	Guan et al., 2023 Polat et al., 2020 Shardell et al., 2019

CFS: clinical frail scale; CHF: chronic heart failure; FP: Frailty Phenotype; FTS: Frailty Trait Scale; HOMA-IR: Homeostasis Model Assessment of Insulin Resistance; IGF-1: insulin growth factor 1; IGFBP: insulin growth factor binding protein; RASM: relative appendicular skeletal muscle mass; SIRT: sirtuin; SNP: single nucleotide polymorphisms.

Prospective studies have shown that diminished testosterone levels in men predict the onset or progression of frailty (Hyde et al., 2010; Swiecicka et al., 2018; Travison et al., 2011). In particular, Chiang and colleagues have shown that a 50% lower concentration of free testosterone was associated with frailty, physical function, sarcopenia, and muscle mass as well as with changes in these outcomes over 12 months (Chiang et al., 2019). Albeit the majority of studies have primarily focused on investigating the association between testosterone and frailty in men, Carcaillon et al. reported the presence of extremely low levels of free testosterone in women with ≥4 Fried's criteria for frailty, being this association confined to obese women (Carcaillon et al., 2012a). A recent systematic review and meta-analysis revealed that low TT and FT were significantly related to frailty in older men but not in women. In contrast, no significant association was reported between SHBG and frailty (Peng et al., 2022).

Cappola and Fried examined the association between serum levels of several anabolic hormones (IGF-1, dehydroepiandrosterone sulfate (DHEAS), and FT) and frailty status in 494 women aged between 70–79 years. They found that those with two or three hormonal deficiencies were significantly more likely to be frail, indicating that frailty might arise from a broader endocrine dysfunction rather than any single hormonal deficiency (Cappola et al., 2009).

Unlike the clear connection between low testosterone and frailty, the link between estradiol levels and frailty is uncertain, with inconsistent study results (Carcaillon et al., 2012b). While some studies have shown a significant negative correlation between estradiol levels and frailty scores in men aged 70 years and older (Travison et al., 2011), others did not detect this association (Cawthon et al., 2009; Eichholzer et al., 2012). In a 4.3-year prospective cohort study of community-dwelling men, higher estradiol levels predicted a lower likelihood of improving frailty phenotype (Swiecicka et al., 2018). Research on obese older adults has not definitively linked estradiol levels with physical performance or strength in either women or men (Aguirre et al., 2014). However, in postmenopausal women, higher estradiol levels have been related to frailty, especially when high-sensitive CRP levels are also elevated (Carcaillon et al., 2012b).

On the other hand, thyroid hormones (THs) change significantly in the elderly and they have also been considered potential contributors to sarcopenia and frailty (Clegg and Hassan-Smith, 2018; Kamwa et al., 2022). Numerous studies have examined the link between thyroid function and frailty, showing inconsistent results. Consequently, the role of THs in frailty remains a subject of debate. One study found that low levels of free triiodothyronine (FT3), but not free thyroxine (FT4), might be a marker of frailty. FT3 levels were significantly associated with frailty scores and could distinguish frail/prefrail individuals from non-frail ones (Bertoli et al., 2017). A recent cross-sectional study conducted among community-dwelling oldest-old, found that subclinical hypothyroidism, higher thyroid stimulating hormone (TSH) level, lower FT3 level, and decreased FT3/FT4 ratio were all associated with frailty assessed by the FRAIL scale (Guan et al., 2022). By contrast, Virgini and colleagues found that subclinical hyperthyroidism, but not subclinical hypothyroidism, was cross-sectionally linked to higher frailty in men, particularly in those under 74 years at baseline. However, after 5 years of follow-up, neither basal condition was consistently associated with overall frailty status or components (Virgini et al., 2015).

The impact of THs on aged skeletal muscle remains unclear. In euthyroid older adults, higher FT4 was related to lower leg lean mass and higher sarcopenia odds (Ibad et al., 2023), while the lowest quartile of FT4 showed a statistically significant higher handgrip strength in a cohort of community-dwelling aged over 90 years (Iorio et al., 2021). A U-shaped association between sarcopenia and TSH, the main regulator of TH synthesis and secretion, was found with either high or low TSH levels correlating to low body muscle mass in older adults without overt thyroid dysfunction (Szlejf et al., 2020). While some authors have described an association between TSH and muscle strength in cross-sectional studies (Kim et al., 2018), this association was absent in a cohort of old euthyroid adults (Gu et al., 2019).

In summary, available evidence indicates an association between the endocrine system and frailty (Table 3.4). But, again, they predominantly arise from cross-sectional studies or studies not assessing their usefulness in terms of clinical diagnostic and prognostic, thus limiting their applicability in clinical practice.

TABLE 3.4
Description of Endocrine Biomarkers Related to Frailty

Endocrine Biomarkers	Study Design	Sample Information	Frailty Assessment	Main Findings	Reference
			Androgens and estrogens		
FT	Cross-sectional and longitudinal	440 men (mean age: 56.1 ± 14.2 years)	FP	• A 50% reduction in FT levels was linked to: • 1.40-fold higher odds of being frail and of becoming frail over 12 months • 1.55-fold higher probability of having sarcopenia and a 1.72-fold higher risk of developing sarcopenia • Lower muscle mass and a decrease in muscle mass over 12 months	Chiang et al., 2019
TT, FT, DHT, estradiol	Longitudinal	3369 men aged 40 to 79 from eight European centers	FP, FI	• Higher T, FT, and DHT were associated with a decreased risk of FI worsening after adjustment for baseline frailty, age, center, and smoking • After additional adjustment by body mass index, only FT remained a significant predictor of FI change • Higher estradiol levels were associated with a reduced probability of experiencing an improvement in FP	Swiecicka et al., 2018
TT, FT	Cross-sectional	1287 (mean age: 74.4 years; 42.1% prefrail and 8.3% frail)	FP	• A U-shaped association between FT levels and frailty was detected in women • Women with four or more frailty criteria exhibited extremely low levels of FT and this association seems to be limited to obese women • In men, the risk of frailty levels linearly decreased with testosterone	Carcaillon et al., 2012a
Estradiol	Cross-sectional	702 postmenopausal women (mean age: 74.5 ± 5.6 years; 309 robust, 219 prefrail, and 37 frail)	FP	• Higher estradiol levels were related to frailty prevalence among women younger than 79 years, but not in the oldest group • The association between estradiol and frailty was intensified in the presence of both higher estradiol and hsCRP compared with women with low levels of both estradiol and hsCRP	Carcaillon et al., 2012b

TT, FT, estradiol, SHBG	Cross-sectional	461 men aged 60 years and older (2.6% frail)	FP	• TT and total and free estradiol serum levels were not statistically significantly related to frailty • Low FT and high SHBG serum levels were associated with a significantly increased odds of frailty after adjustment for age and race/ethnicity	Eichholzer et al., 2012
T, FT, DHT, estradiol, estrone	Cross-sectional and longitudinal	1705 men (aged 70 years or older)	FP and SOF index	• Individuals within the lowest quintile of T, DHT, estradiol, estrone, and calculated FT had a higher likelihood of displaying more pronounced FP/SOF frailty compared to those in the highest T quintile • One standard deviation decrease in TT, calculated FT over a two-year period, was linked to a 1.2- to 1.3-fold increase in the likelihood of frailty progression (increased severity)	Travison et al., 2011
TT, FT, SHBG	Cross-sectional and longitudinal	3616 community-dwelling men aged 70–88 years at baseline (548 frail) Frailty was reassessed in 1586 men aged 76–93 years at follow-up (364 frail)	FRAIL scale	• Each standard deviation in T and FT was cross-sectionally related to increased odds of frailty • Only lower FT levels predicted frailty at follow-up	Hyde et al., 2010
TT, FT, SHBG	Cross-sectional	108 adults 65 years old or older (17 robust, 67 prefrail, and 24 frail)	FP	• Median (range) TT and FT were lower in frail than in prefrail and robust men and women • After adjusting for confounders, low serum TT and FT levels were independently associated with frailty in the elderly	Wu et al., 2010
FT, DHEAS, IGF-1	Cross-sectional	494 women aged 70–79 years (11.9% robust, 61.9% prefrail, and 26.1% frail)	FP	In adjusted models, when compared to women without hormonal deficiencies, those with one hormone deficiency were not likely to be frail. However, women with two or three deficiencies exhibited a notably elevated probability of being frail	Cappola et al., 2009
T, estradiol, SHBG	Cross-sectional and longitudinal	1469 community-dwelling men at least 65 years old 1245 men had their frailty status reassessed 4.1 years later	FP	• Low concentrations of bioavailable T were independently associated with worse baseline frailty status • Estradiol, bioavailable estradiol, T, and SHBG were not associated either cross-sectionally or longitudinally with frailty status after adjustment	Cawthon et al., 2009

(Continued)

TABLE 3.4 (*Continued*)
Description of Endocrine Biomarkers Related to Frailty

Endocrine Biomarkers	Study Design	Sample Information	Frailty Assessment	Main Findings	Reference
			Androgens and estrogens		
TT, FT, SHBG	Cross-sectional	646 men aged 50 to 86 years	FP	• Total and free T levels were not associated with FP • A significant correlation between SHBG and FP was observed yielding an adjusted odds ratio of 1.25 (95% confidence interval=1.06–1.46) for every 10-nM increase in SHBG levels	Mohr et al., 2007
			Thyroid hormones		
TSH, FT3, FT4	Cross-sectional	487 community-dwelling oldest-old (mean age: 86.0 ± 2.9 years)	FRAIL scale	• Subclinical hypothyroidism, higher TSH level, lower FT3 level, and decreased FT3/FT4 ratio were all related to frailty • FT4 levels were not associated to frailty	Guan et al., 2022
TSH, FT3, FT4	Cross-sectional	112 elderly aged 65 years or above	SHARE-FI	• FT3, but not FT4, was significantly correlated with frailty score, both in patients with hip fracture and in patients from the control group • FT3 under normal limits is effective in discriminating frail/prefrail subjects from non-frail subjects	Bertoli et al., 2017

DHEAS: dehydroepiandrosterone sulfate; DHT: dihydrotestosterone; FI: Frailty Index; FP: Frailty Phenotype; FT3: free triiodothyronine; FT4: free thyroxine; SHBG: sex hormone-binding globulin; SOF: study of osteoporotic fractures; SHARE-FI: Survey of Health, Ageing and Retirement in Europe Frailty Instrument; TSH: thyroid stimulating hormone.

Other Laboratory Biomarkers

Concluding this section of the review on laboratory biomarkers of frailty, we will briefly touch upon four additional categories with potential significance for future research: miRNAs, senescence markers, serum bone markers, and vascular function markers.

MiRNAs, short single-strand RNA molecules, are promising candidates for early frailty diagnosis in older individuals (Rusanova et al., 2019) due to their role in regulating gene expression and modulating mechanisms involved in muscle degeneration, mitochondrial dysfunction, and inflammaging (Lozupone et al., 2024). A cross-sectional study on healthy older participants found that miR-21 expression correlated with frailty (Rusanova et al., 2018). Additionally, miR-125b-5p, which inhibits apoptosis and the expression of inflammatory genes induced by Il-1b (Rasheed et al., 2019), has been identified as a protective factor against frailty in both disabled and non-disabled frail individuals (Gomez-Cabrero et al., 2021). Furthermore, Carini and colleagues' analysis of the whole miRNome found that miR-101-3p and miR-142-5p are specifically downregulated in frail subjects. The target genes of these miRNAs are involved in processes and signaling pathways related to frailty pathophysiology (Carini et al., 2022). Further studies are needed to determine if these miRNA alterations precede the onset of frailty or contribute to its progression.

Cellular senescence, a state of permanent growth arrest triggered by various stresses, is characterized by the emergence of a complex SASP consisting of pro-inflammatory cytokines, chemokines, proteases, and other factors (Birch and Gil, 2020). Circulating proteins secreted by senescent cells including vascular endothelial growth factor A (VEGFA), GDF15, osteopontin, and other senescence biomarkers were linked to declines in Short Physical Performance Battery (SPPB) scores in older subjects. Moreover, senescent biomarkers like VEGFA, tumor necrosis factor receptor 1 (TNFR1), and matrix metallopeptidase 7 (MMP7) were significantly related to the onset of major mobility disability and persistent mobility in older subjects (Fielding et al., 2024), reinforcing cellular senescence as a contributing mechanism of age-related functional decline.

Taking into account the recent concept of "osteosarcopenia" and its contribution to age-related functional decline (Kirk and Duque, 2022), several biomarkers linked to bone turnover have been assessed. Elevated levels of N-terminal propeptide of type I procollagen, in addition to low levels of 25-hydroxyvitamin D, have been shown to be associated with frailty (Alvarez-Ríos et al., 2015). At the same time, circulating factors associated with inflammation, like homocysteine, have been shown to increase bone resorption in frail and prefrail older people (Álvarez-Sánchez et al., 2018), emphasizing once more that the generation or clinical manifestation of frailty involves the concurrent alterations of several systems.

Vascular system has also been implicated in frailty pathophysiology and some of its risk factors and diseases are also closely related to frailty through different mechanisms: diabetes (Assar et al., 2016; García-Esquinas et al., 2015), subclinical cardiovascular disease (Chaves et al., 2004), obesity (Strandberg et al., 2012), lipids (Yin et al., 2023), heart fatty-acid binding proteins (Liu et al., 2023), ischemic heart disease (Strandberg et al., 2012), heart failure (Kleipool et al., 2018), and peripheral artery disease (Alonso-Bouzón et al., 2014; Jakubiak et al., 2020). While the roles of these conditions in frailty, as well as the influence of frailty on their outcomes and prognosis, are clear, the role of frailty in predisposing to many of them is controversial.

In the TSHA cohort, an association between endothelial dysfunction, assessed by circulating ADMA levels, and frailty was shown, with an interaction with inflammation (Alonso-Bouzón et al., 2014). In the FRAILOMIC study, which evaluated over 35,000 frailty biomarkers using machine learning, the three most significant biomarkers associated with a high risk of frailty were of vascular origin: Pro-BNP, cardiac troponin-T, and sRAGE (Gomez-Cabrero et al., 2021). Additionally, sRAGE showed to be a predictive factor for total and cardiovascular death in older people with frailty, but not in those who were robust (Butcher et al., 2021, 2019).

IMAGING BIOMARKERS

Anthropometric changes can indicate the onset of specific health events in older adults, such as the development of frailty, with sarcopenia often recognized as a key biological factor in this condition (Rodriguez-Mañas et al., 2020). Research has demonstrated a strong association between sarcopenia and both prevalent (Davies et al., 2018) and incident frailty (Álvarez-Bustos et al., 2022). This evidence suggests that imaging biomarkers assessing muscle mass can be utilized to evaluate frailty. The most appropriate technique depends on various factors, including patient characteristics, functional status, access to technical resources, and the purpose of the assessment.

Dual Energy X-Ray Absorptiometry (DXA)

DXA is a widely available, non-invasive technique used to assess total body lean mass and appendicular lean mass (ALM). It is employed in both clinical practice and research, offering a reproducible estimate of ALM within minutes (Masanés et al., 2017; Minetto et al., 2021). Due to its accuracy, reproducibility, and established sex-specific cut-off values (Cruz-Jentoft and Sayer, 2019), DXA is the preferred method for assessing lean skeletal muscle mass in routine clinical practice. Important considerations when using DXA include adjusting measurements according to body size (height squared, weight, or body mass index – BMI) (Kim et al., 2016) and ensuring stable measurement conditions, such as good hydration status, to avoid errors in assessing the various body compartments.

Bioelectrical Impedance Analysis (BIA)

BIA is a quick, accurate, and reliable screening technique for estimating muscle mass based on whole-body electrical conductivity, using a conversion equation calibrated against reference values of DXA-measured lean mass from a defined population (Gonzalez and Heymsfield, 2017; Sergi et al., 2015). A prospective study of community-dwelling older adults established frailty as an independent risk factor for lean mass index (LMI) decline, assessed by BIA, particularly in those with lower hand-grip strength (Jung et al., 2014). BIA has also been used in hospitalized older adults, as demonstrated by a Chinese cross-sectional study (Xu et al., 2020) which found that frail older inpatients had low skeletal muscle mass and high body fat mass.

BIA-derived phase angle assesses both muscle quantity and quality (Heymsfield et al., 2015). Lower phase angle values are linked to frailty, sarcopenia, malnutrition, morbidity, longer hospital stays, and higher mortality at 1 and 12 months in cardiac surgery patients (Mullie et al., 2018). A lower phase angle value triples the risk of frailty, as shown in a nationally representative NHANES study (Wilhelm-Leen et al., 2014), with a fourfold increased likelihood of worsening frailty for each one-degree increment in phase angle over a one-year period (Norman et al., 2023; Zanforlini et al., 2019). Phase angle values are related not only to incident frailty but also to changes in frailty status in patients with rheumatoid arthritis (Matsumoto et al., 2022; Norman et al., 2023). Additionally, BIA can assess visceral fat obesity, a risk factor for frailty (Li et al., 2022).

Several studies have shown a high correlation between DXA-measured and BIA-measured skeletal muscle mass (Cheng et al., 2021), making BIA a good alternative when DXA is not available (Sbrignadello et al., 2022). However, some studies suggest that BIA may overestimate lean muscle mass compared to DXA, potentially misclassifying one out of six patients (Reiss et al., 2016). Other disadvantages of BIA include its sensitivity to hydration status and the fact that its standardization equation is based on an older European population, which means results should be interpreted with caution in different age groups, races, ethnicities, and other demographic parameters (Gonzalez and Heymsfield, 2017).

Computed Tomography (CT) and Magnetic Resonance Imaging (MRI)

Computed tomography (CT) and magnetic resonance imaging (MRI) serve as the gold standards for non-invasive muscle mass assessments (Cruz-Jentoft and Sayer, 2019). These advanced imaging techniques provide high-resolution visualization, allowing for precise quantification of muscle mass even in the early stages of sarcopenia. Various body regions can be assessed using the following modalities:

A. Skeletal muscle cross-sectional area can be measured at two distinct levels: (1) the 3rd lumbar (L3) vertebra, encompassing muscles such as the psoas, erector spinae, quadratus lumborum, transversus abdominis, external and internal obliques, and rectus abdominis, and (2) the 3rd cervical vertebra, including muscles like the sternocleidomastoid and paravertebral muscles. Studies demonstrate a strong correlation between cross-sectional skeletal muscle area at the latter level of L3 (de Bree et al., 2022). Independent associations between skeletal muscle mass and frailty have been observed in head and neck cancer patients (Alwani et al., 2020; de Bree et al., 2022; Meerkerk et al., 2022; Zwart et al., 2019).

B. Mid-thigh muscle measurement serves as a reliable predictor of whole-body skeletal muscle mass, exhibiting a stronger correlation with total body muscle volume than lumbar muscle areas. Recent research has linked mid-thigh muscle volume (assessed via MRI) with frailty indicators such as the Frailty Index and gait speed (Farrow et al., 2021).

Both CT and MRI techniques allow for the evaluation of muscle quality, enabling the assessment of microscopic and macroscopic changes in muscle architecture and composition, as well as muscle function per unit of muscle mass (McGregor et al., 2014), including the determination of fat infiltration into muscle tissue. However, due to their high cost, lack of portability, requirement for highly trained personnel, and absence of defined cut-off points for low muscle mass, these tools are not widely utilized in routine clinical settings, although they remain extensively employed for research purposes (Beaudart et al., 2016).

Ultrasound (US)

Ultrasound, a reliable, cost-effective, and valid imaging modality, is increasingly utilized in clinical settings to assess muscle quantity (thickness, length, area, and pennation angle) and quality (degree of atrophy, muscle fibrosis, fatty infiltration, and vascularization), as well as muscle wasting (Bauer et al., 2019). While most studies focus on ultrasound evaluations of the quadriceps femoris, other muscles such as the gastrocnemius, forearm, or biceps can also be scanned. Ultrasound can effectively detect decreases in muscle thickness and cross-sectional area over short periods with good intra- and inter-observer reliability (Martín et al., 2017). Another parameter assessable via ultrasound is echo intensity, which identifies structural muscle changes due to degeneration. Using these methods to assess quantitative muscle changes, studies have demonstrated an association between reduced muscle mass and frailty in hospitalized patients (Bencivenga et al., 2022), as well as an inverse correlation between echo intensity and muscle strength, and a direct correlation with frailty (Mombiela et al., 2017).

The European Geriatric Medicine Society (EuGMS) Sarcopenia Special Interest Group and the SARCUS Working Group have proposed a consensus protocol for ultrasound use in muscle assessment, encompassing the evaluation of several parameters: muscle thickness, cross-sectional area, fascicle length, pennation angles, and echogenicity (related to muscle quality) (Perkisas et al., 2019).

Considering that muscle ultrasound is a convenient, safe, affordable, and portable tool, capable of being performed at the bedside without exposing patients to radiation, its potential as a predictor of frailty outcomes in daily clinical practice is becoming increasingly recognized, with a good intra-observer reproducibility (Álvarez et al., 2021; Cho et al., 2017; Sions et al., 2014). Nevertheless, it is essential to note the lack of established cut-off points, with variations across studies depending on the muscle assessed and the population studied. For instance, a Spanish study established a cut-off point for low muscle mass of ≤1.85 cm (transverse plane) and 1.73 cm (longitudinal plane) when assessing the gastrocnemius muscle (Álvarez et al., 2021). Italian studies suggested cut-off points based on rectus femoris thickness of ≤2 cm in men and ≤1.6 cm in women (Minetto et al., 2016), while another study proposed ≤0.9 cm and ≤0.7 cm in men and women, respectively (Rustani et al., 2019). Meanwhile, different cut-off points have been proposed in Asian populations, such as ≤1.51 cm in men and ≤1.43 in women according to a Japanese study (Fukumoto et al., 2021), and a recent Thai study suggested ≤1.1 cm for men and ≤1 cm for women (Sri-On et al., 2022).

D3 Creatine Dilution (DCD)

The DCD test has emerged as a promising method for assessing muscle mass. While lean body mass (LBM) or fat-free mass (FFM) has often been used interchangeably with muscle mass, this assumption has led to inconsistent results regarding the relationship between LBM/FFM and falls, disability, or functional capacity in older individuals (Evans et al., 2019). The DCD test offers an indirect, accurate, and non-invasive means to evaluate skeletal muscle mass (specifically myofiber contractile mass) by estimating whole-body muscle mass based on the renal excretion rate of creatinine. Moreover, it has demonstrated efficacy in assessing changes in muscle mass in response to resistance exercise training (Balachandran et al., 2023).

This technique relies on three key assumptions: (1) 98% of the total body creatine pool resides in skeletal muscle, (2) creatine undergoes turnover in muscle via conversion to creatinine, and (3) an orally administered tracer dose of deuterated creatine (D3-Cr) is 100% bioavailable, transported across the sarcolemma, and sequestered in sarcomeres (Evans et al., 2019). Unlike other methods, such as 24-hour creatinine excretion, DCD measurement does not necessitate dietary control or the collection of 24-hour urine samples; instead, a single spot urine sample taken 48–96 hours after dosing is sufficient.

Various studies conducted in animals and humans have underscored the utility of the DCD method in assessing muscle mass. For instance, a cross-sectional study in rats demonstrated a robust correlation between DCD muscle mass and lean mass assessed by MRI, as well as muscle mass of the lower limbs (Stimpson et al., 2012). Another recent study comparing the DCD method to DXA-determined LBM in aged mice found that DXA consistently overestimated muscle mass with age, potentially reflecting non-muscle tissue as part of LBM (Wimer et al., 2023).

In human studies, particularly those involving older male populations like the MrOS research group, the DCD method has emerged as a valuable predictor of disability, physical performance, falls, incident fractures, mobility limitation, and mortality (Cawthon et al., 2021, 2019; Orwoll et al., 2022; Rodriguez-Mañas et al., 2020; Zanker et al., 2020). Additionally, associations have been found between DCD muscle mass and nutrient biomarkers, particularly those related to sarcopenia (Hetherington-Rauth et al., 2023).

Though fewer studies have examined DCD as a muscle mass measurement in women, an inverse relationship has been noted between muscle mass according to DCD and insulin resistance in this population (Banack et al., 2022). Comparisons between DCD and DXA have revealed moderate correlations with DXA LBM ($r = 0.50$) and DXA ALM ($r = 0.50$). Furthermore, higher DCD values in older women have been associated with better physical function assessed by the SPPB (Zhu et al., 2021).

However, the DCD method does have limitations. First, the formula used for calculating creatine pool size and skeletal muscle mass assumes 100% type II fibers, leading to a false assumption regarding the creatine uptake rates of type I fibers. Second, as a non-direct measure of total skeletal muscle mass, potential sources of measurement error exist, such as variability in creatine concentration between muscles, muscle composition, age, and disease states. Moreover, specialized laboratory facilities are required for DCD measurement, which may result in delays in obtaining results (McCarthy et al., 2022).

A description of the different imaging biomarkers is displayed in Table 3.5.

TABLE 3.5
Description of Different Imaging Biomarkers

Imaging Technique	Measurement	Advantages	Disadvantages	Utility
Dual Energy X-ray Absorptiometry (DXA)	Total body lean mass and ALM	Availability, non-invasive Good accuracy and repeatability Preferred method to assess lean skeletal muscle mass in daily clinical practice	Adjust by body size Measurements should be done in stable conditions (i.e., hydration status may influence the results)	Clinical practice and research
Bioelectrical impedance analysis (BIA)	ASM Muscle quality through phase angle.	Quick, accurate, and reliable tool Non-invasive High correlation with DXA	Overestimates muscle mass Measurements should be done in stable conditions (i.e., hydration status may influence the results) Standardization equation is based on older European population	Clinical practice and research
Computed Tomography (CT) Magnetic Resonance Imaging (MRI)	Total body skeletal muscle mass Muscle quality	Gold standard Non-invasive High-resolution imaging (quantify muscle mass in early sarcopenia stages)	High cost, lack of portability and defined cut-off points, need of highly trained staff	Research
Ultrasound (US)	Muscle quantity, quality (echogenicity), and wasting	Reliable technique with a good intra- and inter-observer reliability, easy, safe, inexpensive, and portable tool	Lack of established cut-off points	Clinical practice and research
D_3 creatine dilution	Skeletal muscle mass through myofiber contractile mass	Accurate, non-invasive method Assess changes in muscle mass Strong correlation with MRI, moderate with DXA A good predictor of functional outcomes, falls and fractures, and mortality	Non-direct method with potential sources of measurement error It cannot discern muscles with functional impairments Specialized laboratory facilities are needed	Research

ALM: appendicular lean mass; ASM: appendicular skeletal mass. DXA: dual energy X-ray absorptiometry. MRI: magnetic resonance imaging.

CONCLUSIONS

Biomarkers of frailty still have a significant journey ahead before they can be routinely integrated into regular clinical practice. Frailty is caused by a combination of aging-related physiological changes and various subclinical and clinical conditions, creating a complex profile and raising theoretical and practical challenges including: (a) the existence of different clinical profiles or subtypes of frailty, each with its own prognosis, underscores the need to identify distinct biomarkers for each subtype, (b) poor agreement among clinical diagnostic tools, (c) assessing biomarkers in terms of sensitivity, specificity, and predictive values for diagnosis and prognosis of frailty outcomes instead of OR or HR, which are commonly used but only assess associations, (d) their reliability considering the dynamic nature of frailty, and (e) the need of using sets of biomarkers and combining lab, clinical and imaging biomarkers, rather than relying on isolated biomarkers, due to the multifactorial nature of frailty.

In addition, frailty biomarkers should be easy to determine, widely available, and affordable, given the high prevalence of frailty and prefrailty in the aging population, which significantly impacts Health and Social Systems. Advances in understanding frailty's pathology, transitions, consequences, and treatment prognosis should promote the development of new biomarkers or improvement of existing ones, aiding daily assessment and treatment of frail older adults or those at risk. This represents a crucial yet challenging path in routine clinical practice.

FUNDING

The present work was funded by grants from the Spanish Ministry of Economy, Industry and Competitiveness, cofinanced by the European Regional Development Funds (Instituto de Salud Carlos III, PI23/00877) and the Centro de Investigación Biomédica en Red en Fragilidad y Envejecimiento Saludable (CB16/10/00464),

REFERENCES

Afilalo, J., 2014. Androgen deficiency as a biological determinant of frailty: Hope or hype? J Am Geriatr Soc 62, 1174–1178. https://doi.org/10.1111/JGS.12835

Aguirre, L.E., Jan, I.Z., Fowler, K., Waters, D.L., Villareal, D.T., Armamento-Villareal, R., 2014. Testosterone and adipokines are determinants of physical performance, strength, and aerobic fitness in frail, obese, older adults. Int J Endocrinol 2014. https://doi.org/10.1155/2014/507395

Alcazar, J., Frandsen, U., Prokhorova, T., Kamper, R.S., Haddock, B., Aagaard, P., Suetta, C., 2021. Changes in systemic GDF15 across the adult lifespan and their impact on maximal muscle power: The Copenhagen Sarcopenia Study. J Cachexia Sarcopenia Muscle 12, 1418–1427. https://doi.org/10.1002/jcsm.12823

Alonso-Bouzón, C., Carcaillon, L., García-García, F.J., Amor-Andrés, M.S., El Assar, M., Rodríguez-Mañas, L., 2014. Association between endothelial dysfunction and frailty: The Toledo Study for Healthy Aging. Age (Dordr) 36, 495–505. https://doi.org/10.1007/S11357-013-9576-1

Al Saedi, A., Feehan, J., Phu, S., Duque, G., 2019. Current and emerging biomarkers of frailty in the elderly. Clin Interv Aging 14, 389–398. https://doi.org/10.2147/CIA.S168687

Álvarez, M.N., Ronda, M.A.V., Rangel, L.S., Thuissard-Vasallo, I.J., Andreu-Vazquez, C., Martin, P.M., Lorite, I.R., Martín, G.S.S., 2021. Muscle assessment by ultrasonography: Agreement with dual-energy X-ray absorptiometry (DXA) and relationship with physical performance. J Nutr Health Aging 25, 956–963. https://doi.org/10.1007/S12603-021-1669-4

Álvarez-Bustos, A., Carnicero-Carreño, J.A., Davies, B., Garcia-Garcia, F.J., Rodríguez-Artalejo, F., Rodríguez-Mañas, L., Alonso-Bouzón, C., 2022. Role of sarcopenia in the frailty transitions in older adults: A population-based cohort study. J Cachexia Sarcopenia Muscle 13, 2352–2360. https://doi.org/10.1002/jcsm.13055

Alvarez-Ríos, A.I., Guerrero, J.M., García-García, F.J., Rodríguez-Mañas, L., Medrano-Campillo, P., de la Torre Lanza, M.A., Alvarez-Sánchez, N., Carrillo-Vico, A., 2015. Associations between frailty and serum N-terminal propeptide of type I procollagen and 25-hydroxyvitamin D in older Spanish women: The Toledo Study for Healthy Aging. Exp Gerontol 69, 79–84. https://doi.org/10.1016/j.exger.2015.05.011

Álvarez-Sánchez, N., Álvarez-Ríos, A.I., Guerrero, J.M., García-García, F.J., Rodríguez-Mañas, L., Cruz-Chamorro, I., Lardone, P.J., Carrillo-Vico, A., 2018. Homocysteine levels are associated with bone resorption in pre-frail and frail Spanish women: The Toledo Study for Healthy Aging. Exp Gerontol 108, 201–208. https://doi.org/10.1016/j.exger.2018.04.019

Álvarez-Sánchez, N., Álvarez-Ríos, A.I., Guerrero, J.M., García-García, F.J., Rodríguez-Mañas, L., Cruz-Chamorro, I., Lardone, P.J., Carrillo-Vico, A., 2020. Homocysteine and C-reactive protein levels are associated with frailty in older Spaniards: The Toledo Study for Healthy Aging. J Gerontol A Biol Sci Med Sci 75, 1488–1494. https://doi.org/10.1093/gerona/glz168

Álvarez-Satta, M., Berna-Erro, A., Carrasco-Garcia, E., Alberro, A., Saenz-Antoñanzas, A., Vergara, I., Otaegui, D., Matheu, A., 2020. Relevance of oxidative stress and inflammation in frailty based on human studies and mouse models. Aging 12, 9982–9999. https://doi.org/10.18632/aging.103295

Alwani, M.M., Jones, A.J., Novinger, L.J., Pittelkow, E., Bonetto, A., Sim, M.W., Moore, M.G., Mantravadi, A.V., 2020. Impact of sarcopenia on outcomes of autologous head and neck free tissue reconstruction. J Reconstr Microsurg 36, 369–378. https://doi.org/10.1055/S-0040-1701696

Angioni, D., Lu, W.H., Sourdet, S., Macaron, T., Takeda, C., Guyonnet, S., Mangin, J.F., Rolland, Y., de Souto Barreto, P., Vellas, B., 2022. Biomarkers of age-related frailty and frailty related to diseases: An exploratory, cross-sectional analysis from the MAPT study. J Nutr Health Aging 26, 545–551. https://doi.org/10.1007/s12603-022-1793-9

Angulo, J., Assar, M.E., Rodríguez-Mañas, L., 2016. Frailty and sarcopenia as the basis for the phenotypic manifestation of chronic diseases in older adults. Mol Aspects Med. https://doi.org/10.1016/j.mam.2016.06.001

Anwar, M., Pradhan, R., Dey, S., Kumar, R., 2023. The role of sirtuins in sarcopenia and frailty. Aging Dis 14, 25. https://doi.org/10.14336/AD.2022.0622

Arauna, D., García, F., Rodríguez-Mañas, L., Marrugat, J., Sáez, C., Alarcón, M., Wehinger, S., Espinosa-Parrilla, Y., Palomo, I., Fuentes, E., 2020. Older adults with frailty syndrome present an altered platelet function and an increased level of circulating oxidative stress and mitochondrial dysfunction biomarker GDF-15. Free Radic Biol Med 149, 64–71. https://doi.org/10.1016/j.freeradbiomed.2020.01.007

Arrieta, H., Hervás, G., Rezola-Pardo, C., Ruiz-Litago, F., Iturburu, M., Yanguas, J.J., Gil, S.M., Rodriguez-Larrad, A., Irazusta, J., 2019. Serum myostatin levels are higher in fitter, more active, and non-frail long-term nursing home residents and increase after a physical exercise intervention. Gerontology 65, 229–239. https://doi.org/10.1159/000494137

Assar, M.E., Angulo, J., Rodríguez-Mañas, L., 2016. Diabetes and ageing-induced vascular inflammation. J Physiol 594, 2125–46. https://doi.org/10.1113/JP270841

Balachandran, A.T., Evans, W.J., Cawthon, P.M., Wang, Y., Shankaran, M., Hellerstein, M.K., Qiu, P., Manini, T., 2023. Comparing D3-creatine dilution and dual-energy X-ray absorptiometry muscle mass responses to strength training in low-functioning older adults. J Gerontol A Biol Sci Med Sci 78, 1591–1596. https://doi.org/10.1093/gerona/glad047

Balboa-Castillo, T., Struijk, E.A., Lopez-Garcia, E., Banegas, J.R., Rodríguez-Artalejo, F., Guallar-Castillon, P., 2018. Low vitamin intake is associated with risk of frailty in older adults. Age Ageing 47, 872–879. https://doi.org/10.1093/ageing/afy105

Banack, H.R., LaMonte, M.J., Manson, J.E., Zhu, K., Evans, W.J., Shankaran, M., Wactawski-Wende, J., 2022. Association of muscle mass measured by D3-creatine (D3Cr), sarcopenic obesity, and insulin-glucose homeostasis in postmenopausal women. PLoS One 17, e0278723. https://doi.org/10.1371/journal.pone.0278723

Bauer, J., Morley, J.E., Schols, A.M.W.J., Ferrucci, L., Cruz-Jentoft, A.J., Dent, E., Baracos, V.E., Crawford, J.A., Doehner, W., Heymsfield, S.B., Jatoi, A., Kalantar-Zadeh, K., Lainscak, M., Landi, F., Laviano, A., Mancuso, M., Muscaritoli, M., Prado, C.M., Strasser, F., von Haehling, S., Coats, A.J., Anker, S., 2019. Sarcopenia: A time for action. An SCWD position paper. J Cachexia Sarcopenia Muscle 10, 956–961. https://doi.org/10.1002/JCSM.12483

Baumann, C.W., Kwak, D., Liu, H.M., Thompson, L.V., 2016. Age-induced oxidative stress: How does it influence skeletal muscle quantity and quality? J Appl Physiol (1985) 121, 1047–1052. https://doi.org/10.1152/JAPPLPHYSIOL.00321.2016

Beaudart, C., McCloskey, E., Bruyère, O., Cesari, M., Rolland, Y., Rizzoli, R., de Carvalho, I.A., Thiyagarajan, J.A., Bautmans, I., Bertière, M.C., Brandi, M.L., Al-Daghri, N.M., Burlet, N., Cavalier, E., Cerreta, F., Cherubini, A., Fielding, R., Gielen, E., Landi, F., Petermans, J., Reginster, J.Y., Visser, M., Kanis, J., Cooper, C., 2016. Sarcopenia in daily practice: Assessment and management. BMC Geriatr 16, 1–10. https://doi.org/10.1186/S12877-016-0349-4

Bencivenga, L., Picaro, F., Ferrante, L., Komici, K., Ruggiero, F., Sepe, I., Gambino, G., Femminella, G.D., Vitale, D.F., Ferrara, N., Rengo, C., Rengo, G., 2022. Muscle ultrasound as imaging domain of frailty. Front Med (Lausanne) 9, 922345. https://doi.org/10.3389/fmed.2022.922345

Bernabei, R., Landi, F., Calvani, R., Cesari, M., Del Signore, S., Anker, S.D., Bejuit, R., Bordes, P., Cherubini, A., Cruz-Jentoft, A.J., Di Bari, M., Friede, T., Ayestarán, C.G., Goyeau, H., Jónsson, P.V., Kashiwa, M., Lattanzio, F., Maggio, M., Mariotti, L., Miller, R.R., Rodriguez-Mañas, L., Roller-Wirnsberger, R., Rýznarová, I., Scholpp, J., Schols, A.M.W.J., Sieber, C.C., Sinclair, A.J., Skalska, A., Strandberg, T., Tchalla, A., Topinková, E., Tosato, M., Vellas, B., Von Haehling, S., Pahor, M., Roubenoff, R., Marzetti, E., 2022. Multicomponent intervention to prevent mobility disability in frail older adults: Randomised controlled trial (SPRINTT project). BMJ 377. https://doi.org/10.1136/BMJ-2021-068788

Bertoli, A., Valentini, A., Cianfarani, M.A., Gasbarra, E., Tarantino, U., Federici, M., 2017. Low FT3: A possible marker of frailty in the elderly. Clin Interv Aging 12, 335–341. https://doi.org/10.2147/CIA.S125934

Birch, J., Gil, J., 2020. Senescence and the SASP: Many therapeutic avenues. Genes Dev 34, 1565–1576. https://doi.org/10.1101/gad.343129.120

Ble, A., Cherubini, A., Volpato, S., Bartali, B., Walston, J.D., Windham, B.G., Bandinelli, S., Lauretani, F., Guralnik, J.M., Ferrucci, L., 2006. Lower plasma vitamin E levels are associated with the frailty syndrome: The InCHIANTI study. J Gerontol A Biol Sci Med Sci 61, 278–283. https://doi.org/10.1093/gerona/61.3.278

Bollen, S.E., Bass, J.J., Fujita, S., Wilkinson, D., Hewison, M., Atherton, P.J., 2022. The vitamin D/vitamin D receptor (VDR) axis in muscle atrophy and sarcopenia. Cell Signal 96, 110355. https://doi.org/10.1016/j.cellsig.2022.110355

Buratto, J., Kirk, B., Phu, S., Vogrin, S., Duque, G., 2023. Safety and efficacy of testosterone therapy on musculoskeletal health and clinical outcomes in men: A systematic review and meta-analysis of randomized placebo-controlled trials. Endocr Pract 29, 727–734. https://doi.org/10.1016/j.eprac.2023.04.013

Buta, B., Choudhury, P.P., Xue, Q.-L., Chaves, P., Bandeen-Roche, K., Shardell, M., Semba, R.D., Walston, J., Michos, E.D., Appel, L.J., McAdams-DeMarco, M., Gross, A., Yasar, S., Ferrucci, L., Fried, L.P., Kalyani, R.R., 2017. The association of vitamin D deficiency and incident frailty in older women: The role of cardiometabolic diseases. J Am Geriatr Soc 65, 619–624. https://doi.org/10.1111/jgs.14677

Butcher, L., Carnicero, J.A., Gomez Cabrero, D., Dartigues, J.F., Pérès, K., Garcia-Garcia, F.J., Rodriguez-Mañas, L., Erusalimsky, J.D., 2019. Increased levels of soluble receptor for advanced glycation end-products (RAGE) are associated with a higher risk of mortality in frail older adults. Age Ageing 48, 691–697. https://doi.org/10.1093/AGEING/AFZ073

Butcher, L., Carnicero, J.A., Pérès, K., Colpo, M., Gomez Cabrero, D., Dartigues, J.-F., Bandinelli, S., Garcia-Garcia, F.J., Rodríguez-Mañas, L., Erusalimsky, J.D., FRAILOMIC Consortium, 2021. Higher sRAGE levels predict mortality in frail older adults with cardiovascular disease. Gerontology 67, 202–210. https://doi.org/10.1159/000512287

Cai, Y., Wanigatunga, A.A., Mitchell, C.M., Urbanek, J.K., Miller, E.R., Juraschek, S.P., Michos, E.D., Kalyani, R.R., Roth, D.L., Appel, L.J., Schrack, J.A., 2022. The effects of vitamin D supplementation on frailty in older adults at risk for falls. BMC Geriatr 22, 312. https://doi.org/10.1186/s12877-022-02888-w

Cappola, A.R., Xue, Q.L., Fried, L.P., 2009. Multiple hormonal deficiencies in anabolic hormones are found in frail older women: The Women's Health and Aging studies. J Gerontol A Biol Sci Med Sci 64, 243–248. https://doi.org/10.1093/GERONA/GLN026

Carcaillon, L., Blanco, C., Alonso-Bouzón, C., Alfaro-Acha, A., Garcia-García, F.-J., Rodriguez-Mañas, L., 2012a. Sex differences in the association between serum levels of testosterone and frailty in an elderly population: The Toledo Study for Healthy Aging. PLoS One 7, e32401. https://doi.org/10.1371/journal.pone.0032401

Carcaillon, L., García-García, F.J., Tresguerres, J.A.F., Gutiérrez Avila, G., Kireev, R., Rodríguez-Mañas, L., 2012b. Higher levels of endogenous estradiol are associated with frailty in postmenopausal women from the Toledo Study for Healthy Aging. J Clin Endocrinol Metab 97, 2898–2906. https://doi.org/10.1210/jc.2012-1271

Carini, G., Mingardi, J., Bolzetta, F., Cester, A., Bolner, A., Nordera, G., Via, L.L., Ieraci, A., Russo, I., Maggi, S., Calza, S., Popoli, M., Veronese, N., Musazzi, L., Barbon, A., 2022. miRNome profiling detects miR-101-3p and miR-142-5p as putative blood biomarkers of frailty syndrome. Genes (Basel) 13. https://doi.org/10.3390/GENES13020231

Cawthon, P.M., Blackwell, T., Cummings, S.R., Orwoll, E.S., Duchowny, K.A., Kado, D.M., Stone, K.L., Ensrud, K.E., Cauley, J.A., Evans, W.J., 2021. Muscle mass assessed by the D3-creatine dilution method and incident self-reported disability and mortality in a prospective observational study of community-dwelling older men. J Gerontol A Biol Sci Med Sci 76, 123–130. https://doi.org/10.1093/gerona/glaa111

Cawthon, P.M., Ensrud, K.E., Laughlin, G.A., Cauley, J.A., Dam, T.T.L., Barrett-Connor, E., Fink, H.A., Hoffman, A.R., Lau, E., Lane, N.E., Stefanick, M.L., Cummings, S.R., Orwoll, E.S., 2009. Sex hormones and frailty in older men: The osteoporotic fractures in men (MrOS) study. J Clin Endocrinol Metab 94, 3806–3815. https://doi.org/10.1210/JC.2009-0417

Cawthon, P.M., Orwoll, E.S., Peters, K.E., Ensrud, K.E., Cauley, J.A., Kado, D.M., Stefanick, M.L., Shikany, J.M., Strotmeyer, E.S., Glynn, N.W., Caserotti, P., Shankaran, M., Hellerstein, M., Cummings, S.R., Evans, W.J., 2019. Strong relation between muscle mass determined by D3-creatine dilution, physical performance, and incidence of falls and mobility limitations in a prospective cohort of older men. J Gerontol A Biol Sci Med Sci 74, 844–852. https://doi.org/10.1093/GERONA/GLY129

Chaves, P.H.M., Kuller, L.H., O'Leary, D.H., Manolio, T.A., Newman, A.B., 2004. Subclinical cardiovascular disease in older adults: Insights from the cardiovascular health study. Am J Geriatr Cardiol 13, 137–149. https://doi.org/10.1111/J.1076-7460.2004.02120.X

Checa-Lopez, M., Rodriguez-Laso, A., Carnicero, J.A., Solano-Jaurrieta, J.J., Obermans, O.S., Sinclair, A., Landi, F., Scuteri, A., Álvarez-Bustos, A., Sepúlveda-Loyola, W., Rodriguez-Manas, L., 2023. Differential utility of various frailty diagnostic tools in non-geriatric hospital departments of several countries: A longitudinal study. Eur J Clin Invest 53, e13979. https://doi.org/10.1111/eci.13979

Cheng, K.Y.K., Chow, S.K.H., Hung, V.W.Y., Wong, C.H.W., Wong, R.M.Y., Tsang, C.S.L., Kwok, T., Cheung, W.H., 2021. Diagnosis of sarcopenia by evaluating skeletal muscle mass by adjusted bioimpedance analysis validated with dual-energy X-ray absorptiometry. J Cachexia Sarcopenia Muscle 12, 2163–2173. https://doi.org/10.1002/JCSM.12825

Chew, J., Tay, L., Lim, J.P., Leung, B.P., Yeo, A., Yew, S., Ding, Y.Y., Lim, W.S., 2019. Serum myostatin and IGF-1 as gender-specific biomarkers of frailty and low muscle mass in community-dwelling older adults. J Nutr Health Aging 23, 979–986. https://doi.org/10.1007/S12603-019-1255-1

Chiang, J.M., Kaysen, G.A., Segal, M., Chertow, G.M., Delgado, C., Johansen, K.L., 2019. Low testosterone is associated with frailty, muscle wasting and physical dysfunction among men receiving hemodialysis: A longitudinal analysis. Nephrol Dial Transplant 34, 802–810. https://doi.org/10.1093/NDT/GFY252

Cho, K.H., Lee, H.J., Lee, W.H., 2017. Intra- and inter-rater reliabilities of measurement of ultrasound imaging for muscle thickness and pennation angle of tibialis anterior muscle in stroke patients. Top Stroke Rehabil 24, 368–373. https://doi.org/10.1080/10749357.2017.1285745

Cleasby, M.E., Jamieson, P.M., Atherton, P.J., 2016. Insulin resistance and sarcopenia: Mechanistic links between common co-morbidities. J Endocrinol 229, R67–R81. https://doi.org/10.1530/JOE-15-0533

Clegg, A., Hassan-Smith, Z., 2018. Frailty and the endocrine system. Lancet Diabetes Endocrinol 6, 743–752. https://doi.org/10.1016/S2213-8587(18)30110-4

Cruz-Jentoft, A.J., Sayer, A.A., 2019. Sarcopenia. The Lancet. https://doi.org/10.1016/S0140-6736(19)31138-9

Davies, B., García, F., Ara, I., Artalejo, F.R., Rodriguez-Mañas, L., Walter, S., 2018. Relationship between sarcopenia and frailty in the Toledo Study of Healthy Aging: A population based cross-sectional study. J Am Med Dir Assoc 19, 282–286.

de Bree, R., Meerkerk, C.D.A., Halmos, G.B., Mäkitie, A.A., Homma, A., Rodrigo, J.P., López, F., Takes, R.P., Vermorken, J.B., Ferlito, A., 2022. Measurement of sarcopenia in head and neck cancer patients and its association with frailty. Front Oncol 12. https://doi.org/10.3389/FONC.2022.884988

Doi, T., Makizako, H., Tsutsumimoto, K., Hotta, R., Nakakubo, S., Makino, K., Suzuki, T., Shimada, H., 2018. Association between insulin-like growth factor-1 and frailty among older adults. J Nutr Health Aging 22, 68–72. https://doi.org/10.1007/S12603-017-0916-1

Drummond, M.J., Addison, O., Brunker, L., Hopkins, P.N., McClain, D.A., Lastayo, P.C., Marcus, R.L., 2014. Downregulation of E3 ubiquitin ligases and mitophagy-related genes in skeletal muscle of physically inactive, frail older women: A cross-sectional comparison. J Gerontol A Biol Sci Med Sci 69, 1040–1048. https://doi.org/10.1093/GERONA/GLU004

Eichholzer, M., Barbir, A., Basaria, S., Dobs, A.S., Feinleib, M., Guallar, E., Menke, A., Nelson, W.G., Rifai, N., Platz, E.A., Rohrmann, S., 2012. Serum sex steroid hormones and frailty in older American men of the Third National Health and Nutrition Examination Survey (NHANES III). Aging Male 15, 208–215. https://doi.org/10.3109/13685538.2012.705366

El Assar, M., Álvarez-Bustos, A., Sosa, P., Angulo, J., Rodríguez-Mañas, L., 2022. Effect of physical Activity/Exercise on oxidative stress and inflammation in muscle and vascular aging. Int J Mol Sci 23, 8713. https://doi.org/10.3390/ijms23158713

El Assar, M., Angulo, J., Carnicero, J.A., Walter, S., García-García, F.J., López-Hernández, E., Sánchez-Puelles, J.-M., Rodríguez-Mañas, L., 2017. Frailty is associated with lower expression of genes involved in cellular response to stress: Results from the Toledo Study for Healthy Aging. J Am Med Dir Assoc 18. https://doi.org/10.1016/j.jamda.2017.04.019. https://doi.org/10.1016/j.jamda.2017.04.019

El Assar, M., Angulo, J., Rodríguez-Mañas, L., 2020. Frailty as a phenotypic manifestation of underlying oxidative stress. Free Radic Biol Med 149, 72–77. https://doi.org/10.1016/j.freeradbiomed.2019.08.011

El Assar, M., Angulo, J., Walter, S., Carnicero, J.A., García-García, F.J., Sánchez-Puelles, J.-M., Sánchez-Puelles, C., Rodríguez-Mañas, L., 2018. Better nutritional status is positively associated with mRNA expression of SIRT1 in community-dwelling older adults in the Toledo Study for Healthy Aging. J Nutr 148, 1408–1414. https://doi.org/10.1093/jn/nxy149

El Assar, M., Carnicero, J.A., Angulo, J., Cámara-Hernández, V., García-García, F.J., Mañas, L.R., 2023. Fat mass accounts for insulin resistance impact on functional decline and mortality in nondiabetic older adults. J Am Med Dir Assoc. https://doi.org/10.1016/j.jamda.2023.09.015

Evans, W.J., Hellerstein, M., Orwoll, E., Cummings, S., Cawthon, P.M., 2019. D3-creatine dilution and the importance of accuracy in the assessment of skeletal muscle mass. J Cachexia Sarcopenia Muscle 10, 14–21. https://doi.org/10.1002/jcsm.12390

Faller, J.W., Pereira, D.N., de Souza, S., Nampo, F.K., Orlandi, F.D.S., Matumoto, S., 2019. Instruments for the detection of frailty syndrome in older adults: A systematic review. PLoS One 14, e0216166. https://doi.org/10.1371/journal.pone.0216166

Farrow, M., Biglands, J., Tanner, S.F., Clegg, A., Brown, L., Hensor, E.M.A., O'Connor, P., Emery, P., Tan, A.L., 2021. The effect of ageing on skeletal muscle as assessed by quantitative MR imaging: An association with frailty and muscle strength. Aging Clin Exp Res 33, 291. https://doi.org/10.1007/S40520-020-01530-2

Fielding, R.A., Atkinson, E.J., Aversa, Z., White, T.A., Heeren, A.A., Mielke, M.M., Cummings, S.R., Pahor, M., Leeuwenburgh, C., LeBrasseur, N.K., 2024. Biomarkers of cellular senescence predict the onset of mobility disability and are reduced by physical activity in older adults. J Gerontol A Biol Sci Med Sci 79. https://doi.org/10.1093/gerona/glad257

Fukumoto, Y., Ikezoe, T., Taniguchi, M., Yamada, Y., Sawano, S., Minani, S., Asai, T., Kimura, M., Ichihashi, N., 2021. Cut-off values for lower limb muscle thickness to detect low muscle mass for sarcopenia in older adults. Clin Interv Aging 16, 1215–1222. https://doi.org/10.2147/CIA.S304972

Fukunaga, T., Mori, S., Omura, T., Noda, Y., Fujita, Y., Ohsawa, I., Shigemoto, K., 2021. Muscle fiber type specific alterations of mitochondrial respiratory function and morphology in aged female mice. Biochem Biophys Res Commun 540, 116–122. https://doi.org/10.1016/J.BBRC.2020.11.071

García-Esquinas, E., Graciani, A., Guallar-Castillón, P., López-García, E., Rodríguez-Mañas, L., Rodríguez-Artalejo, F., 2015. Diabetes and risk of frailty and its potential mechanisms: A prospective cohort study of older adults. J Am Med Dir Assoc 16, 748–754. https://doi.org/10.1016/j.jamda.2015.04.008

Gomez-Cabrero, D., Walter, S., Abugessaisa, I., Miñambres-Herraiz, R., Palomares, L.B., Butcher, L., Erusalimsky, J.D., Garcia-Garcia, F.J., Carnicero, J., Hardman, T.C., Mischak, H., Zürbig, P., Hackl, M., Grillari, J., Fiorillo, E., Cucca, F., Cesari, M., Carrie, I., Colpo, M., Bandinelli, S., Feart, C., Peres, K., Dartigues, J.-F., Helmer, C., Viña, J., Olaso, G., García-Palmero, I., Martínez, J.G., Jansen-Dürr, P., Grune, T., Weber, D., Lippi, G., Bonaguri, C., Sinclair, A.J., Tegner, J., Rodriguez-Mañas, L., FRAILOMIC initiative, 2021. A robust machine learning framework to identify signatures for frailty: A nested case-control study in four aging European cohorts. Geroscience 43, 1317–1329. https://doi.org/10.1007/s11357-021-00334-0

Gonçalves, R.S.D.S.A., Maciel, ÁC.C., Rolland, Y., Vellas, B., de Souto Barreto, P., 2022. Frailty biomarkers under the perspective of geroscience: A narrative review. Ageing Res Rev 81, 101737. https://doi.org/10.1016/j.arr.2022.101737

Gonzalez, M.C., Heymsfield, S.B., 2017. Bioelectrical impedance analysis for diagnosing sarcopenia and cachexia: What are we really estimating? J Cachexia Sarcopenia Muscle 8, 187. https://doi.org/10.1002/JCSM.12159

Gu, Y., Meng, G., Wu, H., Zhang, Q., Liu, L., Bao, X., Wang, Y., Zhang, S., Sun, S., Wang, X., Zhou, M., Jia, Q., Song, K., Niu, K., 2019. Thyroid function as a predictor of handgrip strength among middle-aged and older euthyroid adults: The TCLSIH cohort study. J Am Med Dir Assoc 20, 1236–1241. https://doi.org/10.1016/j.jamda.2019.06.014

Guan, B., Luo, J., Huang, X., Tian, F., Sun, S., Ma, Y., Yu, Y., Liu, R., Cao, J., Fan, L., 2022. Association between thyroid hormone levels and frailty in the community-dwelling oldest-old: A cross-sectional study. Chin Med J (Engl) 135, 1962–1968. https://doi.org/10.1097/CM9.0000000000002208

Guan, Z., Ma, L., Wu, C., 2023. Association between serum klotho and physical frailty in middle-aged and older adults: Finding from the National Health and Nutrition Examination Survey. J Am Med Dir Assoc. https://doi.org/10.1016/j.jamda.2023.02.103

Hafızoğlu, M., Eren, F., Neşelioğlu, S., Şahiner, Z., Karaduman, D., Atbaş, C., Dikmeer, A., İleri, I., Balcı, C., Doğu, B.B., Cankurtaran, M., Erel, Ö, Halil, M.G., 2024. Physical frailty is related to oxidative stress through thiol/disulfide homeostasis parameters. Eur Geriatr Med 15, 423–434. https://doi.org/10.1007/s41999-023-00911-w

Henning, T., Kochlik, B., Ara, I., González-Gross, M., Fiorillo, E., Marongiu, M., Cucca, F., Rodriguez-Artalejo, F., Carnicero Carreño, J.A., Rodriguez-Mañas, L., Grune, T., Weber, D., 2023. Patterns of dietary blood markers are related to frailty status in the FRAILOMIC validation phase. Nutrients 15. https://doi.org/10.3390/nu15051142

Hetherington-Rauth, M., Johnson, E., Migliavacca, E., Parimi, N., Langsetmo, L., Hepple, R.T., Grzywinski, Y., Corthesy, J., Ryan, T.E., Ferrucci, L., Feige, J.N., Orwoll, E.S., Cawthon, P.M., 2023. Nutrient metabolites associated with low D3Cr muscle mass, strength, and physical performance in older men. J Gerontol A Biol Sci Med Sci. https://doi.org/10.1093/GERONA/GLAD217

Heymsfield, S.B., Gonzalez, M.C., Lu, J., Jia, G., Zheng, J., 2015. Skeletal muscle mass and quality: Evolution of modern measurement concepts in the context of sarcopenia. Proc Nutr Soc 74, 355–366. https://doi.org/10.1017/S0029665115000129

Hsu, B., Hirani, V., Cumming, R.G., Naganathan, V., Blyth, F.M., Wright, F.C., Waite, L.M., Seibel, M.J., Handelsman, D.J., Le Couteur, D.G., 2019. Cross-sectional and longitudinal relationships between inflammatory biomarkers and frailty in community-dwelling older men: The Concord Health and Ageing in Men Project. J Gerontol A Biol Sci Med Sci 74, 835–841. https://doi.org/10.1093/gerona/glx142

Hwang, A.-C., Liu, L.-K., Lee, W.-J., Chen, L.-Y., Peng, L.-N., Lin, M.-H., Chen, L.-K., 2015. Association of frailty and cardiometabolic risk among community-dwelling middle-aged and older people: Results from the I-Lan Longitudinal Aging Study. Rejuvenation Res 18, 564–572. https://doi.org/10.1089/rej.2015.1699

Hyde, Z., Flicker, L., Almeida, O.P., Hankey, G.J., McCaul, K.A., Chubb, S.A.P., Yeap, B.B., 2010. Low free testosterone predicts frailty in older men: The health in men study. J Clin Endocrinol Metab 95, 3165–3172. https://doi.org/10.1210/JC.2009-2754

Ibad, H.A., Mammen, J.S., Simonsick, E.M., Kwoh, C.K., Guermazi, A., Demehri, S., 2023. Higher thyroid hormone has a negative association with lower limb lean body mass in euthyroid older adults: Analysis from the Baltimore Longitudinal study of aging. Front Aging 4. https://doi.org/10.3389/fragi.2023.1150645

Inglés, M., Gambini, J., Carnicero, J.A., García-García, F.J., Rodríguez-Mañas, L., Olaso-González, G., Dromant, M., Borrás, C., Viña, J., 2014. Oxidative stress is related to frailty, not to age or sex, in a geriatric population: Lipid and protein oxidation as biomarkers of frailty. J Am Geriatr Soc 62, 1324–1328. https://doi.org/10.1111/JGS.12876

Iorio, A.D, Paganelli, R., Abate, M., Barassi, G., Ireland, A., Macchi, C., Molino-Lova, R., Cecchi, F., 2021. Thyroid hormone signaling is associated with physical performance, muscle mass, and strength in a cohort of oldest-old: Results from the Mugello study. Geroscience 43, 1053–1064. https://doi.org/10.1007/s11357-020-00302-0

Jakubiak, G.K., Pawlas, N., Cieślar, G., Stanek, A., 2020. Chronic lower extremity ischemia and its association with the frailty syndrome in patients with diabetes. Int J Environ Res Public Health 17, 1–17. https://doi.org/10.3390/IJERPH17249339

Jang, Y.C., Rodriguez, K., Lustgarten, M.S., Muller, F.L., Bhattacharya, A., Pierce, A., Choi, J.J., Lee, N.H., Chaudhuri, A., Richardson, A.G., Van Remmen, H., 2020. Superoxide-mediated oxidative stress accelerates skeletal muscle atrophy by synchronous activation of proteolytic systems. Geroscience 42, 1579–1591. https://doi.org/10.1007/s11357-020-00200-5

Ju, S.Y., Lee, J.Y., Kim, D.H., 2018. Low 25-hydroxyvitamin D levels and the risk of frailty syndrome: A systematic review and dose-response meta-analysis. BMC Geriatr 18, 206. https://doi.org/10.1186/s12877-018-0904-2

Jung, H.W., Kim, S.W., Lim, J.Y., Kim, K.W., Jang, H.C., Kim, C.H., Kim, K.I., 2014. Frailty status can predict further lean body mass decline in older adults. J Am Geriatr Soc 62, 2110–2117. https://doi.org/10.1111/JGS.13107

Kamwa, V., Welch, C., Hassan-Smith, Z.K., 2022. The endocrinology of sarcopenia and frailty. Minerva Endocrinol 46, 453–468. https://doi.org/10.23736/S2724-6507.20.03198-3

Kane, A.E., Sinclair, D.A., 2019. Frailty biomarkers in humans and rodents: Current approaches and future advances. Mech Ageing Dev 180, 117–128. https://doi.org/10.1016/J.MAD.2019.03.007

Kim, B.-J., Lee, S.H., Isales, C.M., Hamrick, M.W., Kwak, M.K., Koh, J.-M., 2018. Association of serum TSH with handgrip strength in community-dwelling euthyroid elderly. J Clin Endocrinol Metab 103, 3986–3992. https://doi.org/10.1210/jc.2018-01095

Kim, K.M., Jang, H.C., Lim, S., 2016. Differences among skeletal muscle mass indices derived from height-, weight-, and body mass index-adjusted models in assessing sarcopenia. Korean J Intern Med 31, 643–650. https://doi.org/10.3904/KJIM.2016.015

Kirk, B., Duque, G., 2022. Muscle and bone: An indissoluble union. J Bone Miner Res 37, 1211–1212. https://doi.org/10.1002/jbmr.4626

Kitaoka, Y., Tamura, Y., Takahashi, K., Takeda, K., Takemasa, T., Hatta, H., 2019. Effects of Nrf2 deficiency on mitochondrial oxidative stress in aged skeletal muscle. Physiol Rep 7, e13998. https://doi.org/10.14814/phy2.13998

Kleipool, E.E.F.E., Hoogendijk, E.O., Trappenburg, M.C., Handoko, M.L., Huisman, M., Peters, M.J.L.J., Muller, M., 2018. Frailty in older adults with cardiovascular disease: Cause, effect or both? Aging Dis 9, 489–497. https://doi.org/10.14336/AD.2017.1125

Ko, Y.J., Ko, I.-G., 2021. Voluntary wheel running exercise improves aging-induced sarcopenia via activation of peroxisome proliferator-activated receptor gamma coactivator-1α/fibronectin type III domain-containing protein 5/adenosine monophosphate-activated protein kinase signalin. Int Neurourol J 25, S27–34. https://doi.org/10.5213/inj.2142170.085

Kochlik, B., Franz, K., Henning, T., Weber, D., Wernitz, A., Herpich, C., Jannasch, F., Aykaç, V., Müller-Werdan, U., Schulze, M.B., Grune, T., Norman, K., 2023. Frailty is characterized by biomarker patterns reflecting inflammation or muscle catabolism in multi-morbid patients. J Cachexia Sarcopenia Muscle 14, 157–166. https://doi.org/10.1002/jcsm.13118

Kochlik, B., Stuetz, W., Pérès, K., Pilleron, S., Féart, C., García García, F.J., Bandinelli, S., Gomez-Cabrero, D., Rodriguez-Mañas, L., Grune, T., Weber, D., 2019. Associations of fat-soluble micronutrients and redox biomarkers with frailty status in the FRAILOMIC initiative. J Cachexia Sarcopenia Muscle 10, 1339–1346. https://doi.org/10.1002/jcsm.12479

Kumar, R., Mohan, N., Upadhyay, A.D., Singh, A.P., Sahu, V., Dwivedi, S., Dey, A.B., Dey, S., 2014. Identification of serum sirtuins as novel noninvasive protein markers for frailty. Aging Cell 13, 975–980. https://doi.org/10.1111/acel.12260

Kuro-O, M., 2018. Molecular mechanisms underlying accelerated aging by defects in the FGF23-klotho system. Int J Nephrol 2018, 1–6. https://doi.org/10.1155/2018/9679841

Landino, K., Tanaka, T., Fantoni, G., Candia, J., Bandinelli, S., Ferrucci, L., 2021. Characterization of the plasma proteomic profile of frailty phenotype. Geroscience 43, 1029–1037. https://doi.org/10.1007/S11357-020-00288-9

Langmann, G.A., Perera, S., Ferchak, M.A., Nace, D.A., Resnick, N.M., Greenspan, S.L., 2017. Inflammatory markers and frailty in long-term care residents. J Am Geriatr Soc 65, 1777–1783. https://doi.org/10.1111/jgs.14876

Lantier, L., Williams, A.S., Williams, I.M., Yang, K.K., Bracy, D.P., Goelzer, M., James, F.D., Gius, D., Wasserman, D.H., 2015. SIRT3 is crucial for maintaining skeletal muscle insulin action and protects against severe insulin resistance in high-fat–fed mice. Diabetes 64, 3081–3092. https://doi.org/10.2337/db14-1810

LeBrasseur, N.K., de Cabo, R., Fielding, R., Ferrucci, L., Rodriguez-Manas, L., Viña, J., Vellas, B., 2021. Identifying biomarkers for biological age: Geroscience and the ICFSR task force. J Frailty Aging 10, 196–201. https://doi.org/10.14283/jfa.2021.5

Lee, W.J., Chen, L.K., Liang, C.K., Peng, L.N., Chiou, S.T., Chou, P., 2016. Soluble ICAM-1, independent of IL-6, is associated with prevalent frailty in community-dwelling elderly Taiwanese people. PLoS One 11. https://doi.org/10.1371/JOURNAL.PONE.0157877

Li, B., Li, Y., Zhang, Y., Liu, P., Song, Y., Zhou, Y., Ma, L., 2022. Visceral fat obesity correlates with frailty in middle-aged and older adults. Diabetes Metab Syndr Obes 15, 2877–2884. https://doi.org/10.2147/DMSO.S383597

Liang, J., Zhang, H., Zeng, Z., Wu, L., Zhang, Y., Guo, Y., Lv, J., Wang, C., Fan, J., Chen, N., 2021. Lifelong aerobic exercise alleviates sarcopenia by activating autophagy and inhibiting protein degradation via the AMPK/PGC-1α signaling pathway. Metabolites 11, 323. https://doi.org/10.3390/metabo11050323

Liu, C.K., Lyass, A., Larson, M.G., Massaro, J.M., Wang, N., D'Agostino, R.B., Benjamin, E.J., Murabito, J.M., 2016. Biomarkers of oxidative stress are associated with frailty: The Framingham Offspring Study. Age (Omaha) 38, 1. https://doi.org/10.1007/s11357-015-9864-z

Liu, D., Fan, Y.-B., Tao, X.-H., Pan, W.-L., Wu, Y.-X., Wang, X.-H., He, Y.-Q., Xiao, W.-F., Li, Y.-S., 2021. Mitochondrial quality control in sarcopenia: Updated overview of mechanisms and interventions. Aging Dis 12, 2016–2030. https://doi.org/10.14336/AD.2021.0427

Liu, F., Austin, T.R., Schrack, J.A., Chen, J., Walston, J., Mathias, R.A., Grams, M., Odden, M.C., Newman, A., Psaty, B.M., Ramonfaur, D., Shah, A.M., Windham, B.G., Coresh, J., Walker, K.A., 2023. Late-life plasma proteins associated with prevalent and incident frailty: A proteomic analysis. Aging Cell. https://doi.org/10.1111/ACEL.13975

Liu, P., Li, Y.Y., Li, S., Zhang, Y., Song, Y., Ji, T., Li, Y.Y., Ma, L., 2022. Serum progranulin as a potential biomarker for frailty in Chinese older adults. Aging Clin Exp Res 35, 399–406. https://doi.org/10.1007/s40520-022-02318-2

López-Armada, M.J., Riveiro-Naveira, R.R., Vaamonde-García, C., Valcárcel-Ares, M.N., 2013. Mitochondrial dysfunction and the inflammatory response. Mitochondrion 13, 106–118. https://doi.org/10.1016/j.mito.2013.01.003

Lozupone, M., Solfrizzi, V., Sardone, R., Dibello, V., Castellana, F., Zupo, R., Lampignano, L., Bortone, I., Daniele, A., Panza, F., 2024. The epigenetics of frailty. Epigenomics 16, 189–202. https://doi.org/10.2217/EPI-2023-0279

Ma, L., Niu, H., Sha, G., Zhang, Y., Liu, P., Li, Y., 2019. Serum SIRT1 is associated with frailty and adipokines in older adults. J Nutr Health Aging 23, 246–250. https://doi.org/10.1007/S12603-018-1149-7

Mankhong, S., Kim, S., Moon, S., Kwak, H.-B., Park, D.-H., Kang, J.-H., 2020. Experimental models of sarcopenia: Bridging molecular mechanism and therapeutic strategy. Cells 9, 1385. https://doi.org/10.3390/cells9061385

Martín, C.A.G., Zepeda, E.M., Méndez, O.A.L., 2017. Bedside ultrasound measurement of rectus femoris: A tutorial for the nutrition support clinician. J Nutr Metab 2017, 2767232. https://doi.org/10.1155/2017/2767232

Marzetti, E., Calvani, R., Cesari, M., Buford, T.W., Lorenzi, M., Behnke, B.J., Leeuwenburgh, C., 2013. Mitochondrial dysfunction and sarcopenia of aging: From signaling pathways to clinical trials. Int J Biochem Cell Biol 45, 2288–2301. https://doi.org/10.1016/j.biocel.2013.06.024

Marzetti, E., Picca, A., Marini, F., Biancolillo, A., Coelho-Junior, H.J., Gervasoni, J., Bossola, M., Cesari, M., Onder, G., Landi, F., Bernabei, R., Calvani, R., 2019. Inflammatory signatures in older persons with physical frailty and sarcopenia: The frailty "cytokinome" at its core. Exp Gerontol 122, 129–138. https://doi.org/10.1016/J.EXGER.2019.04.019

Masanés, F., i Luque, X.R., Salvà, A., Serra-Rexach, J.A., Artaza, I., Formiga, F., Cuesta, F., Soto, A.L., Ruiz, D., Cruz-Jentoft, A.J., 2017. Cut-off points for muscle mass – not grip strength or gait speed – determine variations in sarcopenia prevalence. J Nutr Health Aging 21, 825–829. https://doi.org/10.1007/S12603-016-0844-5

Matsumoto, Y., Tada, M., Yamada, Y., Mandai, K., Hidaka, N., Koike, T., 2022. The bioimpedance phase angle is more useful than sarcopenia as a predictor of falls in patients with rheumatoid arthritis: Results from a 2-y prospective cohort study. Nutrition 102, 111729. https://doi.org/10.1016/j.nut.2022.111729

McCarthy, C., Schoeller, D., Brown, J.C., Gonzalez, M.C., Varanoske, A.N., Cataldi, D., Shepherd, J., Heymsfield, S.B., 2022. D3-creatine dilution for skeletal muscle mass measurement: Historical development and current status. J Cachexia Sarcopenia Muscle 13, 2595–2607. https://doi.org/10.1002/JCSM.13083

McGregor, R.A., Cameron-Smith, D., Poppitt, S.D., 2014. It is not just muscle mass: A review of muscle quality, composition and metabolism during ageing as determinants of muscle function and mobility in later life. Longev Healthspan 3. https://doi.org/10.1186/2046-2395-3-9

Meerkerk, C.D.A., Chargi, N., de Jong, P.A., van den Bos, F., de Bree, R., 2022. Low skeletal muscle mass predicts frailty in elderly head and neck cancer patients. Eur Arch Otorhinolaryngol 279, 967–977. https://doi.org/10.1007/S00405-021-06835-0

Minetto, M.A., Busso, C., Lalli, P., Gamerro, G., Massazza, G., 2021. DXA-derived adiposity and lean indices for management of cardiometabolic and musculoskeletal frailty: Data interpretation tricks and reporting tips. Front Rehabil Sci 2. https://doi.org/10.3389/fresc.2021.712977

Minetto, M.A., Caresio, C., Menapace, T., Hajdarevic, A., Marchini, A., Molinari, F., Maffiuletti, N.A., 2016. Ultrasound-based detection of low muscle mass for diagnosis of sarcopenia in older adults. PM R 8, 453–462. https://doi.org/10.1016/J.PMRJ.2015.09.014

Mohr, B.A., Bhasin, S., Kupelian, V., Araujo, A.B., O'Donnell, A.B., McKinlay, J.B., 2007. Testosterone, sex hormone-binding globulin, and frailty in older men. J Am Geriatr Soc 55, 548–555. https://doi.org/10.1111/J.1532-5415.2007.01121.X

Mombiela, R.M., de Castro, F.F., Moreno, P., Borras, C., 2017. Ultrasonic echo intensity as a new noninvasive in vivo biomarker of frailty. J Am Geriatr Soc 65, 2685–2690. https://doi.org/10.1111/JGS.15002

Mullie, L., Obrand, A., Bendayan, M., Trnkus, A., Ouimet, M.C., Moss, E., Chen-Tournoux, A., Rudski, L.G., Afilalo, J., 2018. Phase angle as a biomarker for frailty and postoperative mortality: The BICS Study. J Am Heart Assoc 7. https://doi.org/10.1161/JAHA.118.008721

Namioka, N., Hanyu, H., Hirose, D., Hatanaka, H., Sato, T., Shimizu, S., 2017. Oxidative stress and inflammation are associated with physical frailty in patients with Alzheimer's disease. Geriatr Gerontol Int 17, 913–918. https://doi.org/10.1111/ggi.12804

Norman, K., Herpich, C., Müller-Werdan, U., 2023. Role of phase angle in older adults with focus on the geriatric syndromes sarcopenia and frailty. Rev Endocr Metab Disord 24, 429–437. https://doi.org/10.1007/S11154-022-09772-3

Oba, K., Ishikawa, J., Tamura, Y., Fujita, Y., Ito, M., Iizuka, A., Fujiwara, Y., Kodera, R., Toyoshima, K., Chiba, Y., Tanaka, M., Araki, A., 2024. Serum growth differentiation factor 15 levels predict the incidence of frailty among patients with cardiometabolic diseases. Gerontology 70, 517–525. https://doi.org/10.1159/000536150

O'Connell, M., Wu, F., 2014. Androgen effects on skeletal muscle: Implications for the development and management of frailty. Asian J Androl 16, 203. https://doi.org/10.4103/1008-682X.122581

Orwoll, E.S., Blackwell, T., Cummings, S.R., Cauley, J.A., Lane, N.E., Hoffman, A.R., Burghardt, A.J., Evans, W.J., Cawthon, P.M., 2022. CT muscle density, D3Cr muscle mass, and body fat associations with physical performance, mobility outcomes, and mortality risk in older men. J Gerontol A Biol Sci Med Sci 77, 790–799. https://doi.org/10.1093/gerona/glab266

Oviedo-Briones, M., Laso, ÁR., Carnicero, J.A., Cesari, M., Grodzicki, T., Gryglewska, B., Sinclair, A., Landi, F., Vellas, B., Checa-López, M., Rodriguez-Mañas, L., 2021. A comparison of frailty assessment instruments in different clinical and social care settings: The frailtools project. J Am Med Dir Assoc 22. https://doi.org/10.1016/J.JAMDA.2020.09.024. https://doi.org/10.1016/J.JAMDA.2020.09.024

Oviedo-Briones, M., Rodríguez-Laso, Á, Carnicero, J.A., Gryglewska, B., Sinclair, A.J., Landi, F., Vellas, B., Artalejo, F.R., Checa-López, M., Rodriguez-Mañas, L., 2022. The ability of eight frailty instruments to identify adverse outcomes across different settings: The FRAILTOOLS project. J Cachexia Sarcopenia Muscle 13, 1487–1501. https://doi.org/10.1002/jcsm.12990

Peng, X., Hou, L., Zhao, Y., Lin, T., Wang, H., Gao, L., Yue, J., 2022. Frailty and testosterone level in older adults: A systematic review and meta-analysis. Eur Geriatr Med 13, 663–673. https://doi.org/10.1007/S41999-022-00614-8

Perazza, L.R., Brown-Borg, H.M., Thompson, L.V., 2022. Physiological systems in promoting frailty. In Comprehensive Physiology. Wiley, pp. 3575–3620. https://doi.org/10.1002/cphy.c210034

Perkisas, S., Baudry, S., Bauer, J., Beckwée, D., de Cock, A.M., Hobbelen, H., Jager-Wittenaar, H., Kasiukiewicz, A., Landi, F., Marco, E., Merello, A., Piotrowicz, K., Sanchez, E., Sanchez-Rodriguez, D., Scafoglieri, A., Cruz-Jentoft, A., Vandewoude, M., 2019. The SARCUS project: Evidence-based muscle assessment through ultrasound. Eur Geriatr Med 10, 157–158. https://doi.org/10.1007/S41999-018-0141-4

Picca, A., Beli, R., Calvani, R., Coelho-Júnior, H.J., Landi, F., Bernabei, R., Bucci, C., Guerra, F., Marzetti, E., 2020. Older adults with physical frailty and sarcopenia show increased levels of circulating small extracellular vesicles with a specific mitochondrial signature. Cells 9, 973. https://doi.org/10.3390/cells9040973

Polat, Y., Yalcin, A., Yazihan, N., Bahsi, R., Mut Surmeli, D., Akdas, S., Aras, S., Varli, M., 2020. The relationship between frailty and serum alpha klotho levels in geriatric patients. Arch Gerontol Geriatr 91. https://doi.org/10.1016/J.ARCHGER.2020.104225

Pronsato, L., Milanesi, L., Vasconsuelo, A., 2020. Testosterone induces up-regulation of mitochondrial gene expression in murine C2C12 skeletal muscle cells accompanied by an increase of nuclear respiratory factor-1 and its downstream effectors. Mol Cell Endocrinol 500, 110631. https://doi.org/10.1016/j.mce.2019.110631

Rasheed, Z., Rasheed, N., Abdulmonem, W.A., Khan, M.I., 2019. MicroRNA-125b-5p regulates IL-1β induced inflammatory genes via targeting TRAF6-mediated MAPKs and NF-κB signaling in human osteoarthritic chondrocytes. Sci Rep 9, 6882. https://doi.org/10.1038/s41598-019-42601-3

Razi, S., Cogger, V.C., Kennerson, M., Benson, V.L., McMahon, A.C., Blyth, F.M., Handelsman, D.J., Seibel, M.J., Hirani, V., Naganathan, V., Waite, L., de Cabo, R., Cumming, R.G., Couteur, D.G.L., 2017. SIRT1 polymorphisms and serum-induced SIRT1 protein expression in aging and frailty: The CHAMP Study. J Gerontol A Biol Sci Med Sci 72, 870–876. https://doi.org/10.1093/gerona/glx018

Reiss, J., Iglseder, B., Kreutzer, M., Weilbuchner, I., Treschnitzer, W., Kässmann, H., Pirich, C., Reiter, R., 2016. Case finding for sarcopenia in geriatric inpatients: Performance of bioimpedance analysis in comparison to dual x-ray absorptiometry. BMC Geriatr 16. https://doi.org/10.1186/S12877-016-0228-Z

Rodríguez-Mañas, L., Angulo, J., Carnicero, J.A., el Assar, M., García-García, F.J., Sinclair, A.J., 2022. Dual effects of insulin resistance on mortality and function in non-diabetic older adults: Findings from the Toledo Study of Healthy Aging. Geroscience 44, 1095–1108. https://doi.org/10.1007/s11357-021-00384-4

Rodriguez-Mañas, L., de Carvalho, I.A., Bhasin, S., Bischoff-Ferrari, H.A., Cesari, M., Evans, W., Hare, J.M., Pahor, M., Parini, A., Rolland, Y., Fielding, R.A., Walston, J., Vellas, B., 2020. ICFSR task force perspective on biomarkers for sarcopenia and frailty. J Frailty Aging 9, 4–8. https://doi.org/10.14283/JFA.2019.32

Rodriguez-Mañas, L., Fried, L.P., 2015. Frailty in the clinical scenario. Lancet 385, e7–e9. https://doi.org/10.1016/S0140-6736(14)61595-6

Rodriguez-Mañas, L., Laosa, O., Vellas, B., Paolisso, G., Topinkova, E., Oliva-Moreno, J., Bourdel-Marchasson, I., Izquierdo, M., Hood, K., Zeyfang, A., Gambassi, G., Petrovic, M., Hardman, T.C., Kelson, M.J., Bautmans, I., Abellan, G., Barbieri, M., Peña-Longobardo, L.M., Regueme, S.C., Calvani, R., De Buyser, S., Sinclair, A.J., 2019. Effectiveness of a multimodal intervention in functionally impaired older people with type 2 diabetes mellitus. J Cachexia Sarcopenia Muscle 10, 721–733. https://doi.org/10.1002/JCSM.12432

Rodriguez-Mañas, L., Rodríguez-Artalejo, F., Sinclair, A.J., 2017. The third transition: The clinical evolution oriented to the contemporary older patient. J Am Med Dir Assoc 18, 8–9. https://doi.org/10.1016/j.jamda.2016.10.005

Rodríguez-Mañas, L., Rodriguez-Sánchez, I., 2021. Research on frailty: Where we stand and where we need to go. J Am Med Dir Assoc 22, 520–523. https://doi.org/10.1016/j.jamda.2021.01.061

Romanello, V., 2020. The interplay between mitochondrial morphology and myomitokines in aging sarcopenia. Int J Mol Sci 22, 91. https://doi.org/10.3390/ijms22010091

Rusanova, I., Diaz-Casado, M.E., Fernández-Ortiz, M., Aranda-Martínez, P., Guerra-Librero, A., García-García, F.J., Escames, G., Mañas, L., Acuña-Castroviejo, D., 2018. Analysis of plasma microRNAs as predictors and biomarkers of aging and frailty in humans. Oxid Med Cell Longev 2018, 7671850. https://doi.org/10.1155/2018/7671850

Rusanova, I., Fernández-Martínez, J., Fernández-Ortiz, M., Aranda-Martínez, P., Escames, G., García-García, F.J., Mañas, L., Acuna-Castroviejo, D., 2019. Involvement of plasma miRNAs, muscle miRNAs and mitochondrial miRNAs in the pathophysiology of frailty. Exp Gerontol 124. https://doi.org/10.1016/J.EXGER.2019.110637

Rustani, K., Kundisova, L., Capecchi, P.L., Nante, N., Bicchi, M., 2019. Ultrasound measurement of rectus femoris muscle thickness as a quick screening test for sarcopenia assessment. Arch Gerontol Geriatr 83, 151–154. https://doi.org/10.1016/J.ARCHGER.2019.03.021

Samson, L.D., Buisman, A.M., Ferreira, J.A., Picavet, H.S.J., Verschuren, W.M.M., Boots, A.M.H., Engelfriet, P., 2022. Inflammatory marker trajectories associated with frailty and ageing in a 20-year longitudinal study. Clin Transl Immunology 11. https://doi.org/10.1002/CTI2.1374

Sanders, J.L., Ding, V., Arnold, A.M., Kaplan, R.C., Cappola, A.R., Kizer, J.R., Boudreau, R.M., Cushman, M., Newman, A.B., 2014. Do changes in circulating biomarkers track with each other and with functional changes in older adults? J Gerontol A Biol Sci Med Sci 69, 174–181. https://doi.org/10.1093/GERONA/GLT088

Saum, K.-U., Dieffenbach, A.K., Jansen, E.H.J.M., Schöttker, B., Holleczek, B., Hauer, K., Brenner, H., 2015. Association between oxidative stress and frailty in an elderly German population: Results from the ESTHER Cohort Study. Gerontology 61, 407–415. https://doi.org/10.1159/000380881

Sbrignadello, S., Göbl, C., Tura, A., 2022. Bioelectrical impedance analysis for the assessment of body composition in sarcopenia and type 2 diabetes. Nutrients 14. https://doi.org/10.3390/NU14091864

Sepúlveda, M., Arauna, D., García, F., Albala, C., Palomo, I., Fuentes, E., 2022. Frailty in aging and the search for the optimal biomarker: A review. Biomedicines 10, 1426. https://doi.org/10.3390/biomedicines10061426

Sergi, G., Rui, M., De, Veronese, N., Bolzetta, F., Berton, L., Carraro, S., Bano, G., Coin, A., Manzato, E., Perissinotto, E., 2015. Assessing appendicular skeletal muscle mass with bioelectrical impedance analysis in free-living Caucasian older adults. Clin Nutr 34, 667–673. https://doi.org/10.1016/J.CLNU.2014.07.010

Shardell, M., Semba, R.D., Kalyani, R.R., Bandinelli, S., Prather, A.A., Chia, C.W., Ferrucci, L., 2019. Plasma klotho and frailty in older adults: Findings from the InCHIANTI Study. J Gerontol A Biol Sci Med Sci 74, 1052–1057. https://doi.org/10.1093/gerona/glx202

Shou, J., Chen, P.J., Xiao, W.H., 2020. Mechanism of increased risk of insulin resistance in aging skeletal muscle. Diabetol Metab Syndr 12. https://doi.org/10.1186/s13098-020-0523-x

Sions, J.M., Velasco, T.O., Teyhen, D.S., Hicks, G.E., 2014. Ultrasound imaging: Intraexaminer and interexaminer reliability for multifidus muscle thickness assessment in adults aged 60 to 85 years versus younger adults. J Orthop Sports Phys Ther 44, 425–434. https://doi.org/10.2519/JOSPT.2014.4584

Soysal, P., Isik, A.T., Carvalho, A.F., Fernandes, B.S., Solmi, M., Schofield, P., Veronese, N., Stubbs, B., 2017. Oxidative stress and frailty: A systematic review and synthesis of the best evidence. Maturitas 99, 66–72. https://doi.org/10.1016/j.maturitas.2017.01.006

Soysal, P., Stubbs, B., Lucato, P., Luchini, C., Solmi, M., Peluso, R., Sergi, G., Isik, A.T., Manzato, E., Maggi, S., Maggio, M., Prina, A.M., Cosco, T.D., Wu, Y.-T., Veronese, N., 2016. Inflammation and frailty in the elderly: A systematic review and meta-analysis. Ageing Res Rev 31, 1–8. https://doi.org/10.1016/j.arr.2016.08.006

Sri-On, J., Rueanthip, S., Vanichkulbodee, A., Paksopis, T., Chetanasilpin, C., 2022. The validity of ultrasonographic measurements of the rectus femoris muscle in older adults with sarcopenia in Thai population. Clin Interv Aging 17, 1249–1259. https://doi.org/10.2147/CIA.S375419

Stimpson, S.A., Turner, S.M., Clifton, L.G., Poole, J.C., Mohammed, H.A., Shearer, T.W., Waitt, G.M., Hagerty, L.L., Remlinger, K.S., Hellerstein, M.K., Evans, W.J., 2012. Total-body creatine pool size and skeletal muscle mass determination by creatine-(methyl-D3) dilution in rats. J Appl Physiol (1985) 112, 1940–1948. https://doi.org/10.1152/japplphysiol.00122.2012

Strandberg, T.E., Sirola, J., Pitkälä, K.H., Tilvis, R.S., Strandberg, A.Y., Stenholm, S., 2012. Association of midlife obesity and cardiovascular risk with old age frailty: A 26-year follow-up of initially healthy men. Int J Obes 36, 1153–1157. https://doi.org/10.1038/ijo.2012.83

Swiecicka, A., Eendebak, R.J.A.H., Lunt, M., O'Neill, T.W., Bartfai, G., Casanueva, F.F., Forti, G., Giwercman, A., Han, T.S., Slowikowska-Hilczer, J., Lean, M.E.J., Pendleton, N., Punab, M., Vanderschueren, D., Huhtaniemi, I.T., Wu, F.C.W., Rutter, M.K., 2018. Reproductive hormone levels predict changes in frailty status in community-dwelling older men: European Male Ageing Study Prospective Data. J Clin Endocrinol Metab 103, 701–709. https://doi.org/10.1210/JC.2017-01172

Szlejf, C., Suemoto, C.K., Janovsky, C.C.P.S., Barreto, S.M., Diniz, M. F.H.S., Lotufo, P.A., Bensenor, I.M., 2020. Thyroid function and sarcopenia: Results from the ELSA-Brasil Study. J Am Geriatr Soc 68, 1545–1553. https://doi.org/10.1111/jgs.16416

Tonkin, J., Villarroya, F., Puri, P.L., Vinciguerra, M., 2012. SIRT1 signaling as potential modulator of skeletal muscle diseases. Curr Opin Pharmacol 12, 372–376. https://doi.org/10.1016/J.COPH.2012.02.010

Tran Van Hoi, E., De Glas, N.A., Portielje, J.E.A., Van Heemst, D., Van Den Bos, F., Jochems, S.P., Mooijaart, S.P., 2023. Biomarkers of the ageing immune system and their association with frailty – A systematic review. Exp Gerontol 176. https://doi.org/10.1016/J.EXGER.2023.112163

Travison, T.G., Nguyen, A.H., Naganathan, V., Stanaway, F.F., Blyth, F.M., Cumming, R.G., Le Couteur, D.G., Sambrook, P.N., Handelsman, D.J., 2011. Changes in reproductive hormone concentrations predict the prevalence and progression of the frailty syndrome in older men: The concord health and ageing in men project. J Clin Endocrinol Metab 96, 2464–2474. https://doi.org/10.1210/JC.2011-0143

Trombetti, A., Hars, M., Hsu, F.-C., Reid, K.F., Church, T.S., Gill, T.M., King, A.C., Liu, C.K., Manini, T.M., McDermott, M.M., Newman, A.B., Rejeski, W.J., Guralnik, J.M., Pahor, M., Fielding, R.A., LIFE Study Investigators, 2018. Effect of physical activity on frailty: Secondary analysis of a randomized controlled trial. Ann Intern Med 168, 309–316. https://doi.org/10.7326/M16-2011

Turkel, I., Ozerklig, B., Yılmaz, M., Ulger, O., Kubat, G.B., Tuncer, M., 2023. Mitochondrial transplantation as a possible therapeutic option for sarcopenia. J Mol Med (Berl) 101. https://doi.org/10.1007/S00109-023-02326-3

Vergara, I., Mateo-Abad, M., Saucedo-Figueredo, M.C., Machón, M., Montiel-Luque, A., Vrotsou, K., Nava Del Val, M.A., Díez-Ruiz, A., Güell, C., Matheu, A., Bueno, A., Núñez, J., Rivas-Ruiz, F., 2019. Description of frail older people profiles according to four screening tools applied in primary care settings: A cross sectional analysis. BMC Geriatr 19. https://doi.org/10.1186/S12877-019-1354-1

Veronesi, F., Borsari, V., Cherubini, A., Fini, M., 2021. Association of Klotho with physical performance and frailty in middle-aged and older adults: A systematic review. Exp Gerontol 154, 111518. https://doi.org/10.1016/j.exger.2021.111518

Viña, J., 2019. The free radical theory of frailty: Mechanisms and opportunities for interventions to promote successful aging. Free Radic Biol Med 134, 690–694. https://doi.org/10.1016/J.FREERADBIOMED.2019.01.045

Virgini, V.S., Rodondi, N., Cawthon, P.M., Harrison, S.L., Hoffman, A.R., Orwoll, E.S., Ensrud, K.E., Bauer, D.C., 2015. Subclinical thyroid dysfunction and frailty among older men. J Clin Endocrinol Metab 100, 4524–4532. https://doi.org/10.1210/jc.2015-3191

Wang, Q., Wang, H., Tian, H., Niu, S., Xu, R., 2023. A prospective case-control study examining the relationship between frailty and serum myostatin in older persons with chronic heart failure. Risk Manag Healthc Policy 16, 1343–1349. https://doi.org/10.2147/RMHP.S412725

Welstead, M., Muniz-Terrera, G., Russ, T.C., Corley, J., Taylor, A.M., Gale, C.R., Luciano, M., 2020. Inflammation as a risk factor for the development of frailty in the Lothian Birth Cohort 1936. Exp Gerontol 139. https://doi.org/10.1016/J.EXGER.2020.111055

Wilhelm-Leen, E.R., Hall, Y.N., Horwitz, R.I., Chertow, G.M., 2014. Phase angle, frailty and mortality in older adults. J Gen Intern Med 29, 147–154. https://doi.org/10.1007/S11606-013-2585-Z

Wimer, L., Goncharova, E., Galkina, S., Nyangau, E., Shankaran, M., Davis, A., Prado, L., Munoz, M.C., Epstein, S., Patterson, C., Shaum, N., Hellerstein, M., Evans, W., Melov, S., 2023. The D3-creatine dilution method non-invasively measures muscle mass in mice. Aging Cell 22, e13897. https://doi.org/10.1111/acel.13897

Wu, I.-C., Lin, X.-Z., Liu, P.-F., Tsai, W.-L., Shiesh, S.-C., 2010. Low serum testosterone and frailty in older men and women. Maturitas 67, 348–352. https://doi.org/10.1016/j.maturitas.2010.07.010

Xu, L., Zhang, J., Shen, S., Hong, X., Zeng, X., Yang, Y., Liu, Z., Chen, L., Chen, X., 2020. Association between body composition and frailty in elder inpatients. Clin Interv Aging 15, 313–320. https://doi.org/10.2147/CIA.S243211

Xu, Y.S., Wang, M.M., Chen, D., Jiang, X., Xiong, Z.F., 2022. Inflammatory biomarkers in older adults with frailty: A systematic review and meta-analysis of cross-sectional studies. Aging Clin Exp Res 34, 971–987. https://doi.org/10.1007/S40520-021-02022-7

Yeap, B.B., Paul Chubb, S.A., Lopez, D., Ho, K.K.Y., Hankey, G.J., Flicker, L., 2013. Associations of insulin-like growth factor-I and its binding proteins and testosterone with frailty in older men. Clin Endocrinol (Oxf) 78, 752–759. https://doi.org/10.1111/cen.12052

Yeo, D., Kang, C., Gomez-Cabrera, M.C., Vina, J., Ji, L.L., 2019. Intensified mitophagy in skeletal muscle with aging is downregulated by PGC-1alpha overexpression in vivo. Free Radic Biol Med 130, 361–368. https://doi.org/10.1016/j.freeradbiomed.2018.10.456

Yin, M., Zhang, X., Zheng, X., Chen, C., Tang, H., Yu, Z., He, X., Jing, W., Tang, X., Xu, X., Ni, J., 2023. Cholesterol alone or in combination is associated with frailty among community-dwelling older adults: A cross-sectional study. Exp Gerontol 180. https://doi.org/10.1016/J.EXGER.2023.112254

Zanforlini, B.M., Trevisan, C., Bertocco, A., Piovesan, F., Dianin, M., Mazzochin, M., Alessi, A., Zoccarato, F., Manzato, E., Sergi, G., 2019. Phase angle and metabolic equivalents as predictors of frailty transitions in advanced age. Exp Gerontol 122, 47–52. https://doi.org/10.1016/j.exger.2019.04.016

Zanker, J., Patel, S., Blackwell, T., Duchowny, K., Brennan-Olsen, S., Cummings, S.R., Evans, W.J., Orwoll, E.S., Scott, D., Vogrin, S., Cauley, J.A., Duque, G., Cawthon, P.M., 2020. Walking speed and muscle mass estimated by the D3-creatine dilution method are important components of sarcopenia associated with incident mobility disability in older men: A classification and regression tree analysis. J Am Med Dir Assoc 21, 1997–2002.e1. https://doi.org/10.1016/j.jamda.2020.03.017

Zhang, H., Liu, X., Wang, X., Jiang, Y., 2024. Association of two novel systemic inflammatory biomarkers and frailty based on NHANES 2007–2018. Front Public Health 12. https://doi.org/10.3389/fpubh.2024.1377408

Zhang, H., Qi, G., Wang, K., Yang, J., Shen, Y., Yang, X., Chen, X., Yao, X., Gu, X., Qi, L., Zhou, C., Sun, H., 2023. Oxidative stress: Roles in skeletal muscle atrophy. Biochem Pharmacol 214, 115664. https://doi.org/10.1016/j.bcp.2023.115664

Zhang, H.-H., Ma, X.-J., Wu, L.-N., Zhao, Y.-Y., Zhang, P.-Y., Zhang, Y.-H., Shao, M.-W., Liu, F., Li, F., Qin, G.-J., 2015. SIRT1 attenuates high glucose-induced insulin resistance via reducing mitochondrial dysfunction in skeletal muscle cells. Exp Biol Med 240, 557–565. https://doi.org/10.1177/1535370214557218

Zhang, L., Zeng, X., He, F., Huang, X., 2023. Inflammatory biomarkers of frailty: A review. Exp Gerontol 179, 112253. https://doi.org/10.1016/j.exger.2023.112253

Zhu, K., Wactawski-Wende, J., Ochs-Balcom, H.M., LaMonte, M.J., Hovey, K.M., Evans, W., Shankaran, M., Troen, B.R., Banack, H.R., 2021. The association of muscle mass measured by D3-creatine dilution method with dual-energy X-ray absorptiometry and physical function in postmenopausal women. The Journals of Gerontology: Series A 76, 1591–1599. https://doi.org/10.1093/gerona/glab020

Zwart, A.T., van der Hoorn, A., van Ooijen, P.M.A., Steenbakkers, R.J.H.M., de Bock, G.H., Halmos, G.B., 2019. CT-measured skeletal muscle mass used to assess frailty in patients with head and neck cancer. J Cachexia Sarcopenia Muscle 10, 1060–1069. https://doi.org/10.1002/JCSM.12443

4 Physical Activity and Exercise in the Prevention and Treatment of Frailty

Hélio José Coelho-Júnior, Riccardo Calvani, and Emanuele Marzetti

INTRODUCTION

Frailty is a state of multisystem derangement and poor psychosocial support [1, 2]. The prevalence of frailty increases with age and is highest among those hospitalized or institutionalized [3–5]. Frailty progression increases the vulnerability to many negative events, including falls and fractures, disability, hospitalization, nursing home placement, and death [6–11]. Such a scenario requires a massive utilization of healthcare services, making frailty a costly condition [12–14]. As such, frailty is recognized as a major public health problem and the search for potential therapies to manage this condition remains a pressing topic in medical science.

Physical activity (PA) and exercise training (ET) have been mentioned as candidate therapeutic strategies for the management of frailty [15–17]. PA refers to any body movement produced by skeletal muscles that increases energy expenditure above baseline levels [18]. In contrast, ET refers to a structured and repetitive form of PA specifically designed to improve or maintain one or more health aspects [18]. In the next paragraphs, the main findings regarding the capacity of both movement strategies to prevent and reduce frailty in older adults will be reviewed.

PREVENTION

PA and ET might contribute to the maintenance and improvement of numerous health parameters associated with the development of frailty, likely promoting its prevention during old age. Notably, these effects are not universal, given that specific patterns of PA and characteristics of ET are necessary to reach these beneficial impacts, particularly in individuals at higher risk of frailty (e.g., sarcopenic) [19].

Physical Activity

Numerous studies concluded that the frequent and continuous practice of moderate-to-high intensity PA might reduce the risk of frailty after accompanying middle-aged and old adults over varying follow-up periods. For instance, Kehler et al. [20] examined data from the NHANES database, which included adults aged 50+ years, and found that prolonged sedentary behavior was associated with an increased likelihood of being frail, whereas those who met PA recommendations were less frail. These results were independent of the intensity (i.e., low, moderate, and vigorous) and pattern (i.e., continuous [≥10 min] or sporadic [<10 min]) of PA. Similar findings are reported by many other investigations employing different study designs (e.g., longitudinal, case-control) and examining different PA assessment methods (e.g., steps per day, accelerometer) [21].

These effects might reflect the adherence to healthy lifestyle habits established early in life. Indeed, although there is not much evidence on this theme, the possibility that early life PA levels

 DOI: 10.1201/9781003398295-4

might reduce the risk of frailty during old age exists. In the work of Kolehmainen et al. [22], for example, Finns middle-aged and older adults were followed by a mean follow-up period of 13.6 years (ranging from 1.8 to 39.6 years). Authors found that the practice of PA significantly reduced the risk of pre-frailty and frailty, regardless of participants' age.

A notable result of many of these investigations is the dose-response relationship between PA intensity and the risk of frailty. Intensity refers to the effort exerted by the body to sustain a determined motor task. Low intensity PA involves those performed without causing significant increases in heart rate (HR) or substantial changes in breathing frequency (BF), such as walking and gardening [23]. With the increase of body demands, as in the case of moderate intensity PA, there is an evident rise in HR and BF to reach physical requirements, but individuals are still able to talk and maintain a conversation [23]. Examples of moderate PA are dancing, light jogging, and playing recreational sports. High intensity or vigorous PA, in turn, typically involves short-duration efforts due to the high level of exertion required, which substantially increases heart rate (HR) and breathing frequency (BF) [23]. During these activities, individuals frequently cannot sustain a conversation.

Peterson et al. [24] analyzed well-functioning older adults (70+ years) participants of the health ABC study and found that those with the highest PA levels had the lowest risk of frailty. Rogers et al. [25] expanded these results by reporting that low intensity PA did not reduce the risk of frailty in robust middle-aged and older community-dwelling adults, whereas significant results were found in those who reported moderate PA at least once a week.

Notably, in some investigations [24, 26], the practice of high-intensity PA was not only important but also essential. Ma et al. [26] found that vigorous PA consistently reduced the risk of frailty in older adults in different age groups (i.e., 60–64, 65–69, 70–74, etc.), whereas the effects of moderate intensity PA were restricted to specific age groups and sex-dependent, and no protective effects were found with low intensity PA. In another investigation [24], the risk of severe frailty increased by up to three times in individuals with higher levels of sedentary behavior and moderate-to-high PA. In contrast, it was significantly reduced in those with the highest levels of PA [24].

Adequate PA levels cannot be confined to a specific period of life, but should be maintained or intensified throughout the aging process to minimize the risk of adverse outcomes. Failing to maintain such practices may hinder the potential of PA to reduce frailty risk. In a study [27] that involved community-dwelling older adults aged 70+ years from different European countries, reductions in the frequency of PA over one year increased the risk of frailty, in comparison with the regular practice of moderate intensity PA. These findings are in accordance with the results of Lin et al. [28], in which increases in PA levels over 20 years were associated with a 55% reduced risk of frailty. More specifically, Blodgett et al. [29] found that adding 1 hour/day of moderate-to-high PA might significantly reduce the risk of frailty.

Occupational PA, which refers to the motor tasks performed as part of an individual's job or work responsibilities, also plays a role in influencing the risk of frailty. Van der Valk et al. [30] examined more than 8000 adults who participated in a large cohort of European community-dwelling retirees and reported that those individuals who had physically demanding main lifetime occupations had a higher risk of frailty. On the other hand, moderate to vigorous PA at retirement was associated with a lower risk of frailty.

To provide more specific recommendations on this matter, investigations have conducted a hypothetical analysis applying the isotemporal replacements method, which examines the effects of substituting one type or intensity of PA with another while keeping the total time constant. Mañas et al. [31] analyzed data from the Toledo Health and Aging study and found that replacing 30 minutes per day of sedentary time with 30 minutes of moderate intensity PA, but not low intensity PA, reduced the risk of frailty. Martins et al. [32] expanded on this knowledge by demonstrating that longer durations of replacement yield greater protective effects. Specifically, replacing 60 minutes per day of sedentary behavior or sleep with 60 minutes per day of moderate-intensity PA resulted in a 48% reduction in frailty risk.

Exercise Training

ET might postpone the development of frailty. These assumptions are primarily based on the fact that older people who adhere to ET, principally those practicing resistance training (RT), exhibit greater muscle power, strength, and functional capacity compared to sedentary individuals and those following other ET modalities (e.g., aerobic training) [33, 34] (see Figure 4.1).

The extension of the beneficial effects of RT is unknown. However, it is possible to suggest that lifelong interventions are necessary to maintain adequate physical performance levels, given that the short-term detraining periods, even those involving a few weeks, are associated with significant reductions in physical performance. Although those declines are not sufficient to promote frailty and can be rapidly reversed by resuming the practice of RT, long periods of sedentary behavior are concerning.

The benefits of RT are particularly important in people at high risk of frailty. In this population, non-supervised and not systematically structured RT programs (e.g., without load progression) commonly have limited effects in frailty-related parameters (e.g., physical performance, depressive symptoms, cognitive impairment), emphasizing the need for specific interventions [35].

Robust and pre-frail community-dwelling older adults commonly experience improvements in muscle strength in response to RT programs performed at moderate-to-high intensities. Kalapotharakos et al. [36, 37] found that a high-intensity RT program performed three times a week, composed of three sets of eight repetitions at 80% of participants' maximum capacity, caused greater improvements in muscle strength in comparison to a moderate-intensity RT. Similar findings were found by Taaffe et al. [38], who observed improvements of 60% and 41% in muscle strength in response to high-intensity and moderate-intensity RT programs, respectively. These findings were supported by a pooled analysis of 25 studies that examined more than 800 participants [39].

Findings of the pooled analysis also indicated that the proper management of timer under tension is vital for improving muscle strength. Time under the tension refers to the period that a muscle is under strain during contractions. According to Borde et al. [39], healthy older adults are more likely to experience improvements in muscle strength when RT programs involve muscle contractions

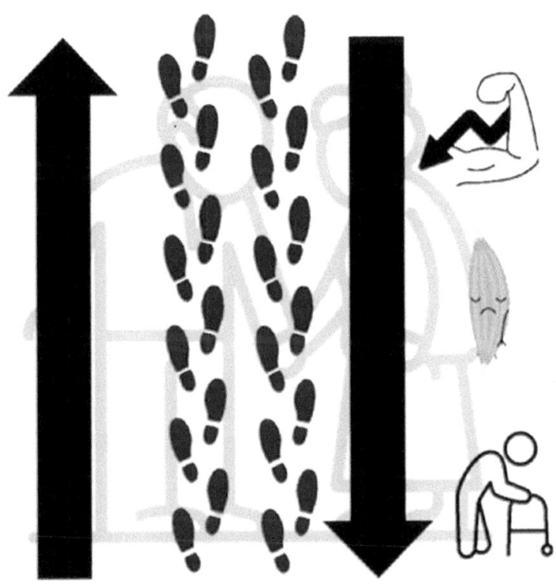

FIGURE 4.1 Bidirectional association between strength, physical exercise, and functional status.

with a mean duration of approximately 5.5 seconds, encompassing nearly 2.5 seconds for concentric actions, when muscles shorten, and nearly 3.0 seconds during eccentric actions, when muscle lengthens while still generating force.

Another ET variable that might impact muscle strength is training volume. It refers to the number of exercises, sets, and repetitions performed during exercise sessions. Galvão and Taaffe [40] found that both low- (one set) and high (three sets)-volume RT programs promoted significant improvements in muscle strength and physical performance, with greater gains obtained with higher-volume regimes. These results were challenged by the observations of Cannon and Marino [41] and Radaelli et al. [42], who reported similar enhancements in response to these variations of RT.

Differences among the results might occur in function of the duration of the interventions. Indeed, low-volume RT might be sufficient to promote neuromuscular gains in untrained older adults, mainly in the first weeks of training, when individuals are learning exercise techniques and gaining resistance, whereas higher physical challenges, as those generated by higher-volume RT, might be necessary to break homeostasis and promote gains in more trained older adults.

Improvements in muscle strength and function might also be reached by performing concentric muscle contractions as fast as possible. This modality of ET is called power training (PT) and has been widely mentioned as an effective strategy to counteract frailty in older adults [43–45]. Particularly, PT is expected to promote greater improvements in physical performance, in comparison to RT [46, 47]. These premises are based on the fact that muscle power, the capacity to generate muscle fast, is more associated with physical performance than other well-established physical capacities, such as muscle strength [48–50].

Empirical evidence on this topic has yielded conflicting results. Bean et al. [51] compared the effects of RT and PT on the short physical performance battery (SPPB) scores of community-dwelling older adults. Results indicated that functional parameters were similarly improved in response to both ET protocols. On the other hand, Ramírez-Campillo et al. [52] found that a 12-week PT composed of six exercises, with three sets of eight repetitions at 75% of participants' maximum capacity, promoted greater improvements in the upper-limb muscle power and walking speed of community-dwelling older women in comparison to RT. In one of the few studies that compared the effects of RT and PT on frailty status, Coelho-Júnior and Uchida [53] noted that 54.5% of the pre-frail participants who took part of a RT program returned to a robust condition after the ET program, while 18.1% participants became robust after the PT program. These improvements were accompanied by greater enhancements in muscle strength in response to RT, but similar increases in other physical performance measures (i.e., 5-time sit-to-stand test [5STS], Timed "Up-and-Go" [TUG]).

Another interesting physical intervention that might reduce the risk of frailty in older adults is multicomponent exercise training (MCET). This type of ET combines many exercise modalities (e.g., RT, aerobic training, balance) in the same exercise session, in an attempt to cause improvements in many physical capabilities simultaneously. Dun et al. [54] investigated the effects of a 12-week MCEP (multicomponent exercise programs) composed of RT, aerobic training, and balance exercises in the pre-frailty status of Chinese older adults who lived in the community. Authors found a reduction of nearly 80% in the prevalence of pre-frailty after the MCET program. These reductions were accompanied by significant enhancements in aerobic endurance, muscle strength, flexibility, agility, and balance.

Although other studies to confirm the effects of MCET on pre-frailty status are still required, many investigations have confirmed its effects on muscle strength and physical performance in older adults with varying conditions [55–59].

Taken as a whole, these findings indicate that RT, PT, and MCET might effectively reduce the risk of frailty in older adults. These results are more evident when ET protocols are designed with specific characteristics (e.g., intensity, volume, velocity of muscle contraction) taken into consideration (see Figure 4.2).

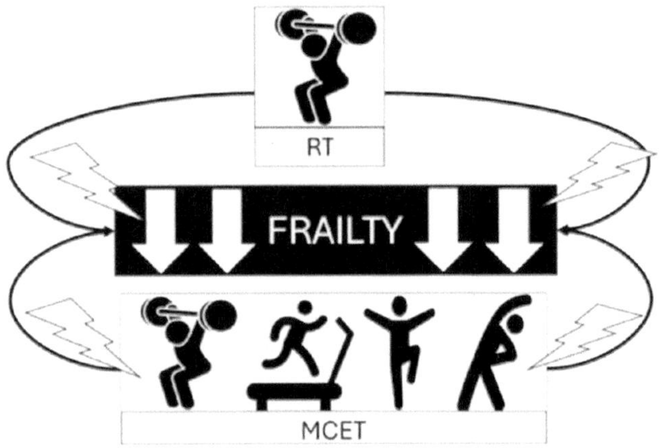

FIGURE 4.2 Schematic representation of the type of physical exercises to be implemented in older people with frailty.

TREATMENT

ET programs are mentioned as a first-line therapy to reduce frailty in older people. Studies have found that RT, PT, and MCEP might effectively enhance physical performance, reduce frailty status, and improve health outcomes in frail older adults. These results were obtained from large randomized clinical trials, systematic reviews, and meta-analyses. Furthermore, individual-based, low-cost, and easy-to-apply ET programs aimed at improving frailty in older adults have been created and show potential to be operationalized in clinical practice.

Since the 1990s, randomized clinical trials have supported the effectiveness of RT protocols in improving physical performance in frail older adults. An influential and elegantly crafted paper was published by Fiatarone et al. [60]. In this study, frail institutionalized nonagenarians showed gains in muscle strength higher than 170% after participating in an 8-week RT program composed of a unique single-joint lower-limb exercise performed at high intensity. The RT program also had direct clinical effects, given that some of the participants, who were not able to walk without a cane or rise from a chair without using their hands at the beginning of the interventions, became able to do so after the RT program. This study paved the way for a series of investigations that supported and expanded its findings.

Meta-analyses on the theme indicate that RT programs promoted significant improvements of 58.1% in muscle strength, 116% in muscle power, and 20.4% in mobility of frail older adults [61, 62]. These benefits are dependent on the specific organizations of RT programs. According to Nagata et al. [62], greater improvements in physical performance are observed in RT programs performed at least three times a week, whereas higher training volume may potentially reduce gains in performance. The fact that frail participants commonly exhibit fatigue is a potential explanation for these findings and indicates that more frequent sessions with short duration might be more recommended for this population.

Studies examining changes in frailty status in response to RT programs are still scarce. A small clinical trial compared the effects of 16-week RT and PT programs on the frailty status and physical performance of nursing home residents with frailty [53]. Authors found that most of the frail participants (~50%) returned to a pre-frail condition after the ET programs. Improvements in frailty status were accompanied by increases in lower-limb muscle strength and power, but no significant changes were observed in balance or mobility. Similar results were found by García Díaz et al. [63] after examining the effects of a 6-month RT protocol with elastic bands on frailty patients with type II diabetes. Nagai et al. [64] proposed a complementary approach examining the effects of a 24-week intervention program combining high-intensity RT and a 10% reduction in sedentary time every 2 weeks. Results indicated that frailty scores and frailty components (i.e., involuntary weight loss and poor grip) were significantly reduced after the intervention.

Reversion in frailty status has also been observed in response to MCET programs. Studies have tested several MCET protocols, which differ in the modalities of exercises included, intensity, volume, and duration. For instance, Tarazona-Santabalbina et al. [65] found that an MCET program performed 5 days a week, for 24 weeks, and composed of endurance, strength, coordination, balance, and flexibility exercises lowered frailty scores in Spanish older adults who attended primary care centers. These results were accompanied by significant improvements in physical performance and cognitive function. Sadjapong et al. [66] observed that a 24-week MCET intervention, 12 weeks of supervised sessions conducted in a health service center, and 12 weeks of home-based exercises, reduced the prevalence of frailty in community-dwelling older adults. Arrieta et al. [67] found similar results after examining the effects of a 24-week MCET intervention on frail older adults living in long-term institutions located in the Basque Country, Spain.

Large multicentric clinical trials have been conducted to examine the effects of ET on the treatment of frailty. The *Lifestyle Interventions and Independence for Elders* (LIFE) study [68, 69] was a randomized clinical trial, conducted in a high-income country in North America, that investigated the effects of a MCET aimed at improving physical function and reducing disability in sedentary older adults with mobility limitations. The program involved aerobic, resistance, balance, and flexibility exercises, which were performed twice a week in a center for 24 months and then at home for more than 7 months. Results of the LIFE study [70] indicated that the MCET significantly reduces the mean number of frailty criteria and the prevalence of frailty. These results were significant even when the analysis was conducted exclusively with younger participants, those with frailty, and those with multimorbidity.

The Sarcopenia and Physical fRailty IN older people: multi-componenT Treatment strategies (SPRINTT) study was a randomized controlled trial designed to investigate the effects of a multicomponent intervention, encompassing MCET and nutritional counselling, on the risk of mobility disability in older adults with sarcopenia and frailty [71]. Results indicated that the SPRINTT MCET program was sufficient to promote significant reductions in the incidence of disability in this population.

The VIVIFRAIL program is a home-based MCET protocol organized according to the specific needs of older adults. This program involves many advantages in comparison to other public health ET programs, given its patient-oriented and hierarchical nature. Moreover, the VIVIFRAIL protocol includes the prescription of PT, which is rarely contemplated, in addition to resistance, power, balance, flexibility, and cardiovascular endurance exercises. The performance of the VIVIFRAIL protocol 3 days a week significantly increased physical function in pre-frail/frail older adults with mild cognitive impairment or mild dementia [72]. Beneficial effects were also observed in acute hospitalized patients, reducing disability and improving physical performance [73].

PRACTICAL VIEW

PA and ET might prevent the incidence of frailty, reduce its progression, and improve related parameters (e.g., physical performance) in older adults. These results have been particularly found in studies that investigated RT, PT, and MCET programs prescribed considering adjustments in specific exercise variables (e.g., intensity, volume, velocity of muscle contraction). As such, these ET programs should be prioritized in public health programs aimed at reducing the frailty burden in older adults.

These premises do not imply that other modalities of ET cannot be used in this context. Actually, using other ET modalities might be necessary to reach participants' desires and increase adherence. However, attention is needed to adequately implement the intervention programs according to the characteristics required to improve physical performance (e.g., high intensity). Furthermore, better results have been obtained when ET protocols are performed under the supervision of exercise specialists.

ET and PA should not be seen as isolated interventions, but as complementary strategies to reduce frailty. This approach is accompanied by inherent problems (e.g., excessive fatigue, lack of motivation) that can be overcome slowly as individuals experience improvements in their physical performance. In fact, exercise instructors might mimic some activities during ET sessions (e.g., climb stairs, change direction quickly, cross the road) to increase the perception of older adults regarding their current capacity to perform physical tasks, since it might be significantly reduced mainly among those who have suffered falls and/or have been recently hospitalized.

Attention is also required for individuals who perform high-intensity PA at work (occupational PA). Although, in common sense, principally among old people, there is a belief that extenuating occupational PA during early life would be sufficient to preserve neuromuscular health during old age, the results abovementioned challenge these premises and indicate that a detailed anamnesis should be conducted before ET sessions to understand individual needs.

Another important aspect to consider is the progression and continuation of ET programs. Most studies have employed short-term strategies (e.g., 6 months) that do not completely represent real-world interventions. The methods that will be utilized to continue improving physical performance across the months and years of ET without losing participants' adherence are challenges and involve the application of ET techniques (e.g., progressive overload), organization methods (e.g., periodization), and individuals' choices and inclinations.

Finally, health professionals responsible for exercise prescription in older adults should consider that increased PA commonly represents adoption of other healthy lifestyle habits, structured routine, and adequate environment. Then, older adults should be supported by a multiprofessional team, depending on the available resources.

CONCLUSIONS

Frailty is a geriatric condition associated with an increased risk for many negative health events in older adults. PA and ET are feasible and effective strategies to prevent and reduce frailty in this population. Specifically, the frequent practice of high-intensity PA reduces the risk of frailty. Specifically, individuals who consistently engage in PA from early adulthood and those who regularly perform high-intensity PA have larger reductions in frailty risk. Regarding ET, RT, PT, and MCEP seem to be the most effective strategies to reduce frailty burden. Health professionals must be aware of the ET variables (e.g., volume, intensity) that should be adequately prescribed to promote improvements in physical performance.

REFERENCES

1. Hoogendijk EO, Afilalo J, Ensrud KE, Kowal P, Onder G, Fried LP. Frailty: Implications for clinical practice and public health. Lancet. 2019;394:1365–75.
2. Clegg A, Young J, Iliffe S, Rikkert MO, Rockwood K. Frailty in elderly people. Lancet. 2013;381:752–62.
3. Kojima G, Iliffe S, Taniguchi Y, Shimada H, Rakugi H, Walters K. Prevalence of frailty in Japan: A systematic review and meta-analysis. J Epidemiol. 2017;27:347–53.
4. Coelho-Junior HJ, Marzetti E, Picca A, Calvani R, Cesari M, Uchida MC. Prevalence of prefrailty and frailty in South America: A systematic review of observational studies. J Frailty Aging. 2020;9.
5. Kojima G. Prevalence of frailty in nursing homes: A systematic review and meta-analysis. J Am Med Dir Assoc. 2015;16:940–5.
6. Fried LP, Tangen CM, Walston J, Newman AB, Hirsch C, Gottdiener J, et al. Frailty in older adults: Evidence for a phenotype. J Gerontol A Biol Sci Med Sci. 2001;56:M146–56.
7. Ensrud KE. Comparison of 2 frailty indexes for prediction of falls, disability, fractures, and death in older women. Arch Intern Med. 2008;168:382.
8. Kojima G. Frailty as a predictor of nursing home placement among community-dwelling older adults. J Geriatr Phys Ther. 2018;41:42–8.
9. Kojima G. Frailty significantly increases the risk of fractures among middle-aged and older people. Evid Based Nurs. 2017;20:119–20.
10. Kojima G. Frailty as a predictor of disabilities among community-dwelling older people: A systematic review and meta-analysis. Disabil Rehabil. 2017;39:1897–908.
11. Kojima G. Frailty as a predictor of hospitalisation among community-dwelling older people: A systematic review and meta-analysis. J Epidemiol Community Health (1978). 2016;70:722–9.

12. Hajek A, Bock J-O, Saum K-U, Matschinger H, Brenner H, Holleczek B, et al. Frailty and healthcare costs-longitudinal results of a prospective cohort study. Age Ageing. 2018;47:233–41.
13. García-Nogueras I, Aranda-Reneo I, Peña-Longobardo LM, Oliva-Moreno J, Abizanda P. Use of health resources and healthcare costs associated with frailty: The FRADEA study. J Nutr Health Aging. 2017; 21:207–14.
14. Ensrud KE, Kats AM, Schousboe JT, Taylor BC, Cawthon PM, Hillier TA, et al. Frailty phenotype and healthcare costs and utilization in older women. J Am Geriatr Soc. 2018;66:1276–83.
15. Aguirre LE, Villareal DT. Physical exercise as therapy for frailty. Nestle Nutr Inst Workshop Ser. 2015; 83:83–92. doi: 10.1159/000382065.
16. Marzetti E, Calvani R, Tosato M, Cesari M, Di Bari M, Cherubini A, et al. Physical activity and exercise as countermeasures to physical frailty and sarcopenia. Aging Clin Exp Res. 2017;29:35–42.
17. Dent E, Martin FC, Bergman H, Woo J, Romero-Ortuno R, Walston JD. Management of frailty: Opportunities, challenges, and future directions. Lancet. 2019;394:1376–86.
18. Garber CE, Blissmer B, Deschenes MR, Franklin BA, Lamonte MJ, Lee I-M, et al. American College of Sports Medicine position stand. Quantity and quality of exercise for developing and maintaining cardiorespiratory, musculoskeletal, and neuromotor fitness in apparently healthy adults: Guidance for prescribing exercise. Med Sci Sports Exerc. 2011;43:1334–59.
19. Coelho-Júnior HJ, Picca A, Calvani R, Marzetti E. Prescription of resistance training for sarcopenic older adults: Does it require specific attention? Ageing Res Rev. 2022;81:101720.
20. Kehler DS, Clara I, Hiebert B, Stammers AN, Hay JL, Schultz A, et al. The association between bouts of moderate to vigorous physical activity and patterns of sedentary behavior with frailty. Exp Gerontol. 2018;104:28–34.
21. Yuki A, Otsuka R, Tange C, Nishita Y, Tomida M, Ando F, et al. Daily physical activity predicts frailty development among community-dwelling older Japanese adults. J Am Med Dir Assoc. 2019;20:1032–6.
22. Kolehmainen L, Havulinna S, Ngandu T, Strandberg T, Levälahti E, Lehtisalo J, et al. Earlier life leisure-time physical activity in relation to age-related frailty syndrome. Age Ageing. 2021;50:161–8.
23. Bull FC, Al-Ansari SS, Biddle S, Borodulin K, Buman MP, Cardon G, Carty C, Chaput JP, Chastin S, Chou R, Dempsey PC, DiPietro L, Ekelund U, Firth J, Friedenreich CM, Garcia L, Gichu M, Jago R, Katzmarzyk PT, Lambert E, Leitzmann M, Milton K, Ortega FB, Ranasinghe C, Stamatakis E, Tiedemann A, Troiano RP, van der Ploeg HP, Wari V, Willumsen JF. World Health Organization 2020 guidelines on physical activity and sedentary behaviour. Br J Sports Med. 2020 Dec;54(24):1451–1462. doi: 10.1136/bjsports-2020-102955. PMID: 33239350; PMCID: PMC7719906.
24. Peterson MJ, Giuliani C, Morey MC, Pieper CF, Evenson KR, Mercer V, et al. Physical activity as a preventative factor for frailty: The health, aging, and body composition study. J Gerontol A Biol Sci Med Sci. 2009;64:61–8.
25. Rogers NT, Marshall A, Roberts CH, Demakakos P, Steptoe A, Scholes S. Physical activity and trajectories of frailty among older adults: Evidence from the English Longitudinal Study of Ageing. PLoS One. 2017;12(2):e0170878.
26. Ma D, Sun Y, Chen G, Hao S, Jiang Z, Wang R, et al. The association between physical activity intensity and frailty risk among older adults across different age groups and genders: Evidence from four waves of the China Health and Retirement Longitudinal Survey. PLoS One. 2024;19(6):e0305346.
27. Zhang X, Tan SS, Franse CB, Bilajac L, Alhambra-Borrás T, Garcés-Ferrer J, et al. Longitudinal association between physical activity and frailty among community-dwelling older adults. J Am Geriatr Soc. 2020;68:1484–93.
28. Lin YK, Chen CY, Cheung DST, Montayre J, Lee CY, Ho MH. The relationship between physical activity trajectories and frailty: A 20-year prospective cohort among community-dwelling older people. BMC Geriatr. 2022;2219(6):e0305346.
29. Blodgett J, Theou O, Kirkland S, Andreou P, Rockwood K. The association between sedentary behaviour, moderate-vigorous physical activity and frailty in NHANES cohorts. Maturitas. 2015;80:187–91.
30. Van Der Valk AM, Theou O, Wallace LMK, Andrew MK, Godin J. Physical demands at work and physical activity are associated with frailty in retirement. Work. 2022;73:695–705.
31. Mañas A, del Pozo-Cruz B, Guadalupe-Grau A, Marín-Puyalto J, Alfaro-Acha A, Rodríguez-Mañas L, et al. Reallocating accelerometer-assessed sedentary time to light or moderate- to vigorous-intensity physical activity reduces frailty levels in older adults: An isotemporal substitution approach in the TSHA study. J Am Med Dir Assoc. 2018;19:185.e1-185.e6.
32. Martins GS, Galvão LL, Tribess S, Meneguci J, Virtuoso Júnior JS. Isotemporal substitution of sleep or sedentary behavior with physical activity in the context of frailty among older adults: A cross-sectional study. Sao Paulo Med J. 2023;141:12–9.

33. Aagaard P, Magnusson PS, Larsson B, Kjær M, Krustrup P. Mechanical muscle function, morphology, and fiber type in lifelong trained elderly. Med Sci Sports Exerc. 2007;39:1989–96.
34. Klitgaard H, Mantoni M, Schiaffino S, Ausoni S, Gorza L, Laurent-Winter C, et al. Function, morphology and protein expression of ageing skeletal muscle: A cross-sectional study of elderly men with different training backgrounds. Acta Physiol Scand. 1990;140:41–54.
35. Gómez-Redondo P, Valenzuela PL, Morales JS, Ara I, Mañas A. Supervised versus unsupervised exercise for the improvement of physical function and well-being outcomes in older adults: A systematic review and meta-analysis of randomized controlled trials. Sports Med. 2024;54:1877.
36. Kalapotharakos VI, Michalopoulos M, Tokmakidis SP, Godolias G, Gourgoulis V. Effects of a heavy and a moderate resistance training on functional performance in older adults. J Strength Cond Res. 2005;19:652–7.
37. Kalapotharakos VI, Michalopoulou M, Godolias G, Tokmakidis SP, Malliou PV, Gourgoulis V. The effects of high- and moderate-resistance training on muscle function in the elderly. J Aging Phys Act. 2004;12:131–43.
38. Taaffe DR, Pruitt L, Pyka G, Guido D, Marcus R. Comparative effects of high- and low-intensity resistance training on thigh muscle strength, fiber area, and tissue composition in elderly women. Clin Physiol. 1996;16:381–92.
39. Borde R, Hortobágyi T, Granacher U. Dose-response relationships of resistance training in healthy old adults: A systematic review and meta-analysis. Sports Med. 2015;45:1693–720.
40. Galvão DA, Taaffe DR. Resistance exercise dosage in older adults: Single- versus multiset effects on physical performance and body composition. J Am Geriatr Soc. 2005;53:2090–7.
41. Cannon J, Marino FE. Early-phase neuromuscular adaptations to high- and low-volume resistance training in untrained young and older women. J Sports Sci. 2010;28:1505–14.
42. Radaelli R, Botton CE, Wilhelm EN, Bottaro M, Lacerda F, Gaya A, et al. Low- and high-volume strength training induces similar neuromuscular improvements in muscle quality in elderly women. Exp Gerontol. 2013;48:710–6.
43. Cadore EL, Izquierdo M. Muscle power training: A hallmark for muscle function retaining in frail clinical setting. J Am Med Dir Assoc. 2018;19:190–2.
44. Radaelli R, Trajano GS, Freitas SR, Izquierdo M, Cadore EL, Pinto RS. Power training prescription in older individuals: Is it safe and effective to promote neuromuscular functional improvements? Sports Med. 2023;53:569–76.
45. Izquierdo M, Cadore EL. Muscle power training in the institutionalized frail: A new approach to counteracting functional declines and very late-life disability. Curr Med Res Opin. 2014;30:1385–90.
46. Tschopp M, Sattelmayer MK, Hilfiker R. Is power training or conventional resistance training better for function in elderly persons? A meta-analysis. Age Ageing. 2011;40:549–56.
47. da Rosa Orssatto LB, de la Rocha Freitas C, Shield AJ, Silveira Pinto R, Trajano GS. Effects of resistance training concentric velocity on older adults' functional capacity: A systematic review and meta-analysis of randomised trials. Exp Gerontol. 2019;127:110731.
48. Lauretani F, Russo CR, Bandinelli S, Bartali B, Cavazzini C, Di Iorio A, et al. Age-associated changes in skeletal muscles and their effect on mobility: An operational diagnosis of sarcopenia. J Appl Physiol (1985). 2003;95:1851–60.
49. Bean JF, Kiely DK, LaRose S, Goldstein R, Frontera WR, Leveille SG, et al. Are changes in leg power responsible for clinically meaningful improvements in mobility among older adults? J Am Geriatr Soc. 2010;58:2363–8.
50. Cuoco A, Callahan DM, Sayers S, Frontera WR, Bean J, Fielding RA. Impact of muscle power and force on gait speed in disabled older men and women. J Gerontol A Biol Sci Med Sci. 2004;59:1200–6.
51. Bean JF, Kiely DK, LaRose S, O'Neill E, Goldstein R, Frontera WR. Increased velocity exercise specific to task training versus the National Institute on Aging's strength training program: Changes in limb power and mobility. J Gerontol A Biol Sci Med Sci. 2009;64:983–91.
52. Ramírez-Campillo R, Castillo A, de la Fuente CI, Campos-Jara C, Andrade DC, Álvarez C, et al. High-speed resistance training is more effective than low-speed resistance training to increase functional capacity and muscle performance in older women. Exp Gerontol. 2014;58:51–7.
53. Coelho-Júnior HJ, Uchida MC. Effects of low-speed and high-speed resistance training programs on frailty status, physical performance, cognitive function, and blood pressure in prefrail and frail older adults. Front Med (Lausanne). 2021;8:702436.
54. Dun Y, Hu P, Ripley-Gonzalez JW, Zhou N, Li H, Zhang W, et al. Effectiveness of a multicomponent exercise program to reverse pre-frailty in community-dwelling Chinese older adults: A randomised controlled trial. Age Ageing. 2022;51(3):afac026.

55. Coelho HJ, Rodrigues B, Feriani DJ, De Oliveira Gonçalves I, Asano RY, Da Silva Aguiar S, et al. Effects of multicomponent exercise on functional and cognitive parameters of hypertensive patients: A quasi-experimental study. J Aging Res. 2017;2017:1978670.
56. Coelho HJ, Sanches IC, Doro M, Asano RY, Feriani DJ, Brietzke C, et al. Multicomponent exercise improves hemodynamic parameters and mobility, but not maximal walking speed, transfer capacity, and executive function of older type II diabetic patients. Biomed Res Int. 2018;2018:4832851.
57. Taguchi N, Higaki Y, Inoue S, Kimura H, Tanaka K. Effects of a 12-month multicomponent exercise program on physical performance, daily physical activity, and quality of life in very elderly people with minor disabilities: An intervention study. J Epidemiol. 2010;20:21–9.
58. Makizako H, Nakai Y, Tomioka K, Taniguchi Y, Sato N, Wada A, et al. Effects of a multicomponent exercise program in physical function and muscle mass in sarcopenic/pre-sarcopenic adults. J Clin Med. 2020;9(5):1386.
59. Chen B, Li M, Zhao H, Liao R, Lu J, Tu J, et al. Effect of multicomponent intervention on functional decline in Chinese older adults: A multicenter randomized clinical trial. J Nutr Health Aging. 2023;27:1063–75.
60. Fiatarone MA, Marks EC, Ryan ND, Meredith CN, Lipsitz LA, Evans WJ. High-intensity strength training in nonagenarians. JAMA. 1990;263:3029.
61. Lopez P, Pinto RS, Radaelli R, Rech A, Grazioli R, Izquierdo M, et al. Benefits of resistance training in physically frail elderly: A systematic review. Aging Clin Exp Res. 2018;30:889–99.
62. Nagata CA, Garcia PA, Hamu TCDS, Caetano MBD, Costa RR, Leal JC, et al. Are dose-response relationships of resistance training reliable to improve functional performance in frail and pre-frail older adults? A systematic review with meta-analysis and meta-regression of randomized controlled trials. Ageing Res Rev. 2023;91:102079.
63. García Díaz E, Alonso Ramírez J, Herrera Fernández N, Peinado Gallego C, Pérez Hernández DG. Effect of strength exercise with elastic bands and aerobic exercise in the treatment of frailty of the elderly patient with type 2 diabetes mellitus. Endocrinol Diabetes Nutr. 2019;66:563–70.
64. Nagaia K, Miyamato T, Okamae A, Tamaki A, Fujioka H, Wada Y, et al. Physical activity combined with resistance training reduces symptoms of frailty in older adults: A randomized controlled trial. Arch Gerontol Geriatr. 2018;76:41–7.
65. Tarazona-Santabalbina FJ, Gomez-Cabrera MC, Perez-Ros P, Martinez-Arnau FM, Cabo H, Tsaparas K, et al. A multicomponent exercise intervention that reverses frailty and improves cognition, emotion, and social networking in the community-dwelling frail elderly: A randomized clinical trial. J Am Med Dir Assoc. 2016;17:426–33.
66. Sadjapong U, Yodkeeree S, Sungkarat S, Siviroj P. Multicomponent exercise program reduces frailty and inflammatory biomarkers and improves physical performance in community-dwelling older adults: A randomized controlled trial. Int J Environ Res Public Health. 2020;17:3760.
67. Arrieta H, Rezola-Pardo C, Gil SM, Virgala J, Iturburu M, Antón I, et al. Effects of multicomponent exercise on frailty in long-term nursing homes: A randomized controlled trial. J Am Geriatr Soc. 2019;67:1145–51.
68. Pahor M, Blair SN, Espeland M, Fielding R, Gill TM, Guralnik JM, et al. Effects of a physical activity intervention on measures of physical performance: Results of the lifestyle interventions and independence for Elders Pilot (LIFE-P) study. J Gerontol A Biol Sci Med Sci. 2006;61:1157–65.
69. Fielding RA, Rejeski WJ, Blair S, Church T, Espeland MA, Gill TM, et al. The lifestyle interventions and independence for elders study: Design and methods. J Gerontol A Biol Sci Med Sci. 2011;66:1226–37.
70. Cesari M, Vellas B, Hsu F-C, Newman AB, Doss H, King AC, et al. A physical activity intervention to treat the frailty syndrome in older persons-results from the LIFE-p study. J Gerontol A Biol Sci Med Sci. 2015;70:216–22.
71. Bernabei R, Landi F, Calvani R, Cesari M, Del Signore S, Anker SD, et al. Multicomponent intervention to prevent mobility disability in frail older adults: Randomised controlled trial (SPRINTT project). BMJ. 2022;377:e068788.
72. Casas-Herrero Á, Sáez de Asteasu ML, Antón-Rodrigo I, Sánchez-Sánchez JL, Montero-Odasso M, Marín-Epelde I, et al. Effects of Vivifrail multicomponent intervention on functional capacity: A multicentre, randomized controlled trial. J Cachexia Sarcopenia Muscle. 2022;13:884–93.
73. Martínez-Velilla N, Abizanda P, Gómez-Pavón J, Zambom-Ferraresi F, Sáez De Asteasu ML, Fiatarone Singh M, et al. Effect of an exercise intervention on functional decline in very old patients during acute hospitalizations: Results of a multicenter, randomized clinical trial. JAMA Intern Med. 2022; 182:345–7.

5 Crosstalk between Fat, Bone Mineral Density, Muscle, and Frailty

Ana María Ayala Copete, Daniel Rivas, and Gustavo Duque

UNDERSTANDING CELLULAR CROSSTALK TO APPROACH THE SYNDROMIC APPROACH

Aging and senescence are complex processes that lead to changes in body composition and the functioning of body systems. Addressing metabolic changes in the musculoskeletal system is crucial for understanding functional trajectories and phenotypes of frailty, disability, and morbidity and, thus, is relevant for recognizing emerging interrelated diseases (1–3). A better comprehension of the interactions among bone, muscle, and fat infiltration is a matter of growing interest for developing interventions that aim to maintain functional status and independence in a fast-growing aging population. Complex crosstalk of metabolic pathways related to physiological and pathological aging under the scope of cellular and molecular hallmarks of aging is core for understanding the pathophysiology of osteoporosis, sarcopenia, osteosarcopenia, and sarcopenic obesity, among other geriatric syndromes (3, 4); thus, the complex crosstalk between bone mineral density (BMD), fat, muscle, and frailty will be addressed in this chapter.

Muscle and bone represent 55% of our body mass, and there is a physiological connection since early embryonic development and shared origin in mesenchymal cells. Thus, it is unsurprising that there is an age-related loss of bone and muscle quantity and quality through complex geroscience mechanisms. Lower BMD seen in osteopenia and osteoporosis has underlying pathways related to aging, hormonal changes, or inadequate intake of nutrients that lead to the imbalance between bone resorption and formation (5). On the other hand, sarcopenia, a muscle disease and syndrome characterized by loss of muscle mass and function (6), can further affect bone health because of the muscles' role and importance in stimulating bone formation through mechanical stress.

Studies have examined biological signaling molecules like proteins, hormones, bioactive lipids, and antioxidants that could be responsible for decreased mass and strength in aged skeletal muscle, potentially contributing to sarcopenia and other age-related disorders. It has been found that elevated C-reactive protein (CRP), interleukin 6 (IL-6), and tumor necrosis factor (TNF) are correlated with muscle mass and strength (7–12). At the cellular level, the connections between inflammatory cytokines, myokines, adipokines, and osteokines summed to epigenetics, comorbidities, prior medical history, and individual risk factors may further accelerate muscle loss, decreased mobility, frailty, higher risk of falls and fractures, and disability trajectories (11, 12).

Aging is a predisposing factor for free fatty acid (FFA) accumulation in muscles, bones, and other tissues (13), which may accelerate the onset of sarcopenia and osteosarcopenia. Concurrently, obesity promotes the buildup of adipose tissue within skeletal muscle and various organs (13, 14). Lipids in muscle cells can induce lipotoxicity and inflammation, resulting in the dedifferentiation of precursor cells to mesenchymal adipocytes that express genes associated with adipose tissue (15). This dedifferentiation may impair muscle function by disrupting the cellular cascade and transcription factors essential for maintaining muscle integrity and buildup. Furthermore, chronic inflammation can lead to fibrosis and may exacerbate insulin resistance due to the compromised capacity of mitochondria

 DOI: 10.1201/9781003398295-5

to oxidize fatty acids and metabolize fats (16–19). A decrease in mitochondria has been correlated with higher oxidative stress and accumulation of lipids within muscle cells. These are approaches to understanding how this may further be triggered by decreases in proteasome activity due to aging, defects in ubiquitylation and autophagy activity, and deficiencies in clearing deteriorated proteins and end-wasting products (16, 20). When integrating these metabolic crosstalk mechanisms and aging, the emerging concept of sarcopenic obesity rises as a syndrome and condition with simultaneous muscle mass and function reduction, with higher amounts of excess adipose tissue.

Recent evidence supports the role of microRNAs (miRNAs) in the transcription and transduction mechanism between fat and muscle and a role in fat metabolism and adipogenesis, demonstrating how miRNAs are involved in the regulation of muscle growth and atrophy and how miRNAs can modulate inflammatory responses that are linked to muscle degradation. (21–23). Specific miRNAs, such as miR-1, miR-133, and miR-206, influence muscle development and maintenance, and their dysregulation can contribute to muscle loss in sarcopenic obesity (23). Faraldi and collaborators found that miR-21 and miR-146a are associated with inflammation and may impact muscle wasting, while miR-27 and miR-103/107 regulate fat storage and metabolic processes. Thus, dysregulation of these miRNAs can lead to increased fat accumulation, contributing to the obesity component of sarcopenic obesity. miRNAs such as miR-29 and miR-122 are implicated in insulin resistance, a common feature of obesity, and can exacerbate muscle loss. Specific miRNA signatures, hsa-miR-126-5p, hsa-miR-146a-5p, and hsa-miR-425-5p; and hsa-miR-126-5p, hsa-miR-146a-5p, hsa-miR-145-5p, and hsa-miR-25-3p, were specifically associated with muscle mass wasting in post-menopausal osteoporotic women (24). Also, in a recent study by Salamanna et al., only three miRs, miR-155, miR-206, and miR-328, showed the same direction of dysregulation (downregulation) in osteoporosis and sarcopenia (25). miRNAs may serve as potential biomarkers for diagnosing and monitoring sarcopenic obesity. The concentrations of these molecules in blood or tissue samples may indicate the condition of muscle and fat metabolism, as well as inflammatory processes. Additionally, targeting specific miRNAs using inhibitors or mimics may provide innovative therapeutic strategies (23–25).

FAT INFILTRATION, LIPOTOXICITY, OSTEOSARCOPENIA, AGING, AND FRAILTY

The complex metabolic pathways involving fat accumulation and oxidative stress are crucial for understanding the pathological changes in bone and muscle with age. Relationships between muscle atrophy and bone loss are linked to genetic and epigenetic factors. The shared complex pathophysiology of osteoporosis, sarcopenia, and obesity in aging may further explain the pathogenesis and appearance of osteosarcopenia and sarcopenic obesity as an interrelated syndrome due to the clear connection between the muscle and bone and the similar risk factors and similar adverse outcomes (4, 25). In addition to the shared common pathophysiology previously addressed, fat accumulation into muscle and myofibers, osteoporotic bone, and bone marrow causes cells to be more prone to premature cell death, catabolic stress, and inflammation (26).

Sarcopenic obesity is becoming increasingly worrisome among older adults due to major health impacts, adverse outcomes, morbimortality, coexisting medical conditions, susceptibility to geriatric syndromes, and frailty. Thresholds for obesity used are BMI ≥30 kg/m^2 or waist circumference in men ≥102 cm and women ≥88 cm (4, 27). Diagnostic criteria for sarcopenic obesity were established in a consensus statement in 2022 where risk factors that include both high body mass index and augmented waist circumference summed to suggestive factors for sarcopenia should involve a further evaluation of skeletal muscle function and body composition (27, 28) since more adipose tissue may predispose to a higher release of cytokines such as adiponectin, IL-6, and TNF among others influencing inflammation, metabolism, and communication of metabolic signals within organs.

The development of sarcopenic obesity is characterized by the interaction between muscle, fat tissue, hormonal fluctuations, inflammation, oxidative stress, and lifestyle (4, 27–29). Several factors may explain the reduction in muscle mass and muscle fiber atrophy seen in sarcopenic obesity, such as the impact on type II fibers, decreased motor neuron count, buildup of collagen, and tissue degradation leading to fiber deterioration (28, 29). Cells undergoing apoptosis contribute to a catabolic state, and the toxicity from fat impacts bone and muscle mass and function, explaining possible underlying mechanisms of why osteokines and myokines increase. In contrast, stem cells decrease, leading to more fat cells and fewer bone and muscle cells (29).

Additional medical conditions related to metabolic syndrome spectrum worsen the loss of muscle and bone with aging. In aged muscles, type II fiber shrinkage, insulin resistance, lipotoxicity, reduced glycogen production, and mitochondrial dysfunction are observed. Additionally, diabetes disrupts anabolic signaling and is a contributing factor to osteoporosis and higher chances of fractures in individuals with osteopenia (30). The reasons behind these issues include prolonged high blood sugar levels, advanced glycation end products, and oxidative stress, all of which can hinder bone formation. Advanced glycation end products also impede the growth of muscle genes and lead to cell death. At the same time, research indicates that inhibiting osteoglycin affects the proliferation of muscle cells, hinting at a common mechanism involving muscle-bone interactions (31, 32). Figure 5.1 shows the comprehensive scope and metabolic hallmarks of LBD, sarcopenia, FFA infiltration, osteosarcopenia, sarcopenic obesity, and frailty.

The complex interactions between muscle loss and fat accumulation underlie the paradigm of how chronic inflammation, insulin resistance, and metabolic alterations contribute to the development of sarcopenic obesity. The interaction of these elements may result in decreased muscle function and higher fat mass, worsening health problems. It can be challenging to diagnose sarcopenic obesity accurately because the symptoms of sarcopenia and obesity overlap, and it is important to enhance diagnostic tools to better distinguish between fat and muscle mass (32–34).

Sarcopenic obesity can result in serious adverse outcomes, including disability, metabolic disorders, and overall death, stressing the importance of increased awareness of identifying this frequently underdiagnosed clinical condition and addressing interventions in public health (4, 27, 32–34).

This dual condition complicates the diagnosis and management of obesity and age-related muscle decline; thus, comprehensive management of sarcopenic obesity includes various strategies, such as lifestyle changes, physical activity, diet, and medical treatments (27, 32–34).

Researchers in the PIONEER study looked at how sarcopenia, osteoporosis, and obesity, either on their own or in combination, were linked to frailty in older Asian adults (35). They determined

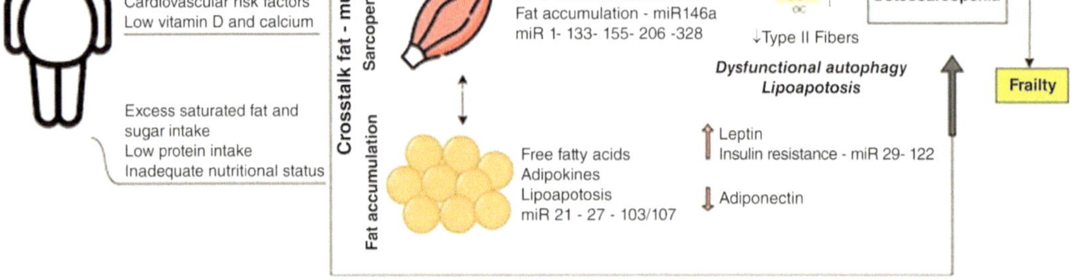

FIGURE 5.1 Underlying mechanism and crosstalk among low bone mineral density, FFA infiltration, muscle, obesity, and frailty

that either a mix of osteoporosis or just osteoporosis on its own was associated with an elevated risk of frailty and, having all three conditions simultaneously, was linked to the greatest risk of frailty. The coexistence of sarcopenia and obesity had a moderating effect on frailty rates, compared to each condition alone, indicating that sarcopenia and obesity play a vital role in the development of frailty, highlighting the need for further research between obesity, frailty, and disability (35).

Another systematic review and meta-analysis conducted by Yuan and Cols explored the link between obesity and three distinct frailty models in 18,764 community-dwelling older adults included in 17 studies. Abdominal circumference and BMI were evaluated. Participants with abdominal obesity were at higher risk of frailty, and participants with a larger waist circumference were found to have a 57% increased risk of frailty compared to those with a normal waist circumference. Furthermore, 12 observational studies involving 37,985 older adults that were incorporated into the meta-analysis exploring the association between BMI and frailty risk found that when compared to the normal BMI group, the risk of frailty was 1.45 times higher in the underweight group and 1.40 times in the obese group, indicating how obesity and underweight were linked to a greater frailty risk among older adults living in the community (36).

Concerning osteosarcopenia, a systematic review and meta-analysis of prospective cohort studies explored osteosarcopenia and an increased mortality risk. This meta-analysis included 14,429 participants with an average age of 70 years, where the prevalence of osteosarcopenia was 12.72%. Following adjustment for four covariates over a follow-up of 6.6 years, osteosarcopenia was associated with an approximately 53% higher likelihood of death (RR: 1.53; 95% CI: 1.28–1.78). The connection between osteosarcopenia and higher mortality risk in older adults emphasizes the importance of addressing osteoporosis and sarcopenia together. This combination of conditions in older individuals can result in complications and, ultimately, mortality (37).

Related to the risk of fractures shared in this complex syndrome, the Osteoporosis Prospective Risk Assessment (OPRA) cohort, a population-based study that included 1604 women older than 75 years of age followed up upon 10 years demonstrated how osteosarcopenia confirmed through anthropometric data prevalence increased from 3.0% to 4.9% to 9.2% between the ages of 75 and 85 and likely enhanced the 10-year risk of major osteoporotic fracture (HR 2.04 [1.27–3.27]), hip fracture (HR 2.67 [1.34–5.32]), and mortality (HR 1.91 [1.21–3.04]) addressing the importance of probable osteosarcopenia as an important risk factor of hip and severe osteoporotic fractures, disability, and mortality (38).

The concept of biochemical interaction between bone, muscle, and fat (Figure 5.1) in a growing older adult population raises a major call to a better understanding of bone and muscle health in aging, aiming a growing need to create new tailored approaches and treatments that may address various conditions like osteoporosis, sarcopenia, and obesity that are directly linked to muscle loss and dysfunction and the phenotypes of osteosarcopenia and sarcopenic obesity.

A COMPREHENSIVE OVERVIEW OF THIS SYNDROME: CURRENT AND FUTURE DIRECTIONS

Osteoporosis, sarcopenia, osteosarcopenia, fat infiltration in the musculoskeletal system, and frailty are interconnected and affect overall physical health and mobility and may be important contributors to frailty, disability, and mortality. An early and comprehensive approach is necessary to mitigate the effects of these conditions and improve the quality of life and functional status of older adults.

The importance of identifying frailty as a predictor of clinical outcomes is recognized in various settings, interventions, and prognostic assessments. The connection between muscle, bone, fat, physical performance, and overall health is reflected in phenotypes of frailty and robustness, highlighting the importance of addressing and identifying these syndromes to predict clinical outcomes in various cardiovascular, oncological, surgical, and decision-making scenarios.

Frailty and its connection to osteosarcopenia was explored in the FRAILTY-AVR cohort study where older adult patients with osteosarcopenia had the highest 1-year mortality (OR, 3.18; 95% CI,

1.54–6.57) and an increased risk of 1-year worsening disability (OR, 2.11; 95% CI, 1.19–3.74) compared to those without this syndrome after transcatheter aortic valve replacement in frail patients. This higher risk of mortality enhances the importance of early detection of syndromes of osteosarcopenia and the need for identifying this syndrome as a risk estimation clinical syndrome that could aid in decision-making and comprehensive rehabilitation efforts in elderly individuals (39).

To address osteosarcopenia and sarcopenic obesity, physical activity, resistance training, and dietary changes focused on protein consumption to improve muscle mass and function are encouraged. Also, an all-encompassing strategy includes tailored lifestyle modifications to lessen the impacts of obesity and muscle loss. By putting these overall and transdisciplinary interventions into practice, older adults may enhance their physical health, performance, and overall well-being (34, 40).

The LIFE study found that engaging in moderate-intensity physical activity led to a significant decrease in long-term movement limitation (HR 0.72, 95% CI 0.57–0.91) compared to an education-based healthy condition program on males aged 60–75 with sarcopenia (41). Currently, the International Clinical Practice Guidelines for Sarcopenia suggest resistance exercise as a main treatment option, while consensus groups advocate for impact exercise to prevent bone loss safely and effectively (42).

Resistance training continues to be the most effective method for addressing osteosarcopenia by improving muscle mass, strength, and function in sarcopenic individuals. Impact exercise has been shown to slightly increase bone density in osteoporotic patients, as addressed in recent Cochrane reviews and research (43–45). The effectiveness of resistance training in reducing falls in older individuals is proven, as 33 studies demonstrated significant improvements in physical abilities among 2,172 participants, indicating an increase in muscle strength. In 73 experiments with 281 participants, there was a mean improvement of 0.84 with a 95% confidence interval of 0.67–1.00, demonstrating improvement in muscle power (43, 45). Maintaining muscle strength also helps protect against overall mortality, such as cancer and heart disease.

Advanced imaging techniques and biomarkers are being investigated as possible answers for addressing these complex syndromes. Effective management of sarcopenic obesity involves a thorough strategy that incorporates resistance training and physical activity for enhancing muscle mass and strength, a diet high in protein and other essential nutrients, and setting tailored treatment approaches under the scope of future direction from a geroscience perspective and combination of nutrition and physical exercise to mitigate frailty (45).

Concerning osteosarcopenia, nutritional approaches, higher protein intake is required to counteract decreased protein synthesis in sarcopenic muscle. Chronic protein supplementation enhances the benefits of resistance exercise, with the PROT-AGE group recommending 1.2–1.5 g/kg/d of protein and 2.5–3 g of leucine per meal for sarcopenic individuals (46). With aging, the body's ability to absorb vitamin D decreases. Thus, the risk of osteosarcopenia is increased due to a combination of factors, including reduced sunlight exposure, dietary habits, and changes in vitamin D receptor expression. Creatine has also been found to improve muscle mass and function, potentially aiding in maintaining bone strength, especially in healthy older adults. Based on the recent evidence, it is advisable to suggest sufficient protein, vitamin D, and creatine consumption as the primary approach to addressing osteosarcopenia. Nonetheless, further research and evidence from clinical trials are needed to evaluate the impact of interventions and to develop strong evidence-based recommendations for patients with osteosarcopenia and sarcopenic obesity (1, 46–48).

Further research aims to further elucidate how miRNAs influence the balance between muscle loss and fat gain. This includes identifying which miRNAs are most significantly involved and understanding their mechanisms of action. Translating miRNA research into clinical practice involves developing reliable assays and creating targeted therapies based on miRNA modulation. Moreover, it is important to investigate new therapeutic targets and incorporate various disciplinary methods to progress the field. Currently, many new approaches are being studied that aim to address both muscle loss and fat accumulation. A comprehensive approach focusing on muscle loss and excess fat is necessary to enhance overall health and quality of life.

Detecting and treating osteosarcopenia and sarcopenic obesity early can reduce morbidity and mortality, as well as improve the quality of life for patients. Although treatments for osteoporosis—such as anabolic medications, bisphosphonates, and hormone replacement therapy—are available, evidence-based clinical practice guidelines often emphasize increased physical activity as the primary treatment for osteosarcopenia.

For sarcopenia, treatment includes upping protein intake, physical activity, resistance training, vitamin D supplementation if needed, and essential amino acids in the diet. Yet no pharmaceutical treatments are approved by the FDA (47, 48). Obesity encompasses complex metabolic pathways that should be addressed individually; nonetheless, common indications for maintaining adequate nutrition, adequate protein intake, and optimal vitamin D levels are essential for older adults to maintain the health of their bones and muscles, as well as their well-being and quality of life. As we have examined in this chapter, age-related metabolic diseases heighten the risk of falls, fractures, disability, and death in older adults.

Further research is needed to explore the combined impact of exercise and nutrition on hormones, growth factors, and inflammatory cytokines that play a role in the development of osteosarcopenia and sarcopenic obesity. The increase in a fast-growing aging population will lead to a higher prevalence of osteosarcopenia, leading to more hospital admissions, fractures, and falls; thus, medical professionals must identify this increasing clinical syndrome, comprehend the common root causes, and implement strategies to promote healthy aging (48).

REFERENCES

1. Kirk, B., & Duque, G. (2020). Osteosarcopenia: A case of geroscience. Aging Medicine, 3(1), 1–6. https://doi.org/10.1002/agm2.12080
2. Fagundes Belchior, G., Kirk, B., Pereira da Silva, E. A. et al. (2020). Osteosarcopenia: Beyond age-related muscle and bone loss. European Geriatric Medicine, 11, 715–724. https://doi.org/10.1007/s41999-020-00355-6
3. Lopez-Otin, C., Blasco, M. A., Partridge, L., Serrano, M., & Kroemer, G. (2013). The hallmarks of aging. Cell, 153(6), 1194–1217. https://doi.org/10.1016/j.cell.2013.05.039
4. Batsis, J. A., & Villareal, D. T. (2018). Sarcopenic obesity in older adults: Aetiology, epidemiology, and treatment strategies. Nature Reviews Endocrinology, 14(9), 513–537. https://doi.org/10.1038/s41574-018-0062-9
5. Cheng, C. H., Chen, L. R., & Chen, K. H. (2022). Osteoporosis due to hormone imbalance: An overview of the effects of estrogen deficiency and glucocorticoid overuse on bone turnover. International Journal of Molecular Sciences, 23(3), 1376. doi: 10.3390/ijms23031376.
6. Cruz-Jentoft, A. J., Bahat, G., Bauer, J., Boirie, Y., Bruyère, O., Cederholm, T., Cooper, C., Landi, F., Rolland, Y., Sayer, A. A., Schneider, S. M., Sieber, C. C., Topinkova, E., Vandewoude, M., Visser, M., & Zamboni, M.; Writing Group for the European Working Group on Sarcopenia in Older People 2 (EWGSOP2), and the Extended Group for EWGSOP2. (2019). Sarcopenia: Revised European consensus on definition and diagnosis. Age Ageing, 48(1), 16–31. doi: 10.1093/ageing/afy169. Erratum in: Age Ageing. 2019 Jul 1;48(4):601. doi: 10.1093/ageing/afz046.
7. Harrison, M. J., & Wood, C. A. (2018). Aging and physical activity: An overview of the current research. Journal of Aging Research, 2018, Article 123456. https://doi.org/10.1155/2018/123456
8. Bano, G., & Kouroumalis, E. (2017). The role of cytokines in inflammatory diseases. Inflammation Research, 66(1), 1–15. https://doi.org/10.1007/s00011-016-1010-7
9. Schaap, L. A., Pluijm, S. M. F., Deeg, D. J. H., et al. (2009). Higher levels of inflammation are associated with lower muscle mass in older adults. Journal of Gerontology: Series A, 64(1), 116–123. https://doi.org/10.1093/gerona/gln034
10. Visser, M., et al. (2002). Muscle mass and muscle strength in relation to physical performance in older men and women. Journal of the American Geriatrics Society, 50(5), 767–773. https://doi.org/10.1046/j.1532-5415.2002.50227.x
11. Peterson, M. D., et al. (2012). Resistance exercise and the prevention of falls: A systematic review and meta-analysis. Journal of Geriatric Physical Therapy, 35(3), 117–126. https://doi.org/10.1519/JPT.0b013e31825b7b87

12. Pirro, M., & Sahebkar, A. (2019). The role of inflammation in sarcopenia: Clinical implications. Journal of Cachexia, Sarcopenia and Muscle, 10(1), 1–12. https://www.ncbi.nlm.nih.gov/pmc/articles/PMC6880711
13. Li, C. W., Yu, K., Shyh-Chang, N., Jiang, Z., Liu, T., Ma, S., Luo, L., Guang, L., Liang, K., Ma, W., Miao, H., Cao, W., Liu, R., Jiang, L. J., Yu, S. L., Li, C., Liu, H. J., Xu, L. Y., Liu, R. J., Zhang, X. Y., & Liu, G. S. (2022). Pathogenesis of sarcopenia and the relationship with fat mass: Descriptive review. Journal of Cachexia Sarcopenia Muscle, 13(2), 781–794. doi: 10.1002/jcsm.12901.
14. Neumann, I., & Sjoberg, J. (2009). Long-term effects of muscle-strengthening activities on older adults. New England Journal of Medicine, 360(14), 1414–1422. https://doi.org/10.1056/NEJMoa0909087
15. Buchowski, M. S., & Sun, M. (2016). Body composition assessment using dual-energy x-ray absorptiometry: A systematic review. Obesity Reviews, 17(5), 394–407. https://doi.org/10.1111/obr.12862
16. Einstein, F. H., Huffman, D. M., Fishman, S., Jerschow, E., Heo, H. J., Atzmon, G., Schechter, C., Barzilai, N., & Muzumdar, R. H. (2010). Aging per se increases the susceptibility to free fatty acid-induced insulin resistance. The Journals of Gerontology: Series A, Biological Sciences and Medical Sciences, 65(8), 800–808.
17. Li, C., Yu, K., Shyh-Chang, N., Jiang, Z., Liu, T., Ma, S., Luo, L., Guang, L., Liang, K., Ma, W., Miao, H., Cao, W., Liu, R., Jiang, L., Yu, S., Li, C., Liu, H., Xu, L., Liu, R., & Liu, G. (2022). Pathogenesis of sarcopenia and the relationship with fat mass: Descriptive review. Journal of Cachexia, Sarcopenia and Muscle, 13(2), 781–794. https://doi.org/10.1002/jcsm.12901
18. Meex, R. C., Hoy, A. J., Morris, A., et al. (2015). Fetuin B is a secreted hepatocyte factor linking steatosis to impaired glucose metabolism. Cell Metabolism, 22(6), 1078–1089. https://doi.org/10.1016/j.cmet.2015.10.001
19. Pal, D., Dasgupta, S., Kundu, R., et al. (2012). Fetuin-A acts as an endogenous ligand of TLR4 to promote lipid-induced insulin resistance. Nature Medicine, 18(8), 1279–1285. https://doi.org/10.1038/nm.2840
20. Kumar, K. G., Trevaskis, J. L., Lam, D. D., et al. (2008). Identification of adropin as a secreted factor linking dietary macronutrient intake with energy homeostasis and lipid metabolism. Cell Metabolism, 8(6), 468–481. https://doi.org/10.1016/j.cmet.2008.09.005
21. Bai, Y., Li, J., & Wang, H. (2021). The role of oxidative stress in the pathogenesis of diabetes mellitus. International Journal of Molecular Sciences, 22(24), 13884. https://doi.org/10.3390/ijms222413884
22. Zhou, Q., Li, J., Wang, L., et al. (2020). Plasma microRNA signature associated with skeletal muscle wasting in post-menopausal osteoporotic women. Journal of Cachexia, Sarcopenia and Muscle, 11(5), 1161–1170. https://doi.org/10.1002/jcsm.12560
23. Iannone, F., Montesanto, A., Cione, E., Crocco, P., Caroleo, M. C., Dato, S., Rose, G., & Passarino, G. (2020). Expression patterns of muscle-specific miR-133b and miR-206 correlate with nutritional status and sarcopenia. Nutrients, 12(2), 297. doi: 10.3390/nu12020297.
24. Faraldi, M., Sansoni, V., Vitale, J., Perego, S., Gomarasca, M., Verdelli, C., Messina, C., Sconfienza, L. M., Banfi, G., Corbetta, S., & Lombardi, G. (2024). Plasma microRNA signature associated with skeletal muscle wasting in post-menopausal osteoporotic women. Journal of Cachexia, Sarcopenia and Muscle, 15(2), 690–701. https://doi.org/10.1002/jcsm.13421
25. Salamanna, F., Contartese, D., Ruffilli, A., Barile, F., Bellavia, D., Marchese, L., Manzetti, M., Viroli, G., Faldini, C., & Giavaresi, G. (2023). Sharing circulating micro-RNAs between osteoporosis and sarcopenia: A systematic review. Life, 13(3), 602. https://doi.org/10.3390/life13030602
26. Zhao, S., Zhang, X., Li, J., Zhang, L., Li, X., & Zhao, W. (2022). Sharing circulating micro-RNAs between osteoporosis and sarcopenia: A systematic review. Journal of Cachexia, Sarcopenia and Muscle, 13(2), 1124–1136. https://doi.org/10.1002/jcsm.12878
27. Prado, C. M., Batsis, J. A., Donini, L. M., et al. (2024). Sarcopenic obesity in older adults: A clinical overview. Nature Reviews Endocrinology, 20, 261–277. https://doi.org/10.1038/s41574-023-00943-z
28. Huang, W., Tan, J., & Lee, J. (2021). Coexistence of osteoporosis, sarcopenia, and obesity in community-dwelling adults: The Yishun Study. Journal of Cachexia, Sarcopenia and Muscle, 12(3), 509–517. https://doi.org/10.1002/jcsm.12628
29. Cameron, J. D., & Bi, L. (2016). Coexistence of osteoporosis, sarcopenia, and obesity in community-dwelling adults: The Yishun Study. Frontiers in Endocrinology, 7, 69. https://doi.org/10.3389/fendo.2016.00069
30. Zhu, L., & Yang, J. (2022). The interplay between obesity, osteoporosis, and sarcopenia in older adults: A review. Clinical Nutrition, 41(3), 699–706. https://doi.org/10.1016/j.clnu.2021.11.014
31. Rivas, A. I., & Martinez-Guerrero, A. (2022). Cell senescence in osteoporosis: A critical review. Cells, 11(1), 160. https://doi.org/10.3390/cells11010160
32. Murray, C. E., & Coleman, C. M. (2019). Impact of diabetes mellitus on bone health. International Journal of Molecular Sciences, 20(19), 4873. https://doi.org/10.3390/ijms20194873

33. Fuchs, S. C., & Cardoso, C. A. (2016). The role of osteocalcin in the interplay between osteoporosis and obesity. Frontiers in Endocrinology, 7, 69. https://doi.org/10.3389/fendo.2016.00069
34. Polyzos, S. A., & Margioris, A. N. (2018). Sarcopenic obesity. Hormones (Athens), 17(3), 321–331. https://doi.org/10.1007/s42000-018-0049-x
35. Heng, M. W. Y., Chan, A. W. D., Man, R. E. K., et al. (2023). Individual and combined associations of sarcopenia, osteoporosis, and obesity with frailty in a multi-ethnic Asian older adult population. BMC Geriatrics, 23, 802. https://doi.org/10.1186/s12877-023-04500-1
36. Yuan, L., Chang, M., & Wang, J. (2021). Abdominal obesity, body mass index and the risk of frailty in community-dwelling older adults: A systematic review and meta-analysis. Age and Ageing, 50(4), 1118–1128. https://doi.org/10.1093/ageing/afab039
37. Veronese, N., Ragusa, F. S., Sabico, S., Dominguez, L. J., Barbagallo, M., Duque, G., & Al-Daghri, N. (2024). Osteosarcopenia increases the risk of mortality: A systematic review and meta-analysis of prospective observational studies. Aging Clinical and Experimental Research, 36(1). https://doi.org/10.1007/s40520-024-02785-9
38. Paulin, T. K., Malmgren, L., McGuigan, F. E., & Akesson, K. E. (2024). Osteosarcopenia: Prevalence and 10-year fracture and mortality risk – A longitudinal, population-based study of 75-year-old women. Calcified Tissue International. https://doi.org/10.1007/s00223-023-01181-1
39. Solla-Suarez, P., Arif, S. G., Ahmad, F., Rastogi, N., Meng, A., Cohen, J. M., Rodighiero, J., Piazza, N., Martucci, G., Lauck, S., Webb, J. G., Kim, D. H., Kovacina, B., & Afilalo, J. (2024). Osteosarcopenia and mortality in older adults undergoing transcatheter aortic valve replacement. JAMA Cardiology, 9(7), 611. https://doi.org/10.1001/jamacardio.2024.0911
40. Hong, A. R., & Kim, S. W. (2018). Effects of resistance exercise on bone health. Endocrinology and Metabolism, 33(4), 435. https://doi.org/10.3803/enm.2018.33.4.435
41. Pahor, M., Guralnik, J. M., Anton, S. D., Ambrosius, W. T., Blair, S. N., Church, T. S., Espeland, M. A., Fielding, R. A., Gill, T. M., Glynn, N. W., Groessl, E. J., King, A. C., Kritchevsky, S. B., Manini, T. M., McDermott, M. M., Miller, M. E., Newman, A. B., & Williamson, J. D. (2020). Impact and lessons from the lifestyle interventions and independence for elders (LIFE) clinical trials of physical activity to prevent mobility disability. Journal of the American Geriatrics Society, 68(4), 872–881. https://doi.org/10.1111/jgs.16365
42. Dent, E., Morley, J. E., Cruz-Jentoft, A. J., Arai, H., Kritchevsky, S. B., Guralnik, J., Bauer, J. M., Pahor, M., Clark, B. C., Cesari, M., Ruiz, J., Sieber, C. C., Aubertin-Leheudre, M., Waters, D. L., Visvanathan, R., Landi, F., Villareal, D. T., Fielding, R., Won, C. W., Theou, O., Martin, F. C., Dong, B., Woo, J., Flicker, L., Ferrucci, L., Merchant, R. A., Cao, L., Cederholm, T., Ribeiro, S. M. L., Rodríguez-Mañas, L., Anker, S. D., Lundy, J., Gutiérrez Robledo, L. M., Bautmans, I., Aprahamian, I., Schols, J. M. G. A., Izquierdo, M., & Vellas, B. (2018). International clinical practice guidelines for sarcopenia (ICFSR): Screening, diagnosis and management, The Journal of Nutrition, Health and Aging, 22(10), 1148–1161. https://doi.org/10.1007/s12603-018-1139-9.
43. Liu, C., & Latham, N. K. (2009). Progressive resistance strength training for improving physical function in older adults. Cochrane Library. https://doi.org/10.1002/14651858.cd002759.pub2
44. Dixon, B. M., & Schmitt, L. R. (2021). The effects of exercise training on muscle strength and endurance in older adults: A systematic review. Journal of Clinical Sleep Medicine, 17(6), 1107–1120. https://doi.org/10.1002/jcsm.12901
45. Hurst, C., Robinson, S. M., Witham, M. D., Dodds, R. M., Granic, A., Buckland, C., De Biase, S., Finnegan, S., Rochester, L., Skelton, D. A., & Sayer, A. A. (2022). Resistance exercise as a treatment for sarcopenia: Prescription and delivery. Age and Ageing, 51(2). https://doi.org/10.1093/ageing/afac003
46. Kirk, B., Zanker, J., & Duque, G. (2020). Osteosarcopenia: Epidemiology, diagnosis, and treatment—Facts and numbers. Journal of Cachexia, Sarcopenia and Muscle, 11(3), 609–618. https://doi.org/10.1002/jcsm.12567
47. Duque, G. (2019). Osteosarcopenia: Bone, muscle, and fat interactions. https://doi.org/10.1007/978-3-030-25890-0
48. Heng, M. W. Y., Chan, A. W. D., Man, R. E. K., Fenwick, E. K., Chew, S. T. H., Tay, L., Sien, N. Y., Ng, D., Koh, F. H., Yong, E., Gupta, P., & Lamoureux, E. L. (2023). Individual and combined associations of sarcopenia, osteoporosis, and obesity with frailty in a multi-ethnic Asian older adult population. BMC Geriatrics, 23(1). https://doi.org/10.1186/s12877-023-04500-1

6 Nutrition and Frailty
Epidemiological and Clinical Facts

Stefano Cacciatore and Francesco Landi

INTRODUCTION

The World Health Organization defines malnutrition as an imbalance in nutrient intake, which can involve deficiencies, excesses, or impaired utilization of essential nutrients.[1] Frailty, malnutrition, and sarcopenia are closely interconnected conditions, each presenting significant health challenges for older adults.[2] The interplay between these conditions can be described as a vicious cycle, where sarcopenia diminishes muscle mass, strength, and performance, increasing frailty risk. In turn, frailty can lead to malnutrition, as frail individuals often experience reduced appetite, difficulties preparing meals, and cognitive or physical impairments that hinder adequate nutrition. Malnutrition further accelerates sarcopenia by depriving the body of essential nutrients, especially proteins and amino acids necessary for muscle repair and maintenance. This cycle results in progressive physical decline, with each condition exacerbating the other.[3] Nutritional interventions play a significant role in breaking this cycle, offering a crucial opportunity to improve health outcomes in older adults.

This chapter presents the key epidemiological studies that establish the connection between nutrition and frailty. It reviews the current evidence while offering practical strategies for nutritional care in frail older adults. Additionally, the chapter highlights potential gaps in both research and clinical practice, providing insights into future directions and areas for further investigation.

EPIDEMIOLOGICAL EVIDENCE ON MALNUTRITION AND FRAILTY

Over the past decades, various observational and interventional studies have explored the relationship between malnutrition and frailty in older adults, consistently underscoring the critical role of nutrition in preventing frailty progression. In particular, large-scale population studies have provided robust evidence on the connection between nutrient intake and frailty development and progression.

Protein intake has been increasingly recognized as a key factor in preventing frailty among older adults. Findings from the Women's Health Initiative demonstrated a strong association between higher dietary protein consumption and a reduced risk of frailty in older women.[4] Moreover, as highlighted by Bollwein et al.,[5] not only the quantity but also the timing and distribution of protein intake throughout the day play significant roles in maintaining muscle mass and physical function. Comprehensive analyses of the available evidence have shown that insufficient protein intake is a major risk factor for frailty,[6] underscoring the importance of adequate protein consumption to preserve muscle strength and support physical function in aging populations.

Beyond protein intake, micronutrient deficiencies have emerged as a crucial factor in frailty progression. The findings from the InCHIANTI Study, for example, revealed a negative association between antioxidant consumption and frailty.[7–12] The Third National Health and Nutrition Examination Survey further identified deficiencies in key micronutrients, such as selenium, carotenoids, and vitamins A, C, and E, as being linked to a heightened risk of frailty in older adults.[13,14] Additionally, the Women's Health and Aging Study found that lower levels of serum carotenoids, vitamin B12, and folate were strongly associated with increased frailty risk.[15–17].

Adherence to dietary patterns such as the Mediterranean Diet has shown protective effects against frailty. Leon-Munoz et al.[18] found that adherence to the Mediterranean diet was associated

 DOI: 10.1201/9781003398295-6

with a reduced risk of incident frailty. Similar results were observed in the Osteoporotic Fractures in Men Study, which showed that high adherence to the Mediterranean diet was protective against frailty over a 4.6-year follow-up period.[19] The Health Aging and Body Composition (ABC) Study further reinforced this, showing that adherence to the Mediterranean diet was linked to a lower risk of mobility decline over a 14-year period.[20] These studies highlight the anti-inflammatory and antioxidant properties of the Mediterranean diet as contributing factors to its protective effects.[21,22]

In addition to nutrient deficiencies, conditions such as anorexia of aging play a critical role in frailty progression. For instance, findings from the ilSIRENTE Study revealed that reduced appetite and food intake accelerate physical decline, thereby significantly contributing to the development and progression of frailty.[23]

Interventional studies aiming to determine whether targeted nutritional interventions can effectively prevent or reverse frailty offer additional insights on the relationship between frailty and nutrition. A meta-analysis by de Moraes et al.[24] synthesized findings from 17 randomized controlled trials focusing on nutritional supplementation in frail older adults. The results were mixed: while protein supplementation led to improvements in muscle mass, the overall impact on frailty status remained limited when used alone. However, interventions combining protein supplementation with resistance exercise, such as the Sarcopenia and Physical fRailty IN older people: multi-componenT Treatment strategies (SPRINTT) trial,[25] demonstrated more significant improvements, particularly in mobility and muscle strength. Notably, this intervention appeared to have a more pronounced effect in those who were frail but not too severely impacted by their condition, emphasizing the importance of early intervention.[25] A meta-analysis by Li et al. further emphasized the importance of leveraging multimodal and multidimensional strategies in nutritional interventions for older adults living with either frailty or pre-frailty.[26]

MALNUTRITION AND SARCOPENIA: KEY DRIVERS OF FRAILTY PROGRESSION

Frailty, malnutrition, and sarcopenia are deeply interconnected, each playing a significant role in accelerating the progression of the others. While aging naturally contributes to the deterioration of muscle function, sarcopenia accelerates this decline, leading to reduced strength and mobility, which is central to the physical manifestations of frailty.[27] Sarcopenia does not arise solely due to aging but is also influenced by a variety of factors, including chronic diseases, inflammations, and hormonal imbalances.[28,29] For example, diseases like diabetes, heart failure, and chronic obstructive pulmonary disease increase muscle catabolism and impair anabolic responses, thus driving muscle loss.[30] Moreover, chronic low-grade inflammation, often referred to as inflammaging, exacerbates muscle degradation through the release of pro-inflammatory cytokines such as TNF-α and IL-6, which activate catabolic pathways like the ubiquitin-proteasome system.[31]

Malnutrition is a pivotal modifiable factor in the development and progression of frailty.[1] The DoMAP model[32] offers a comprehensive perspective on the multifactorial causes of malnutrition in older adults, emphasizing the importance of addressing both direct and indirect causes to prevent and manage nutritional deficits. It identifies three main pathways leading to malnutrition: inadequate intake, heightened nutrient demands, and diminished nutrient bioavailability. These pathways are shaped by both direct and indirect influences, including physical, psychological, and social factors. For instance, difficulties with chewing or swallowing directly impact food intake, while broader determinants like socioeconomic status, food insecurity, and the quality of healthcare indirectly increase the risk of malnutrition.[32]

As highlighted by epidemiological studies, nutritional deficiencies are a significant driver of frailty, particularly in older adults. Poor dietary intake, particularly insufficient protein and energy consumption, accelerates muscle wasting and reduces physical resilience.[33,34] Adequate protein consumption is essential for muscle protein synthesis, which is crucial in maintaining muscle mass and preventing muscle breakdown.[35] In older adults, anabolic resistance, defined as a diminished muscle response to protein intake, can impair MPS, leading to frailty.[36] Consuming higher quantities of protein, especially essential amino acids like leucine, can help overcome anabolic resistance and stimulate muscle protein synthesis.[37–39] Distributing protein intake evenly across meals ("nutritional periodization") can optimize muscle function and prevent muscle loss associated with frailty.[40,41] Inadequate energy intake, often seen in older adults, contributes to muscle wasting and frailty.[42,43] Energy malnutrition can stem from decreased appetite, reduced physical activity, or illness, which collectively lead to insufficient nutrient intake to meet the metabolic demands of muscle repair and growth.[32] This imbalance accelerates muscle loss, decreasing strength and functional capacity, thereby increasing the risk of frailty.[44] Oxidative stress is another important factor contributing to frailty. As individuals age, the balance between reactive oxygen species production and the antioxidant defenses of the body becomes disrupted, leading to cellular damage.[45] This oxidative damage, particularly in muscle tissue, accelerates muscle atrophy and contributes to the development of sarcopenia.[46] Frail individuals often exhibit elevated levels of oxidative stress markers, which correlate with declines in physical performance.[47] Adequate intake of antioxidant-rich foods can potentially mitigate the damaging effects of oxidative stress and improve muscle health.[32] Antioxidant deficiencies, on the other hand, can impair muscle regeneration and increase oxidative stress and inflammation, both of which exacerbate muscle degradation and frailty. Additionally, vitamins and minerals like calcium, magnesium, and zinc are essential for maintaining muscle contractility and preventing physical decline.[45,48,49] Also, metabolic dysregulation, particularly insulin resistance, contributes to frailty by impairing muscle protein synthesis and promoting fat accumulation and muscle loss.[50] Addressing oxidative stress and metabolic dysregulation through targeted interventions, such as antioxidant supplementation and improved protein intake, may help slow the progression of frailty.[51]

Emerging research highlights the critical role of the gut microbiota in influencing frailty. Dysbiosis, characterized by an imbalance in gut microbial populations, has been closely associated with frailty in older adults.[51] A diverse and healthy gut microbiota supports overall health, while gut dysbiosis contributes to increased systemic inflammation and metabolic dysregulation, both of which are strongly linked to frailty.[52] Healthy gut bacteria, such as *Faecalibacterium prausnitzii*, produce short-chain fatty acids, such as butyrate, which play anti-inflammatory roles and support muscle health. In frail individuals, reduced short-chain fatty acid production contributes to inflammaging,[53,54] and increased levels of pro-inflammatory cytokines, such as interleukin-6 and tumor necrosis factor-alpha, promote muscle breakdown, contributing to sarcopenia and frailty.[54,55] The gut–muscle axis demonstrates the critical connection between gut health and frailty.[55] Studies suggest that interventions aimed at restoring gut microbiota diversity, such as the use of probiotics and prebiotics, may help reduce inflammation and support muscle health. These interventions could be promising for managing frailty by addressing the gut–muscle axis and improving overall health outcomes in older adults.[51]

INTERVENTIONS FOR FRAILTY PREVENTION AND MANAGEMENT: NUTRITION AND PHYSICAL ACTIVITY

Nutritional Interventions: Dietary Supplements, Protein, and Micronutrient Intake

Nutritional interventions are a cornerstone of frailty management, with a particular focus on increasing protein intake and addressing micronutrient deficiencies. According to the European Society for Clinical Nutrition and Metabolism (ESPEN) guidelines, older adults should aim for a daily protein intake of at least 1.0–1.2 g/kg of body weight. For those facing chronic illness or in recovery, this requirement can rise to 1.2–1.5 g/kg per day, and even up to 2.0 g/kg in cases of severe illness.[56] High-quality proteins, such as whey and casein, are essential for stimulating muscle protein synthesis.[57] Increased protein intake is critical for muscle maintenance and repair in older adults, who often struggle to consume adequate amounts of dietary protein. Studies show that protein supplementation, particularly when combined with resistance training, can significantly improve muscle strength, functional capacity, and overall body composition.[58,59] Branched-chain amino acids (BCAAs), particularly leucine, are known to activate the mTORC1 pathway, which is key for stimulating muscle protein synthesis.[60]

Beta-hydroxy-beta-methylbutyrate (HMB), a metabolite of leucine, is another powerful supplement that promotes MPS while decreasing muscle protein breakdown. The anti-inflammatory properties of HMB, achieved by inhibiting the NF-κB signaling pathway, help preserve muscle mass and improve function in older adults.[61,62]

Creatine, known for its role in ATP production and promotion of muscle hydration, supports increased training capacity and muscle strength. By enhancing bioenergetics, creatine indirectly reduces inflammation and fosters muscle regeneration through satellite cell proliferation and muscle fiber repair.[63,64] This is particularly useful when combined with resistance training to boost overall muscle mass and performance.[63]

Arginine, an amino acid and precursor of nitric oxide, helps promote MPS via the mTORC1 pathway. It also provides vasodilatory and anti-inflammatory benefits, improving blood flow to muscles and aiding in recovery and regeneration.[65,66] Dietary nitrates, found in foods like beetroot juice, improve mitochondrial efficiency and indirectly support muscle protein synthesis. They also reduce blood pressure and inflammation, while enhancing blood flow to muscles, contributing to better endurance and muscle performance.[67,68]

Antioxidants, including vitamins C and E, and Coenzyme Q10, play a significant role in protecting muscle cells from oxidative damage. They help reduce oxidative stress and inflammation, indirectly supporting muscle protein synthesis and promoting muscle recovery.[69,70]

Combined Approaches: Integrating Nutrition and Exercise for Frailty Prevention and Management

Physical activity, especially resistance training, is critical in preventing and managing frailty. The Lifestyle Interventions and Independence for Elders (LIFE) study demonstrated that structured physical activity programs focused on aerobic and resistance training reduced the risk of major mobility disability in frail older adults.[71] However, there are challenges in prescribing physical activity to older adults. Barriers such as a lack of motivation, fear of injury, and limited access to exercise facilities often reduce participation in structured physical activity programs. Older adults may feel reluctant to engage in exercise due to a lack of confidence or concerns about injury, particularly if they have pre-existing mobility limitations.[72] To address these concerns, physicians must consider the patient's baseline functional status, comorbidities, and preferences when prescribing exercise.[73] A comprehensive pre-exercise assessment is essential to ensure safety and design an appropriate program tailored to the individual's capabilities.[72] Addressing these barriers through education and resources for healthcare providers is necessary. Regular follow-up and monitoring are essential to ensure that older adults are adhering to their exercise programs and achieving the desired outcomes without injury.[74]

Evidence supports the synergy between nutritional interventions and physical activity in improving frailty outcomes. Studies like the SPRINTT trial highlight the significant impact of combining exercise with nutritional support, including protein supplementation.[25] Similarly, a study by Elliot and Hamlin demonstrated that those who received both interventions showed greater improvements in mobility, muscle strength, and lean mass retention compared to those receiving either intervention alone.[75] In clinical practice, the challenge lies in tailoring these interventions to meet individual needs. Frailty manifests differently in each person, and interventions must be personalized based on factors like comorbidities, functional capacity, and nutritional status.[74] Long-term adherence to both nutritional and exercise programs is critical for sustained benefits, although this can be challenging to achieve. Programs such as VIVIFRAIL, which offer personalized exercise regimens based on frailty levels, have successfully improved functional capacity and reduced falls in frail older adults.[76] Despite these challenges, the integration of nutrition and physical activity remains one of the most effective strategies for managing frailty in older adults. By addressing both malnutrition and physical decline, healthcare providers can significantly improve health outcomes, quality of life, and functional independence in this vulnerable population.

ADAPTING NUTRITIONAL CARE FOR FRAIL PATIENTS: PRACTICAL CONSIDERATIONS

Adapting nutritional care for frail older adults requires a personalized approach that addresses the complex interplay between malnutrition, frailty, and age-related conditions. Frailty is often accompanied by multi-system decline, making nutritional optimization both essential and challenging. Effective interventions must focus on each patient's specific needs, prioritizing the prevention of muscle mass loss, correcting micronutrient deficiencies, and managing related conditions like sarcopenia and chronic diseases. To manage malnutrition in frail older adults, strategies must target both modifiable factors (e.g., physical limitations, depression, social isolation, and financial difficulties) and non-modifiable factors (e.g., physiological changes due to aging, such as decreased taste and smell sensitivity and hormonal fluctuations that influence appetite). Additionally, incorporating comprehensive assessments that consider the individual's medical history, lifestyle, and preferences is crucial for developing effective nutritional interventions.

Nutritional Assessment and Screening

A critical first step in managing frailty in older adults is the thorough assessment and regular screening of nutritional status. The ESPEN guidelines emphasize the importance of early identification of individuals at risk of malnutrition, advocating for initial screening at the first medical encounter or within 24 hours of hospital admission. In long-term care facilities, it is recommended that screening be repeated at least monthly and annually in outpatient settings or whenever significant changes are observed in the patient's clinical conditions.[56] Several screening tools are commonly used to assess nutritional risk in older adults. The Mini Nutritional Assessment (MNA) is one of the most effective tools specifically designed for older populations. It allows for early detection of malnutrition risk, starting with a quick screening version (MNA–short form) that can be extended to a more detailed assessment if necessary.[77] The Malnutrition Universal Screening Tool (MUST)[78] and the Short Nutritional Assessment Questionnaire (SNAQ)[79] are also widely used for screening. After identifying patients at risk through screening, a comprehensive nutritional assessment should follow, including anthropometric measurements, physical examination, body composition assessment through direct (i.e., dual-energy X-ray absorptiometry) or indirect methods (i.e., calf circumference), and laboratory testing for markers of malnutrition (i.e., serum albumin).[80,81] The aim is not only to confirm malnutrition but also to identify contributing factors such as anorexia of aging, dysphagia, or gastrointestinal issues, enabling the implementation of targeted interventions that address these underlying causes.[56] Given the high susceptibility of frail older adults to malnutrition, particularly during acute illness or exacerbation of chronic diseases, ongoing monitoring and re-evaluation are essential components of their care.

Another relevant framework for identifying and treating malnutrition is provided by the Global Leadership Initiative on Malnutrition (GLIM). The GLIM criteria combine phenotypic and etiologic components to enhance the accuracy and consistency of nutritional assessments across different clinical settings.[82] They emphasize the importance of identifying both the severity and the underlying causes of malnutrition, facilitating targeted interventions for frail older adults.

Protein and Energy Requirements and the Role of Oral Nutritional Supplements

A key aspect of managing frailty is ensuring adequate protein and energy intake to prevent further muscle mass loss and support recovery. ESPEN guidelines recommend a daily protein intake of at least 1.0–1.2 g/kg of body weight. For those facing chronic illness or in recovery, this requirement can rise to 1.2–1.5 g/kg per day, and even up to 2.0 g/kg in cases of severe illness. Energy intake should also be individualized, typically aiming for 25–30 kcal/kg body weight per day, depending on the patient's condition and energy needs.[56]

High-protein oral nutritional supplements (ONS) are valuable for patients who struggle with poor appetite or swallowing difficulties, especially when dietary interventions alone fail to meet nutritional needs.[83] ONS are available in various formulations to address specific clinical requirements, such as high-protein formulas for those experiencing muscle loss, high-calorie formulas for patients needing concentrated energy intake, or fiber-enriched formulas to support gastrointestinal health. The goal of ONS is to bridge the gap between a patient's nutritional needs and their current intake, ensuring they receive the necessary nutrients to maintain or improve health. It is crucial to note that ONS should supplement regular meals rather than replace them. They are often recommended between meals or before bedtime to increase total caloric and protein intake without overwhelming patients who may already have limited appetite.[56] Tailoring ONS to the patient's clinical condition is essential. For example, in patients with diabetes, nutritional supplements designed to minimize postprandial glycemic spikes should be considered; in patients with chronic kidney disease, lower-protein supplements may be needed to prevent further kidney stress, while for patients undergoing dialysis, higher protein intake might be necessary to compensate for protein loss during dialysis sessions.[84,85] By integrating ONS strategically into a patient's dietary plan, healthcare providers can effectively support nutritional requirements while adapting to the individual's medical and dietary needs.

Micronutrient Supplementation

Micronutrient deficiencies are common in frail older adults and can exacerbate frailty symptoms. Vitamin D deficiency, for example, is widespread and contributes to muscle weakness and an increased risk of fractures. Supplementation with vitamin D, along with calcium, is recommended to maintain bone health and muscle function.[86] Other micronutrients, such as B vitamins, zinc, and omega-3 fatty acids, are important for maintaining cognitive function, immune health, and reducing inflammation.[87] However, caution must be exercised in using supplements, particularly in older adults who may be susceptible to polypharmacy (i.e., the use of multiple medications and supplements), which increases the risk of adverse interactions. The concept of "dietary supplement polypharmacy" is becoming an unrecognized public health concern, as highlighted in research examining the use of multiple supplements without proper medical guidance.[88] This can lead to complications such as liver toxicity or interactions with prescribed medications.[88] Therefore, while micronutrient supplementation offers clear benefits, healthcare providers should closely monitor and individualize supplementation to avoid excessive or inappropriate use. Regular assessments, along with the careful introduction of supplements based on evidence-based guidelines, can ensure that older adults receive the nutrients they need without the risks associated with overuse.

Addressing Poor Appetite

Poor appetite, a key determinant of malnutrition in frail older adults, can significantly impact nutrient intake and overall diet quality. Studies highlight how older adults with poor appetite tend to consume fewer essential nutrients, particularly protein, folate, and zinc, while increasing their intake of carbohydrates. This imbalance is further exacerbated by factors such as chewing difficulties, polypharmacy, and social isolation. As a result, these individuals often have a lower intake of vegetables and other nutrient-rich foods, leading to a decline in diet quality.[89] To address these challenges in clinical practice, regular assessment of appetite using tools like the Simplified Nutritional Appetite Questionnaire (SNAQ) is crucial.[90] Interventions that focus on enhancing meal palatability, offering a variety of foods, and adjusting meal frequency can help improve appetite. For example, providing smaller, nutrient-dense meals more frequently throughout the day can be beneficial for older adults experiencing anorexia of aging. Additionally, the use of ONS can be effective in ensuring adequate intake of protein and calories, particularly in those with limited appetites.[91]

Obesity and Sarcopenic Obesity in Older Adults: Is the "Obesity Paradox" Real?

An important consideration in the nutritional evaluation of older and frail patients is the use of body mass index (BMI), which can be problematic due to age-related changes in body composition. Currently, no universally established cut-off points for BMI exist specifically for older adults, and WHO reference values for adults of all ages are typically employed to define obesity.[92] Interestingly, higher BMI values in older adults have been associated with a lower mortality risk, a phenomenon known as the "obesity paradox." However, despite this paradox, obesity in older adults still presents significant health concerns due to associated cardiometabolic, mechanical, and psychosocial complications, which are as impactful on frailty as undernutrition.[93–95] In older adults, the combination of muscle loss and fat redistribution can make BMI less reliable for classifying obesity. The BMI cut-offs used for younger populations may not adequately reflect the nutritional status of older adults and are often inaccurate for identifying cardiometabolic complications in this age group.[96] While BMI can help assess nutritional risk, both low and high values may indicate malnutrition or health concerns. Optimal BMI ranges for older adults tend to be higher, generally between 24 and 29 kg/m^2, which are associated with better survival rates and reduced frailty risks.[97] This discrepancy, however, underscores a critical gap in clinical practice, as standardized BMI reference ranges specifically for older adults are still lacking.

Sarcopenic obesity, a condition where excess fat coexists with reduced muscle mass and function, presents a significant diagnostic challenge in older adults.[98] The excess fat can mask muscle loss, making the BMI appear within a normal range. This condition is particularly concerning because it is associated with increased risks of metabolic disorders, cardiovascular diseases, frailty, and mortality.[98] Managing sarcopenic obesity requires a comprehensive approach that focuses on reducing fat mass while preserving or improving muscle mass. Weight loss interventions must focus not only on reducing fat but also on preventing further muscle decline. This can be achieved through a combination of dietary modifications and physical exercise. Nutritional interventions should include diets rich in protein and essential nutrients to support muscle maintenance.[99] Strength-based exercises can help increase muscle strength and mass while promoting fat loss, and aerobic exercises like walking or cycling further contribute to a reduction in fat mass and improvements in cardiovascular health.[99] Additionally, psychological counseling can be crucial in managing sarcopenic obesity, particularly in addressing behavioral challenges related to diet and physical activity. Many older adults face barriers such as reduced motivation, depression, or fear of injury, which can hinder adherence to exercise programs. Counseling provides support, strategies, and encouragement for maintaining a healthy lifestyle.[100]

Intrinsic Capacity, Nutrition, and Healthy Ageing

One of the novel aspects discussed during the 2024 ESPEN Symposium was the need to address intrinsic capacity in dietary management of older adults.[101] The WHO introduced the concept of intrinsic capacity to reflect the total physical and mental capabilities an individual can draw upon, emphasizing a holistic, person-centered approach to care. Intrinsic capacity consists of six domains: mobility, cognition, vision, hearing, mood, and vitality. The main objective of the WHO Healthy Aging Framework is to maintain or improve these capacities to promote healthy aging and sustain functional independence in older adults.[102]

Frailty and intrinsic capacity serve complementary roles in healthcare. Both frailty and intrinsic capacity can be analyzed longitudinally. However, frailty assessments typically address short-term vulnerability to stressors, helping clinicians identify immediate risks and personalize interventions. On the other hand, intrinsic capacity takes a proactive, long-term approach, focusing on optimizing and preserving functional abilities across multiple domains to prevent declines in autonomy.[103]

Among the domains of intrinsic capacity, vitality is closely linked to nutrition and is pivotal in delaying frailty and supporting functional independence. Vitality, which includes energy metabolism, neuromuscular function, and immune response, is strongly influenced by an individual's nutritional status.[104] Malnutrition and reduced appetite have been observed as major contributors to frailty and diminished intrinsic capacity in older adults. This is supported by data from the INSPIRE ICOPE-Care Program, which found that appetite and weight loss, hallmarks of malnutrition, were common in older populations and closely associated with impairments in cognition, mood, mobility, and sensory domains.[105] The WHO's Integrated Care for Older People (ICOPE) framework prioritizes early detection of nutritional decline, focusing on key indicators such as unintentional weight loss and reduced appetite, to prevent disability and maintain vitality.[106] To preserve or improve intrinsic capacity, particularly vitality, personalized nutrition interventions are critical. Ensuring adequate protein intake, addressing micronutrient deficiencies, and promoting balanced diets can enhance muscle mass, energy levels, and overall physiological function.[105] Moreover, current research is focusing on the contribution of dietary patterns to sustain and improve vitality capacity in older adults.[107] From this perspective, implementing nutritional interventions may not only reduce the risk of frailty but also support long-term health and independence in older adults.

CONCLUSION AND FUTURE PERSPECTIVES IN NUTRITION FOR FRAIL OLDER ADULTS

The management and prevention of frailty in older adults through personalized nutrition and physical activity are key areas of focus in research and clinical practice. However, several challenges exist in implementing effective nutritional care for frail patients. The individualized nature of frailty requires tailored approaches that account for individual variability, resource limitations, and patient adherence due to reduced appetite, social isolation, and comorbidities.[108,109] It should be noted that compliance with dietary interventions is often low, particularly among socially isolated or cognitively impaired patients.[32,100] To overcome these barriers, a holistic, patient-centered approach is crucial. Personalized care plans that consider each individual's socioeconomic status, nutritional needs, and physical capacity are essential for improving outcomes.[101] Strategies to enhance adherence include involving family members in meal preparation, offering tailored dietary education, and utilizing programs like meals-on-wheels.[56,110] Nutritional care should be implemented as part of a broader multidisciplinary approach to frailty management. Moreover, healthcare systems must prioritize the integration of nutrition and exercise interventions into routine geriatric care.[111,112]

Looking forward, future research should focus on refining these interventions by examining genetic, physiological, and environmental factors that influence individual responses to nutrition and exercise. Personalized strategies based on biomarkers of frailty, such as inflammatory markers or hormonal levels, could improve the precision of interventions, helping to identify frailty earlier and tailor treatments accordingly. Research on the gut microbiome, its interaction with nutrition, and its role in frailty is also an emerging field that may reveal new therapeutic targets.[113]

Furthermore, the integration of novel technologies in frailty management offers promising avenues for improving care. Telemedicine, wearable devices, and digital platforms for tracking dietary intake provide opportunities for enhancing personalized nutrition and monitoring patient progress. These technologies are especially beneficial for older adults living in underserved or rural areas, where access to healthcare may be limited. Digital tools, such as touch-screen apps and smart sensors, have been shown to engage older adults in managing their dietary needs, improving adherence to nutritional interventions.[113,114] Future developments in these technologies could include artificial intelligence-powered dietary monitoring systems that allow for more precise tracking of nutritional intake and offer personalized recommendations based on an individual's health status and biomarkers.[115] Expanding the use of telemedicine for nutritional counseling could further enhance access to professional care, particularly for individuals who cannot attend in-person consultations.[116] These innovations may help overcome the challenges faced by healthcare providers in implementing comprehensive frailty management programs.

Addressing malnutrition and frailty in older adults requires a multifaceted approach, combining personalized nutrition, regular physical activity, and the incorporation of emerging technologies. As research deepens our understanding of the interaction between diet, muscle health, and frailty, clinical practice must evolve to implement these findings more effectively. Ultimately, the goal is to promote healthy aging, reduce frailty incidence, and improve the quality of life for older adults, particularly by leveraging tailored interventions that account for individual variability and accessibility issues.

REFERENCES

1. World Health Organization. Malnutrition. Available at: https://www.who.int/health-topics/malnutrition. Accessed September 11, 2024.
2. Cruz-Jentoft AJ, Kiesswetter E, Drey M, Sieber CC. Nutrition, frailty, and sarcopenia. Aging Clin Exp Res 2017;29(1):43–48. https://doi.org/10.1007/s40520-016-0709-0.
3. Yannakoulia M, Ntanasi E, Anastasiou CA, Scarmeas N. Frailty and nutrition: From epidemiological and clinical evidence to potential mechanisms. Metabolism 2017;68:64–76. https://doi.org/10.1016/j.metabol.2016.12.005.
4. Beasley JM, LaCroix AZ, Neuhouser ML, et al. Protein intake and incident frailty in the Women's Health Initiative observational study. J Am Geriatr Soc 2010;58(6):1063–71. https://doi.org/10.1111/j.1532-5415.2010.02866.x.
5. Bollwein J, Diekmann R, Kaiser MJ, et al. Distribution but not amount of protein intake is associated with frailty: A cross-sectional investigation in the region of Nurnberg. Nutr J 2013;12:109. https://doi.org/10.1186/1475-2891-12-109.
6. Coelho-Junior HJ, Calvani R, Tosato M, Landi F, Picca A, Marzetti E. Protein intake and physical function in older adults: A systematic review and meta-analysis. Ageing Res Rev 2022;81:101731. https://doi.org/10.1016/j.arr.2022.101731.
7. Urpi-Sarda M, Andres-Lacueva C, Rabassa M, et al. The relationship between urinary total polyphenols and the frailty phenotype in a community-dwelling older population: The InCHIANTI Study. J Gerontol A Biol Sci Med Sci 2015;70(9):1141–7. https://doi.org/10.1093/gerona/glv026.
8. Shardell M, D'Adamo C, Alley DE, et al. Serum 25-hydroxyvitamin D, transitions between frailty states, and mortality in older adults: The Invecchiare in Chianti Study. J Am Geriatr Soc 2012;60(2):256–64. https://doi.org/10.1111/j.1532-5415.2011.03830.x.
9. Shardell M, Hicks GE, Miller RR, et al. Association of low vitamin D levels with the frailty syndrome in men and women. J Gerontol A Biol Sci Med Sci 2009;64(1):69–75. https://doi.org/10.1093/gerona/gln007.
10. Lauretani F, Semba RD, Bandinelli S, et al. Carotenoids as protection against disability in older persons. Rejuvenation Res 2008;11(3):557–63. https://doi.org/10.1089/rej.2007.0581.
11. Lauretani F, Semba RD, Bandinelli S, Ray AL, Guralnik JM, Ferrucci L. Association of low plasma selenium concentrations with poor muscle strength in older community-dwelling adults: The InCHIANTI Study. Am J Clin Nutr 2007;86(2):347–52. https://doi.org/10.1093/ajcn/86.2.347.
12. Ble A, Cherubini A, Volpato S, et al. Lower plasma vitamin E levels are associated with the frailty syndrome: The InCHIANTI study. J Gerontol A Biol Sci Med Sci 2006;61(3):278–83. https://doi.org/10.1093/gerona/61.3.278.
13. Smit E, Winters-Stone KM, Loprinzi PD, Tang AM, Crespo CJ. Lower nutritional status and higher food insufficiency in frail older US adults. Br J Nutr 2013;110(1):172–8. https://doi.org/10.1017/S000711451200459X.
14. Wilhelm-Leen ER, Hall YN, Deboer IH, Chertow GM. Vitamin d deficiency and frailty in older Americans. J Intern Med 2010;268(2):171–80. https://doi.org/10.1111/j.1365-2796.2010.02248.x.
15. Semba RD, Ferrucci L, Sun K, et al. Oxidative stress and severe walking disability among older women. Am J Med 2007;120(12):1084–9. https://doi.org/10.1016/j.amjmed.2007.07.028.
16. Semba RD, Varadhan R, Bartali B, et al. Low serum carotenoids and development of severe walking disability among older women living in the community: The women's health and aging study I. Age Ageing 2007;36(1):62–7. https://doi.org/10.1093/ageing/afl122.
17. Michelon E, Blaum C, Semba RD, Xue QL, Ricks MO, Fried LP. Vitamin and carotenoid status in older women: Associations with the frailty syndrome. J Gerontol A Biol Sci Med Sci 2006;61(6):600–7. https://doi.org/10.1093/gerona/61.6.600.

18. Leon-Munoz LM, Garcia-Esquinas E, Lopez-Garcia E, Banegas JR, Rodriguez-Artalejo F. Major dietary patterns and risk of frailty in older adults: A prospective cohort study. BMC Med 2015;13:11. https://doi.org/10.1186/s12916-014-0255-6.
19. Shikany JM, Barrett-Connor E, Ensrud KE, et al. Macronutrients, diet quality, and frailty in older men. J Gerontol A Biol Sci Med Sci 2014;69(6):695–701. https://doi.org/10.1093/gerona/glt196.
20. Shahar DR, Houston DK, Hue TF, et al. Adherence to mediterranean diet and decline in walking speed over 8 years in community-dwelling older adults. J Am Geriatr Soc 2012;60(10):1881–8. https://doi.org/10.1111/j.1532-5415.2012.04167.x.
21. Coelho-Junior HJ, Trichopoulou A, Panza F. Cross-sectional and longitudinal associations between adherence to Mediterranean diet with physical performance and cognitive function in older adults: A systematic review and meta-analysis. Ageing Res Rev 2021;70:101395. https://doi.org/10.1016/j.arr.2021.101395.
22. Dominguez LJ, Donat-Vargas C, Sayon-Orea C, et al. Rationale of the association between Mediterranean diet and the risk of frailty in older adults and systematic review and meta-analysis. Exp Gerontol 2023;177:112180. https://doi.org/10.1016/j.exger.2023.112180.
23. Landi F, Onder G, Bernabei R. Sarcopenia and diabetes: Two sides of the same coin. J Am Med Dir Assoc 2013;14(8):540–1. https://doi.org/10.1016/j.jamda.2013.05.004.
24. de Moraes MB, Avgerinou C, Fukushima FB, Vidal EIO. Nutritional interventions for the management of frailty in older adults: Systematic review and meta-analysis of randomized clinical trials. Nutr Rev 2021;79(8):889–913. https://doi.org/10.1093/nutrit/nuaa101.
25. Bernabei R, Landi F, Calvani R, et al. Multicomponent intervention to prevent mobility disability in frail older adults: Randomised controlled trial (SPRINTT project). BMJ 2022;377:e068788. https://doi.org/10.1136/bmj-2021-068788.
26. Li W, Wu Z, Liao X, et al. Nutritional management interventions and multi-dimensional outcomes in frail and pre-frail older adults: A systematic review and meta-analysis. Arch Gerontol Geriatr 2024;125:105480. https://doi.org/10.1016/j.archger.2024.105480.
27. Bernabei R, Martone AM, Vetrano DL, Calvani R, Landi F, Frailty ME. Physical frailty, sarcopenia: A new conceptual model. Stud Health Technol Inform 2014;203:78–84. https://doi.org/10.3233/978-1-61499-425-1-78.
28. Cruz-Jentoft AJ, Sayer AA. Sarcopenia. Lancet 2019;393(10191):2636–2646. https://doi.org/10.1016/S0140-6736(19)31138-9.
29. Fried LP, Cohen AA, Xue QL, Walston J, Bandeen-Roche K, Varadhan R. The physical frailty syndrome as a transition from homeostatic symphony to cacophony. Nat Aging 2021;1(1):36–46. https://doi.org/10.1038/s43587-020-00017-z.
30. Taylor JA, Greenhaff PL, Bartlett DB, Jackson TA, Duggal NA, Lord JM. Multisystem physiological perspective of human frailty and its modulation by physical activity. Physiol Rev 2023;103(2):1137–91. https://doi.org/10.1152/physrev.00037.2021.
31. Marzetti E, Picca A, Marini F, et al. Inflammatory signatures in older persons with physical frailty and sarcopenia: The frailty "cytokinome" at its core. Exp Gerontol 2019;122:129–38. https://doi.org/10.1016/j.exger.2019.04.019.
32. Volkert D, Kiesswetter E, Cederholm T, et al. Development of a model on determinants of malnutrition in aged persons: A MaNuEL Project. Gerontol Geriatr Med 2019;5:2333721419858438. https://doi.org/10.1177/2333721419858438.
33. Bonnefoy M, Berrut G, Lesourd B, et al. Frailty and nutrition: Searching for evidence. J Nutr Health Aging 2015;19(3):250–7. https://doi.org/10.1007/s12603-014-0568-3.
34. Olaniyan ET, O'Halloran F, McCarthy AL. Dietary protein considerations for muscle protein synthesis and muscle mass preservation in older adults. Nutr Res Rev 2021;34(1):147–57. https://doi.org/10.1017/S0954422420000219.
35. Wang L, Xu Z, Ling D, Li J, Wang Y, Shan T. The regulatory role of dietary factors in skeletal muscle development, regeneration and function. Crit Rev Food Sci Nutr 2022;62(3):764–82. https://doi.org/10.1080/10408398.2020.1828812.
36. Haran PH, Rivas DA, Fielding RA. Role and potential mechanisms of anabolic resistance in sarcopenia. J Cachexia Sarcopenia Muscle 2012;3(3):157–62. https://doi.org/10.1007/s13539-012-0068-4.
37. Guillet C, Masgrau A, Walrand S, Boirie Y. Impaired protein metabolism: Interlinks between obesity, insulin resistance and inflammation. Obes Rev 2012;13 Suppl 2:51–7. https://doi.org/10.1111/j.1467-789X.2012.01037.x.
38. Smeuninx B, McKendry J, Wilson D, Martin U, Breen L. Age-related anabolic resistance of myofibrillar protein synthesis is exacerbated in obese inactive individuals. J Clin Endocrinol Metab 2017;102(9):3535–45. https://doi.org/10.1210/jc.2017-00869.

39. Beaudry KM, Binet ER, Collao N, Lisio D M. Nutritional regulation of muscle stem cells in exercise and disease: The role of protein and amino acid dietary supplementation. Front Physiol 2022;13:915390. https://doi.org/10.3389/fphys.2022.915390.
40. Conlon JA, Newton RU, Tufano JJ, et al. Periodization strategies in older adults: Impact on physical function and health. Med Sci Sports Exerc 2016;48(12):2426–36. https://doi.org/10.1249/MSS.0000000000001053.
41. Oikawa SY, Brisbois TD, van Loon LJC, Rollo I. Eat like an athlete: Insights of sports nutrition science to support active aging in healthy older adults. Geroscience 2021;43(5):2485–95. https://doi.org/10.1007/s11357-021-00419-w.
42. Amorim JA, Coppotelli G, Rolo AP, Palmeira CM, Ross JM, Sinclair DA. Mitochondrial and metabolic dysfunction in ageing and age-related diseases. Nat Rev Endocrinol 2022;18(4):243–58. https://doi.org/10.1038/s41574-021-00626-7.
43. Marco-Bonilla M, Fresnadillo M, Largo R, Herrero-Beaumont G, Mediero A. Energy regulation in inflammatory sarcopenia by the purinergic system. Int J Mol Sci 2023;24(23). https://doi.org/10.3390/ijms242316904.
44. Almeida NS, Rocha R, de Souza CA, et al. Energy and nutrient intake by people with and without sarcopenia diagnosed by the European Working Group on Sarcopenia in Older People: A systematic review and meta-analysis. Nutr Rev 2023. https://doi.org/10.1093/nutrit/nuad154.
45. Soysal P, Isik AT, Carvalho AF, et al. Oxidative stress and frailty: A systematic review and synthesis of the best evidence. Maturitas 2017;99:66–72. https://doi.org/10.1016/j.maturitas.2017.01.006.
46. Fulle S, Protasi F, Di Tano G, et al. The contribution of reactive oxygen species to sarcopenia and muscle ageing. Exp Gerontol 2004;39(1):17–24. https://doi.org/10.1016/j.exger.2003.09.012.
47. El Assar M, Angulo J, Rodriguez-Manas L. Frailty as a phenotypic manifestation of underlying oxidative stress. Free Radic Biol Med 2020;149:72–7. https://doi.org/10.1016/j.freeradbiomed.2019.08.011.
48. van Dronkelaar C, van Velzen A, Abdelrazek M, van der Steen A, Weijs PJM, Tieland M. Minerals and sarcopenia; The role of calcium, iron, magnesium, phosphorus, potassium, selenium, sodium, and zinc on muscle mass, muscle strength, and physical performance in older adults: A systematic review. J Am Med Dir Assoc 2018;19(1):6–11 e3. https://doi.org/10.1016/j.jamda.2017.05.026.
49. Abeywickrama HM, Uchiyama M, Sumiyoshi T, Okuda A, Koyama Y. The role of zinc on nutritional status, sarcopenia, and frailty in older adults: A scoping review. Nutr Rev 2024;82(7):988–1011. https://doi.org/10.1093/nutrit/nuad094.
50. Barzilay JI, Blaum C, Moore T, et al. Insulin resistance and inflammation as precursors of frailty: The cardiovascular health study. Arch Intern Med 2007;167(7):635–41. https://doi.org/10.1001/archinte.167.7.635.
51. Cacciatore S, Calvani R, Esposito I, et al. Emerging targets and treatments for sarcopenia: A narrative review. Nutrients 2024;16(19):3271. https://doi.org/10.3390/nu16193271.
52. Jackson MA, Jeffery IB, Beaumont M, et al. Signatures of early frailty in the gut microbiota. Genome Med 2016;8(1). https://doi.org/10.1186/s13073-016-0262-7.
53. Wang XM, Fan L, Meng CC, et al. Gut microbiota influence frailty syndrome in older adults: Mechanisms and therapeutic strategies. Biogerontology 2024;25(1):107–29. https://doi.org/10.1007/s10522-023-10082-7.
54. Conway JNAD. Ageing of the gut microbiome: Potential influences on immune senescence and inflammageing. Ageing Res Rev 2021;68:101323. https://doi.org/10.1016/j.arr.2021.101323.
55. Liu C, Cheung WH, Li J, et al. Understanding the gut microbiota and sarcopenia: A systematic review. J Cachexia, Sarcopenia Muscle 2021;12(6):1393–407. https://doi.org/10.1002/jcsm.12784.
56. Volkert D, Beck AM, Cederholm T, et al. ESPEN practical guideline: Clinical nutrition and hydration in geriatrics. Clin Nutr 2022;41(4):958–89. https://doi.org/10.1016/j.clnu.2022.01.024.
57. Song Z, Pan T, Tong X, Yang Y, Zhang Z. The effects of nutritional supplementation on older sarcopenic individuals who engage in resistance training: A meta-analysis. Front Nutr 2023;10:1109789. https://doi.org/10.3389/fnut.2023.1109789.
58. Calvani R, Picca A, Coelho-Junior HJ, Tosato M, Marzetti E, Landi F. Diet for the prevention and management of sarcopenia. Metabolism 2023;146:155637. https://doi.org/10.1016/j.metabol.2023.155637.
59. Doza GL, van Heden S, Oliveira Felix F, Singh V, Beaudart C. Impact of interventions on sarcopenia from the perspective of older persons: A systematic literature review. J Frailty Aging 2024;13(3):224–32. https://doi.org/10.14283/jfa.2024.47.
60. Ko CH, Wu SJ, Wang ST, et al. Effects of enriched branched-chain amino acid supplementation on sarcopenia. Aging (Albany NY) 2020;12(14):15091–103. https://doi.org/10.18632/aging.103576.

61. Landi F, Calvani R, Picca A, Marzetti E. Beta-hydroxy-beta-methylbutyrate and sarcopenia: From biological plausibility to clinical evidence. Curr Opin Clin Nutr Metab Care 2019;22(1):37–43. https://doi.org/10.1097/MCO.0000000000000524.
62. Kaczka P, Michalczyk MM, Jastrząb R, Gawelczyk M, Kubicka K. Mechanism of action and the effect of beta-hydroxy-beta-methylbutyrate (HMB) supplementation on different types of physical performance—A systematic review. Journal of Human Kinetics 2019;68(1):211–22. https://doi.org/10.2478/hukin-2019-0070.
63. Candow DG, Chilibeck PD, Forbes SC, Fairman CM, Gualano B, Roschel H. Creatine supplementation for older adults: Focus on sarcopenia, osteoporosis, frailty and cachexia. Bone 2022;162:116467. https://doi.org/10.1016/j.bone.2022.116467.
64. Sobolewski EJ, Thompson BJ, Smith AE, Ryan ED. The physiological effects of creatine supplementation on hydration: A review. Am J Lifestyle Med 2011;5(4):320–27. https://doi.org/10.1177/1559827611406071.
65. Kapil V, Khambata RS, Jones DA, et al. The noncanonical pathway for in vivo nitric oxide generation: The nitrate-nitrite-nitric oxide pathway. Pharmacol Rev 2020;72(3):692–766. https://doi.org/10.1124/pr.120.019240.
66. Lundberg JO, Weitzberg E. Nitric oxide signaling in health and disease. Cell 2022;185(16):2853–78. https://doi.org/10.1016/j.cell.2022.06.010.
67. Chen L, Zhu Y, Hu Z, Wu S, Jin C. Beetroot as a functional food with huge health benefits: Antioxidant, antitumor, physical function, and chronic metabolomics activity. Food Sci Nutr 2021;9(11):6406–20. https://doi.org/10.1002/fsn3.2577.
68. Jones AM, Vanhatalo A, Seals DR, Rossman MJ, Piknova B, Jonvik KL. Dietary nitrate and nitric oxide metabolism: Mouth, circulation, skeletal muscle, and exercise performance. Med Sci Sports Exerc 2021;53(2):280–94. https://doi.org/10.1249/MSS.0000000000002470.
69. Linton C, Wright HH, Wadsworth DP, Schaumberg MA. Dietary inflammatory index and associations with sarcopenia symptomology in community-dwelling older adults. Nutrients 2022;14(24). https://doi.org/10.3390/nu14245319.
70. Chen Q, Huang W, Capanoglu E, Amrouche AT, Lu B. Targeting mitochondrial quality control in muscle aging: Natural dietary products as potential interventions. Food Frontiers 2023;4(3):1206–41. https://doi.org/10.1002/fft2.261.
71. Pahor M, Guralnik JM, Ambrosius WT, et al. Effect of structured physical activity on prevention of major mobility disability in older adults: The LIFE study randomized clinical trial. JAMA 2014; 311(23):2387–96. https://doi.org/10.1001/jama.2014.5616.
72. Bethancourt HJ, Rosenberg DE, Beatty T, Arterburn DE. Barriers to and facilitators of physical activity program use among older adults. Clin Med Res 2014;12(1–2):10–20. https://doi.org/10.3121/cmr.2013.1171.
73. Merchant RA, Morley JE, Izquierdo M. Editorial: Exercise, aging and frailty: Guidelines for increasing function. J Nutr Health Aging 2021;25(4):405–9. https://doi.org/10.1007/s12603-021-1590-x.
74. Izquierdo M, Merchant RA, Morley JE, et al. International Exercise Recommendations in Older Adults (ICFSR): Expert consensus guidelines. J Nutr Health Aging 2021;25(7):824–53. https://doi.org/10.1007/s12603-021-1665-8.
75. Elliot CA, Hamlin MJ. Combined diet and physical activity is better than diet or physical activity alone at improving health outcomes for patients in New Zealand's primary care intervention. BMC Public Health 2018;18(1):230. https://doi.org/10.1186/s12889-018-5152-z.
76. Casas-Herrero A, Saez de Asteasu ML, Anton-Rodrigo I, et al. Effects of vivifrail multicomponent intervention on functional capacity: A multicentre, randomized controlled trial. J Cachexia Sarcopenia Muscle 2022;13(2):884–93. https://doi.org/10.1002/jcsm.12925.
77. Vellas B, Guigoz Y, Garry PJ, et al. The mini nutritional assessment (MNA) and its use in grading the nutritional state of elderly patients. Nutrition 1999;15(2):116–22. https://doi.org/10.1016/s0899-9007(98)00171-3.
78. Stratton RJ, Hackston A, Longmore D, et al. Malnutrition in hospital outpatients and inpatients: Prevalence, concurrent validity and ease of use of the 'malnutrition universal screening tool' ('MUST') for adults. Br J Nutr 2004;92(5):799–808. https://doi.org/10.1079/bjn20041258.
79. Kruizenga HM, Seidell JC, de Vet HC, Wierdsma NJ, van Bokhorst-de van der Schueren MA. Development and validation of a hospital screening tool for malnutrition: The short nutritional assessment questionnaire (SNAQ). Clin Nutr 2005;24(1):75–82. https://doi.org/10.1016/j.clnu.2004.07.015.
80. Prado CM, Landi F, Chew STH, et al. Advances in muscle health and nutrition: A toolkit for healthcare professionals. Clin Nutr 2022;41(10):2244–63. https://doi.org/10.1016/j.clnu.2022.07.041.

81. Martone AM, Ciciarello F, Galluzzo V, et al. A simple medical device development according to normative values of calf circumference across ages: Results from the Italian Longevity Check-up 7+ (Lookup 7+) project. J Cachexia Sarcopenia Muscle 2024;15(1):36–44. https://doi.org/10.1002/jcsm.13286.
82. Cederholm T, Jensen GL, Correia M, et al. GLIM criteria for the diagnosis of malnutrition – A consensus report from the global clinical nutrition community. Clin Nutr 2019;38(1):1–9. https://doi.org/10.1016/j.clnu.2018.08.002.
83. Thomson KH, Rice S, Arisa O, et al. Effectiveness and cost-effectiveness of oral nutritional supplements in frail older people who are malnourished or at risk of malnutrition: A systematic review and meta-analysis. Lancet Healthy Longev 2022;3(10):e654–e666. https://doi.org/10.1016/S2666-7568(22)00171-4.
84. Piccoli GB, Cederholm T, Avesani CM, et al. Nutritional status and the risk of malnutrition in older adults with chronic kidney disease – Implications for low protein intake and nutritional care: A critical review endorsed by ERN-ERA and ESPEN. Clinical Nutr 2023;42(4):443–57. https://doi.org/10.1016/j.clnu.2023.01.018.
85. Huhmann MB, Smith KN, Schwartz SL, Haller SK, Irvin S, Cohen SS. Plasma glucose and insulin response to two oral nutrition supplements in adults with type 2 diabetes mellitus. BMJ Open Diabetes Res Care 2016;4(1):e000240. https://doi.org/10.1136/bmjdrc-2016-000240.
86. Chevalley T, Brandi ML, Cashman KD, et al. Role of vitamin D supplementation in the management of musculoskeletal diseases: Update from an European Society of Clinical and Economical Aspects of Osteoporosis, Osteoarthritis and Musculoskeletal Diseases (ESCEO) working group. Aging Clin Exp Res 2022;34(11):2603–23. https://doi.org/10.1007/s40520-022-02279-6.
87. O'Connor D, Molloy AM, Laird E, Kenny RA, O'Halloran AM. Sustaining an ageing population: The role of micronutrients in frailty and cognitive impairment. Proc Nutr Soc 2023;82(3):315–28. https://doi.org/10.1017/S0029665123002707.
88. Nisly NL, Gryzlak BM, Zimmerman MB, Wallace RB. Dietary supplement polypharmacy: An unrecognized public health problem? Evid Based Complement Alternat Med 2010;7(1):107–13. https://doi.org/10.1093/ecam/nem150.
89. Scheufele P, Rappl A, Visser M, Kiesswetter E, Volkert D. Dietary characteristics of community-dwelling older adults with poor appetite: A cross-sectional analysis. Age Ageing 2024;53(Suppl 2): ii4–ii12. https://doi.org/10.1093/ageing/afae040.
90. Volkert D, Corish CA, Dardevet D, et al. Innovative plAnt Protein fibre and Physical activity solutions to address poor appEtite and prevenT undernutrITion in oldEr adults – APPETITE. Nutrition Bulletin 2021;46(4):486–96. https://doi.org/10.1111/nbu.12529.
91. Aprahamian I, Coats AJ, Morley JE, et al. Anorexia of aging: An international assessment of healthcare providers' knowledge and practice gaps. J Cachexia Sarcopenia Muscle 2023;14(6):2779–92. https://doi.org/10.1002/jcsm.13355.
92. Winter JE, MacInnis RJ, Wattanapenpaiboon N, Nowson CA. BMI and all-cause mortality in older adults: A meta-analysis. Am J Clin Nutr 2014;99(4):875–90. https://doi.org/10.3945/ajcn.113.068122.
93. Yuan L, Chang M, Wang J. Abdominal obesity, body mass index and the risk of frailty in community-dwelling older adults: A systematic review and meta-analysis. Age Ageing 2021;50(4):1118–28. https://doi.org/10.1093/ageing/afab039.
94. Banack HR, Stokes A. The 'obesity paradox' may not be a paradox at all. Int J Obes (Lond) 2017; 41(8):1162–3. https://doi.org/10.1038/ijo.2017.99.
95. Samper-Ternent R, Al Snih S. Obesity in older adults: Epidemiology and implications for disability and disease. Rev Clin Gerontol 2012;22(1):10–34. https://doi.org/10.1017/s0959259811000190.
96. Cacciatore S, Calvani R, Marzetti E, et al. Predictive values of relative fat mass and body mass index on cardiovascular health in community-dwelling older adults: Results from the Longevity Check-up (Lookup) 7. Maturitas 2024;185:108011. https://doi.org/10.1016/j.maturitas.2024.108011.
97. Kiskac M, Soysal P, Smith L, Capar E, Zorlu M. What is the optimal body mass index range for older adults? Ann Geriatr Med Res 2022;26(1):49–57. https://doi.org/10.4235/agmr.22.0012.
98. Donini LM, Busetto L, Bischoff SC, et al. Definition and diagnostic criteria for sarcopenic obesity: ESPEN and EASO consensus statement. Obes Facts 2022;15(3):321–35. https://doi.org/10.1159/000521241.
99. Eglseer D, Traxler M, Schoufour JD, et al. Nutritional and exercise interventions in individuals with sarcopenic obesity around retirement age: A systematic review and meta-analysis. Nutr Rev 2023;81(9):1077–90. https://doi.org/10.1093/nutrit/nuad007.
100. Collazo-Castineira P, Sanchez-Izquierdo M, Reiter LJ, et al. Analysis of behavioral change techniques used in exercise and nutritional interventions targeting adults around retirement age with sarcopenic obesity in a systematic review. Arch Gerontol Geriatr 2024;123:105437. https://doi.org/10.1016/j.archger.2024.105437.

101. Volkert D, Delzenne N, Demirkan K, et al. Nutrition for the older adult – Current concepts. Report from an ESPEN symposium. Clin Nutr 2024;43(8):1815–24. https://doi.org/10.1016/j.clnu.2024.06.020.
102. Beard JR, Officer A, de Carvalho IA, et al. The world report on ageing and health: A policy framework for healthy ageing. Lancet 2016;387(10033):2145–54. https://doi.org/10.1016/S0140-6736(15)00516-4.
103. Belloni G, Cesari M. Frailty and intrinsic capacity: Two distinct but related constructs. Front Med (Lausanne) 2019;6:133. https://doi.org/10.3389/fmed.2019.00133.
104. Bautmans I, Knoop V, Amuthavalli Thiyagarajan J, et al. WHO working definition of vitality capacity for healthy longevity monitoring. Lancet Healthy Longev 2022;3(11):e789–e796. https://doi.org/10.1016/S2666-7568(22)00200-8.
105. Gaussens L, Gonzalez-Bautista E, Bonnefoy M, et al. Associations between vitality/nutrition and the other domains of intrinsic capacity based on data from the INSPIRE ICOPE-care program. Nutrients 2023;15(7) https://doi.org/10.3390/nu15071567.
106. World Health Organization. 2019. Integrated care for older people (ICOPE): Guidance for person-centred assessment and pathways in primary care. World Health Organization.
107. Kim J, Lee Y, Kim M, Won CW, Kim MK, Shim JS. Dietary patterns and intrinsic capacity in older adults: A 6-year prospective cohort study. J Nutr Health Aging 2024;28(8):100314. https://doi.org/10.1016/j.jnha.2024.100314.
108. Chen L, Huang H, Jiang S, et al. Facilitators and barriers to the implementation of dietary nutrition interventions for community-dwelling older adults with physical frailty and sarcopenia: A qualitative meta-synthesis. Int J Nurs Sci 2024;11(1):18–30. https://doi.org/10.1016/j.ijnss.2023.12.007.
109. Lee BY, Ordovas JM, Parks EJ, et al. Research gaps and opportunities in precision nutrition: An NIH workshop report. Am J Clin Nutr 2022;116(6):1877–900. https://doi.org/10.1093/ajcn/nqac237.
110. Dickinson A, Wills W. Meals on wheels services and the food security of older people. Health Soc Care Community 2022;30(6):e6699–e6707. https://doi.org/10.1111/hsc.14092.
111. Lyons OC, Flynn MAT, Corish CA, et al. Nutrition policy: Developing scientific recommendations for food-based dietary guidelines for older adults living independently in Ireland. Proc Nutr Soc 2022;81(1):49–61. https://doi.org/10.1017/S0029665122001008.
112. World Health Organization. 2024. Global Strategy on Diet, Physical Activity and Health. World Health Organization.
113. Govindaraju T, Owen AJ, McCaffrey TA. Past, present and future influences of diet among older adults – A scoping review. Ageing Res Rev 2022;77:101600. https://doi.org/10.1016/j.arr.2022.101600.
114. Izquierdo M, Duque G, Morley JE. Physical activity guidelines for older people: Knowledge gaps and future directions. Lancet Healthy Longev 2021;2(6):e380–e383. https://doi.org/10.1016/S2666-7568(21)00079-9.
115. Abeltino A, Riente A, Bianchetti G, et al. Digital applications for diet monitoring, planning, and precision nutrition for citizens and professionals: A state of the art. Nutr Rev 2024. https://doi.org/10.1093/nutrit/nuae035.
116. Wegener EK, Kayser L. Smart health technologies used to support physical activity and nutritional intake in fall prevention among older adults: A scoping review. Exp Gerontol 2023;181:112282. https://doi.org/10.1016/j.exger.2023.112282.

7 Physical Function-Based Programs in Frailty

Eleftheria Antoniadou

INTRODUCTION

Definition of Physical Function

In the framework of functional medicine, the term "function" is associated with the progressive realization that disease represents a final stage, while function embodies an ongoing process. Functionality is a trajectory that can progress or regress. The trajectory of functional changes or as is known functional ability is influenced by the distinctive interplay between an individual's intrinsic capacity and their surrounding environment, dietary choices, and lifestyle (1) (Figure 7.1). Functional ability is the capacity that allows every person to be and do the things that they have reason to value, such as meeting their basic needs, learning, growing, and making decisions, being mobile, forming and maintaining relationships, and making a positive contribution to society.

Healthy aging is defined as "the process of developing and maintaining the functional ability that enables wellbeing in older ages", as health status is rather defined by the functional status rather than morbidity, as recognised by older people themselves, health-care professionals, and policy makers (2–7). Applying the International Classification of Functioning (ICF) concepts in this age category, in order to create effective interventions, it becomes clear that as the body structure and function components are usually impaired, and the possibility of prolonging life is dramatically reduced (8), the importance must be focused on activity and participation in order to improve the quality of life (3). To this end, we must recognize, according to the World Health Organisation (9) how the relationship between what I am able to do (performance qualifier) and the challenges of the

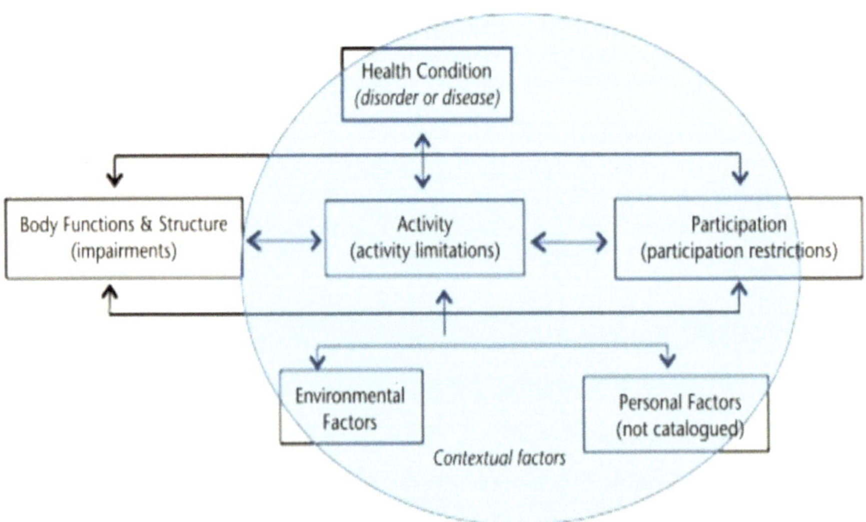

FIGURE 7.1 ICF model of functioning and disability. (3)

DOI: 10.1201/9781003398295-7

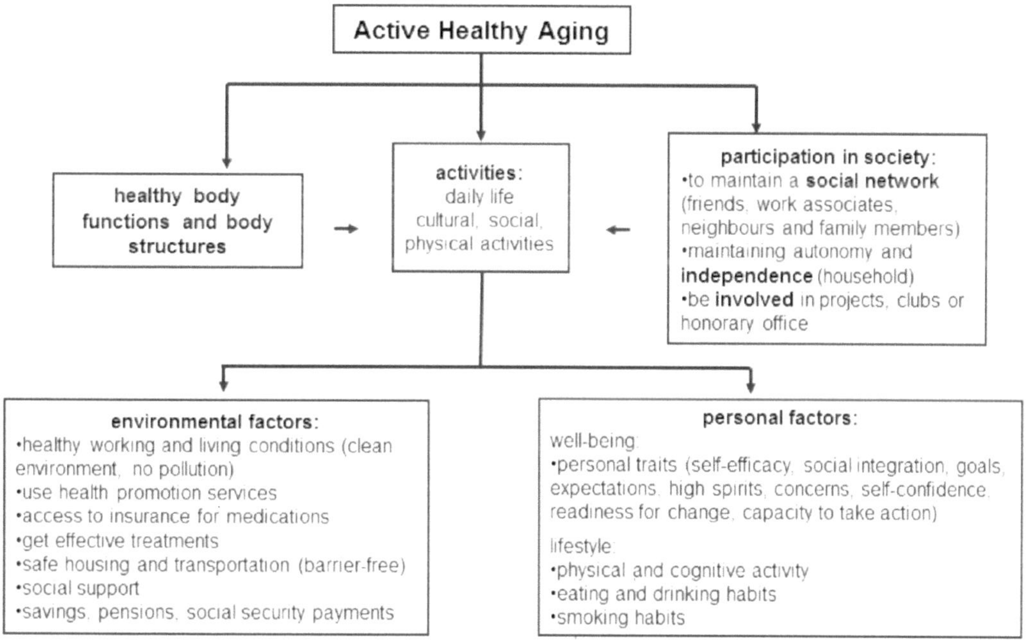

FIGURE 7.2 Interactions of the ICF-components for healthy aging (based on World Health Organization, 2009).

environment (the context where the person exists) (capacity qualifier) marks the presence of functional autonomy or disability (Figure 7.2). In the process of maximizing functional abilities, we can either address the performance by trying to increment physical function or implement compensatory and coping strategies (such as assistive devices) (4).

In this chapter, we will discuss effective interventions to maintain and improve functional performance and, as a consequence, functional ability. To start, we need to uncover what is physical function. With the term physical function, we describe the capacity of the body to carry out a range of physical tasks and activities, typically associated with daily living, work, recreation, or exercise, and is essential for overall functional ability. It encompasses the body's ability to carry out movements, the capacity to exert force against resistance (strength), the ability to sustain physical activity over an extended period without becoming excessively fatigued (endurance), the capability to maintain stability and control while performing various movements (balance and coordination), the range of motion in joints and muscles (flexibility), which affects the ability to perform movements without pain or discomfort, the ability to change direction quickly and effectively (agility), fine motor skills that allow for precise and coordinated movements (dexterity), and the speed at which a person can respond to a stimulus (reaction time). The physical function domain is theoretically composed of subdomains, which are conceptually interrelated but distinct, like mobility related to the lower extremities, dexterity related to the upper extremities, axial or central related to core, and complex activities which involve more than one subdomain and are related to functional abilities (5, 6).

Frailty is a key player when addressing physical function in the older population, as it is the geriatric syndrome that hinders their functional ability and independence (6, 7), and as a consequence, if no intervention is carried out, it can lead to disability. Frailty is a functional-related syndrome, a reversible condition, and taking action against it can minimize adverse outcomes, as it also the case when acting against various age-related processes associated with cognitive decline, and brain aging, which are interconnected with frailty syndrome, and share similar pathophysiological mechanisms (6, 10).

From all the above, we can clearly comprehend that assessing and addressing physical function in frail older people is of great importance.

Assessment of Physical Function

Undertaking a comprehensive evaluation of the various function domains and tailoring the intervention accordingly is the most efficient method of putting physical function programs into practice. A wide range of instruments, including single-domain and multi-domain evaluators, tests, and questionnaires, can be used to assess physical function. There are objective assessments that, regardless of complexity, include predefined, objective evaluation criteria (such as completion time) to measure the results and subjective measurements, like patient questionnaires or others, that are a crucial step in understanding the viewpoints of the patients. Physical tasks such as musculoskeletal strength, power, endurance, range of motion, or basic physical abilities like walking or balance are evaluated by the simple on domain assessors. Complicated performance-based assessments are frequently employed to assess intricate tasks like self-care.

Objective Assessment Tools

- Single-domain evaluation tools are designed to assess only one domain of physical function like:
  - *Strength testing:* Handgrip strength (on dominant arm) and for the lower body the quadriceps Isometric Strength Testing are the most common (11). The one-repetition maximum (1-RM) is defined as the maximal weight an individual can lift for just one repetition with correct technique, and it is considered the gold standard for assessing muscle strength and also determining the intensity of the exercises (12).
  - *Musculoskeletal power* can be measured in muscle lower exercises, leg press, and knee extension, showing good reliability and validity (13). The five times sit-to-stand (5STS) (asking the patient to sit and stand five times quickly) is a valid and clinically relevant tool (14). Power training intensity should range from 30% to 60% of 1-RM load because maximal power output is maximized at these intensities (15).
  - *Endurance testing:* Aerobic capacity is usually measured via VO_2 maximum or peak. To assess it, the cardiopulmonary exercise testing (CPET) is the best evaluation tool, but it needs a special setting, and the equipment is expensive. Two of the conventional submaximal exercise tests commonly used in older people as measures of walking speed and mobility performance are the 6-min walking test and the 400-m walking test (16). *The BORG and mBORG scales can be used to track exertion* (17).
  - *Range of motion (ROM) assessments* using goniometers or visually assessing the degree of joint movement and the presence of amplitude restrictions.
  - *Balance assessments*: Many tools exist in the literature fit for the purpose to evaluate a complex ability like Balance. They range from simple and quick tools, such as the Sharpened Romberg, the Timed Up and Go (TUG), Single Leg Balance Test, to more complex ones where a thorough evaluation of all the aspects of balance is performed such as the Berg Balance Scale, Mini BEST test, the Tinetti, and the mCTSIB (18).
  - *Gait speed* can be assessed in a 3 or 4 m corridor in usual or maximum pace (19). From the meta-analysis of Chou, a clear connection can be made between gait speed and the prediction of functional decline, nursing home placement, and mortality (20).
- Multi-domain tools that can evaluate more than one domain like:
  - *The SPPB (Short Physical Performance Battery)* (21), to evaluate physical performance, assesses gait speed at the individual's usual pace, balance (using the Romberg progressive test), and power in lower limbs (with the 5STS test). It is accepted that every point of incrementation is translated with a difference in function performance (22).
  - *Functional abilities are* usually assessed by assessment of activities of daily living (ADLs). These are tests that evaluate the ability to perform essential daily tasks like dressing, bathing, eating, and toileting called basic activities of daily living BADLs (Katz, Barthel FIM). There are also tools that evaluate Instrumental Activities of Daily Living (IADLs),

meaning more complicated tasks like meal preparation, managing medications, and handling finances (23). Compared to functional performance tests like the SPPB, ADL measurement, in question format only, generally has weak reliability, validity, reproducibility, and sensitivity to change; mostly when we analyze a community-dwelling population that live independently, we can expect a prominent ceiling effect (24). Nevertheless, a variety of such ADL instruments are used as a routine in studies of older adults.

Subjective Assessment Tools

- Questionnaires and self-reports:
  - Self-assessment instruments evaluate physical function and quality of life by using validated questionnaires such as the EuroQol-5D (25), or the SF-36, where a predetermined list of tasks or open-ended questions is used to evaluate one's physical functioning. These tests take into account subjective elements and evaluate how well one feels one is operating or what challenges one has faced (26, 27). One benefit of using questionnaires is that they allow for comparisons both within and between groups and situations. Additionally, open-ended questions can help identify key issues that might not have been evaluated (28).

Specialized Tests

Depending on the specific needs and conditions of the individual, specialized tests like pulmonary function tests for lung function, gait analysis, or balance assessments using force plates may be conducted.

Objective Measurement Tools

Wearable fitness trackers and accelerometers can provide objective data on physical activity levels, step counts, and energy expenditure.

In completing the examination, a key component is also the traditional physical exam where many limiting factors can be unmasked. Problems in older people often develop in areas of the body not covered by "conventional" examination, unless actively looked like the feet or the oral cavity. If a pathology of the upper limbs is present, specific assessment is important to test manual dexterity and fine motor ability (such as tying shoelaces or managing buttons).

There is no wide consensus on the value of clinically meaningful change as a reduction in frailty or functional ability (28–30). In contrast, performance-based physical outcome measures such as the SPPB, mobility, balance, body mass, and activity levels have consistently reported strong associations with future health, functional ability, and service use in older and frail adults (25, 27, 31).

PHYSICAL FUNCTION-BASED PROGRAMS

These are programs aiming to address the physical function domain through a combination of exercise, activities, and therapies with the goal to promote independence, prevent further decline, and maintain functional abilities for as long as possible (32), in order to maintain a sense of independence and identity (33, 34). These programs are designed to address the specific functional needs of frail individuals and improve their overall quality of life (35), also taking into account their socioeconomic status, particularly those in low-income urban communities. They can be a simple one-domain intervention to address a particular physical domain like balance or gait or a more complex intervention that aims to address multiple domains or functional abilities as a whole, some also having a cognitive intervention important for the execution of daily activities. These programs vary widely in complexity, duration, intensity, personnel implementing the intervention, target population, measurable outcomes, components, and setup. But still some conclusions about what is really impacting functional performance and therefore functional ability of older people can be drawn.

Programs Components

Multi-Monodomain Programs

Concerning the components of the intervention, they can be monodomain interventions, usually exercise programs, but also multi-domain, considering that function is a complex, multifaceted condition, and therefore multidimensional interventions must be conducted to manage cognition, mobility, and overall intrinsic capacity (36).

There have been many multi-domain protocols to investigate the effects of multi-domain interventions in (pre)frail elderly on frailty, functional, and cognitive status. From the systematic review of Dedeyne et al. (37), we can deduce that multi-domain interventions improve **functional abilities** (ADL and/or IADL and/or personal activities of daily living dependency or scores) but also **physical functioning** (SPPB) (38–40) in methodologically good studies. As for the components of interventions, they can be a blend of physical activity, nutrition, geriatric assessment, and physical activity; those that incorporated a physical activity component were consistently the most effective at improving frailty status, physical outcomes (e.g., body mass index, muscle mass, strength, gait speed, exhaustion, physical activity), and/or functional ability (37, 41). No consistent effects of multi-domain interventions on falls, psychosocial behavior, or depression were found. However, beneficial effects may have been missed due to a low number of studies examining these outcomes or insufficient power, as several studies did not do a proper sample size calculation for these outcomes or did not include these outcomes as primary outcomes (42, 43). For quality of life, there are inconclusive effects, but evidence is pointing toward a positive effect (20, 43). The effect on preventing related disability is very hard to uncover because the majority of the studies did not include long-term follow-up measurement. The most promising findings were found for welfare technology implementation, CGA followed by treatment and, to a lesser extent, physical exercise programs. Technology, adapted to the needs of the frail older persons, as a means of support, may be very effective in preventing disability in community-dwelling frail older persons, though more research is needed in this area (44).

From the various domains of these studies, exercise has a significant improved effect on ADL and IADL (39). To be more specific, physical exercise seems to play an essential role in the multi-domain intervention, with some improvements by an additional intervention (e.g., nutritional intervention). The positive effects of nutritional supplementation increase when associated with physical exercise, and the positive effects of physical exercise increase when associated with nutritional supplementation (45–47). Also, it could be claimed that the physical exercise component accounts for the greatest improvements, which has also been suggested in reviews discussing frail (44) and sarcopenic older people (45). Cognitive training (CT), when added to a multicomponent program together with nutrition and exercise, as in the work of Ng (48), doesn't improve physical performance but has a positive effect on frailty after 6 to 12 months (48). The intervention consisted in randomized participants with a mean age of 70 years to physical exercise (participant tailored 90 min twice week for 12 weeks which focused on strength and balance, and 12 weeks of home exercise), nutrition (received supplements of iron and folate, vitamins B6 and B12, calcium and vitamin D for 24 weeks), CT (2 h weekly sessions for 12 weeks of cognitive-enhancing activities designed to stimulate short-term memory and enhance attention, information-processing skills, reasoning, and problem solving abilities), or combination treatment program for 6 months to examine the impact on frailty characteristics, with specific focus on mobility and strength performance outcomes. Frailty was assessed and participants classified as pre-frail or frail were included, though no stratified analyses were conducted. Frailty scores were significantly reduced at 6 and 12 months in all groups, with the combination group reporting the greatest because frailty also has a cognitive component, whereas physical performance doesn't (49).

Exploring the durations of the results after a multi-domain programs but also the cost-effectiveness, there is evidence from the MID-Frail project (2) that a multi-domain intervention composed by exercise nutrition and education in pre-frail and frail diabetic patient has positive effects on physical performance that also last 12 months after the completion of the intervention and is cost-effective. The intervention must have produced a sufficient challenge to the body to have this effect. This particular intervention evaluated 964 frail and pre-frail (according to Fried phenotype) older people with diabetes. A three-pronged intervention, including a resistance exercise program, a nutritional and educational program, and a standardized across sites researcher-enhanced training for best practices in diabetes care, was compared with a usual care group, with a duration of 12 months. The physical exercise intervention was a supervised individualized resistance exercise program consisting of a 2-week pre-training and learning phase followed by a 16-week program with participants undergoing two 45-min exercise sessions per week. Dropouts were higher in frail participants and in the intervention group. Although the exercise intervention ceased after 16 weeks, participants did not experience a rapid loss in physical function supposedly because is likely to have provided a sufficient stimulus to large leg muscle groups by using resistance training machines and moderate loads that ranged from 40% to 80% of individual maximal voluntary strength following a program proven to increase muscle mass, strength, and functional ability in older people with T2DM. There were no clear differences between the groups in measures of IADL or Barthel ADL because of the ceiling effect in a population with no clear disability. The study also evaluated as secondary outcomes mortality, hospitalization, institutionalization, quality of life, burden on caregivers, the frequency and severity of hypoglycemia episodes, and the cost-effectiveness of the intervention of the treatment group compared to the usual care group, that were positively impacted, but also it proved to be cost-efficient with average savings of 428.02 euros per participant per year.

Monodomains-Physical Exercise

The most effective monodomain intervention is exercise, whether is part of multi-domain program or is the only one, and it has been proven to delay physical dependence in older people (50–53) Exercise improves physical function domains like mobility, balance, functional abilities, muscle strength, body composition, and frailty as such (54–56) and functional performance based measures of physical function in community-dwelling frail and pre-frail older people as proved from many systematic reviews and meta-analysis (20, 25, 42, 43, 57).

Exercise programs are also beneficial in improving functional abilities as measured by ADL performance in frail older people. By just incrementing gait speed 0.1 m/s, we have a 10% decrease in the ability to perform instrumental ADL and an increase in falls (58, 59), which, in turn, often leads to a loss of independent living and to institutionalization. Although there is no clear evidence, it seems that the improvement in ADLs might be associated with exercise type (55); there is an important argument toward task-oriented or functional practice (39, 44, 60, 61) to maximize performance. Researchers observed that the outcome of task-oriented ADLs training with repetitive practice could motivate subjects and recommended that it should be performed often. Individualized high-intensity training program under supervision was also (62) observed to increase the baseline performance in ADLs (Gait speed, Barthel Index, FIM, and FSQ tools were used to measure performance) within 6 to 9 months in the program (62, 63). In the work of Tarazona-Santabalbina (39) (a supervised-facility multicomponent exercise for frail older people, 65 min per session, performed 5 days per week for 24 weeks), there is proof that exercise also has a positive effect on the IADLs. From this study, we can also deduce another important point, that is, the preservation of performance; people in the control group, but not those in the intervention group, decreased their performance in these functional measurements. This goal was achieved after 6 months and by attending 50% of the sessions. Maintaining performance in a group of people that is heading toward disability can be considered of great value and clinical relevance because a progressive decline in mobility and physical functioning are common and considered to be a major health concern for frail older people (64). Small positive effects were also found in frailty, functionality, cognitive and emotional status, social networking, and biological markers of frailty, when compared with a controlled population that received no training. Newly disabled people are able to regain some physical functioning, but they are at high risk of recurrent disability. Therefore, very large effects should not be expected when studying already physically impaired older adults.

From the secondary analysis of the LIFE-P Study (65), we can also see a positive effect from a non-structured exercise program. This is a two-arm, single blind, multicenter, randomized controlled trial comparing a physical activity intervention versus a successful aging (SA) health education program targeting older people; we can support the hypothesis that a PA program significantly reduced the presence and severity of frailty in a sample of sedentary elders, thus potentially modifying their risk profile. Secondary analyses suggest that participants benefiting most from the intervention included those with frailty and multimorbidity at baseline.

But what type of exercise has been proven to be more effective? From the Vivifrail project, we can uncover that the exercise program that has a major impact is a multicomponent one. In this project, a primary endpoint was the change in functional capacity from baseline (beginning of the intervention) to 12 weeks after intervention, as assessed with the SPPB (a meaningful clinical change was considered a change of ≥1 point in the score), clearly demonstrated a gain in functional performance only in the intervention group (22).

For older individuals who are frail or pre-frail, multicomponent exercise programs have been suggested as the most effective way to support neuromuscular adaptations (66, 67), lower the risk of falls, which would subsequently avoid injury, illness, and mortality (17, 68). This program's elements include strength, power, endurance, flexibility, and range of motion.

During the first few weeks of training, endurance exercises can last 5 to 10 min. Afterward, they can last 15 to 30 min (17). An alternate technique for determining exercise intensity is the Rate of Perceived Exertion scale; an intensity of 12 to 14 on the Borg scale seems to be well tolerated (69).

Strength training, similar to muscle strength workouts, appears to be the most important factor in retaining older people's functional capacity while preventing falls and sarcopenia (46, 68, 70). This could entail performing bodyweight exercises, such as squats, leg lifts, wall push-ups, and seated rows, that target the main muscle groups in the arms, legs, chest, and core. It may additionally involve utilizing resistance bands or weights. It is advised to be performed every other day, particularly when one initially starts out with exercise (71). Two to three times a week and two to three sets of 8–12 repetitions at an intensity that begins at 20–30% and increases to 80% of 1-RM should be the goal of resistance-training regimens (72). Resistance training protocols should incorporate exercises that imitate daily tasks, such as the sit-to-stand exercise, to maximize each person's functional ability.

As the only exercise technique that has been shown to minimize falls and their negative effects, balance training is also essential (68). There are numerous exercises that can help with balance, but the programs should be tailored to the individual needs and should be based on a functional assessment of balance. These could include walking on a balance board, toe and heel walking, tandem gait, single leg stances, semi-tandem and tandem stances, eye-hand or eye-leg coordination, joining the arms to the body or opening them, modifying the surface's stability, or adding cognitive components or sensory stimuli (73). Pre-established protocols like as Tai Chi and Otago can also be utilized (74, 75).

Given that aging has a significant impact on power, older adults should prioritize power training (concentric exercises as fast as possible). The SPPB demonstrated improvements as a consequence of a recent meta-analysis (71). It should be noted that studies have shown that higher gains in muscle power are linked to lower training volumes (76).

Flexibility and range of motion exercises like stretching are incorporated to improve joint flexibility and maintain a full range of motion in order to prevent stiffness by increasing static and dynamic stretching tolerance (66). Static stretching refers to the ability to maintain the position and is preferably incorporated in the end as part of the cool-down phase (66) of the ROM, whereas dynamic stretching refers to achieving, on a repeated gradual transition on any part of the body, a progressive increase in ROM (67), and could be included in warm-up (68).

While the general efficacy of all these programs has been demonstrated, the optimal physical exercise program composition is still unknown. Like any treatment, it needs to be dosed appropriately and raised (or decreased) in accordance with each patient's progress. The SPPB should be used for prescribing physical activity-based therapies, according to the Life-P study (24, 80). The Vivifrail program (81) can serve as a basis for the workout type in the following ways.

Patients with low grip strength, low knee strength, and poor 5STS, as well as those with higher Brachial/Ankle Index (BAI) scores that may be associated with peripheral artery disease, may benefit from strength interventions (82). According to the MID-FRAIL study's findings (as evaluated by SPPB), the pre-frail adult population is likely to benefit the most from strength training in terms of preventing functional decline, provided that the appropriate dose, frequency, and duration are applied (82). For patients with low gait speed, low 5STS, and low balance test scores, power interventions should be prioritized. Patients with low scores on the Romberg or other balance tests should have a focus on balance intervention. Dual tasks, strength training (83), and multicomponent exercise regimens (84, 85) have been shown to improve verbal fluency. Patients who exhibit sedentary behavior, low gait speed, or poor physical condition should prioritize aerobic training (82). An improved quality of life in terms of health is reported by older adults who exercise for more than 180 min a week (82). Also to be highlighted, there is evidence that multicomponent, highly intense physical exercise programs may be beneficial, particularly for older people who live in communities and are moderately physically fragile (42, 45, 58).

An intriguing observation from the Vivifrail program is that, after 3 months of intervention, they found that a multicomponent exercise program of moderate intensity had an impact on the cognition and mood (86) as measured by MOCA and MEC-Lobo, in addition to the physical performance and functional ability of older people with cognitive impairment. This is extremely important, in order to enhance all domains of intrinsic capacity and maintain autonomy as a whole, because functional abilities also have a cognitive component, as a consequence enhancing autonomy means promoting well-being.

Multicomponent exercise regimes work in promoting functional ability, but it is highly challenging to standardize the workout regimes' intensity, duration, and repetition. The studies that included those parameters were heterogeneous in terms of included participants (frailty diagnostic tool), intervention strategies (type of interventions, number of interventions, and combinations of interventions), and intervention duration. We can deduce from the evidence that longer training regimens may be advised because it has been proposed that training duration plays a significant role in the retention of neuromuscular adaptations after training is completed. It is sometimes necessary to begin functional-based exercise programs for frail older adults with shorter sessions than for healthy older adults (31). It is important to remember that high-quality studies (42) have shown evidence of a positive effect for both short- and longer-term interventions (42). The effects on functional capacity in the oldest old, though, can be observed, according to the Vivifrail, after only 1-month intervention (17). Most exercise programs were conducted three times a week; sometimes this frequency can be difficult to observe by frail older people with frequent multiple diseases (69).

Similarly, it is challenging to discuss the persistence of the effects that have been achieved because so few studies have included follow-up measurements in their research. However, there is evidence that, when sufficiently stimulated, the effects of exercise persist long after the intervention is over (69). Additionally, there is evidence that high-intensity exercise therapy has a positive impact on quality of life when compared to low-intensity exercise therapy (42).

Acute wards, secondary and primary health care facilities, community settings, and home-based protocols were among the varied but beneficial settings in which the protocols were conducted (62, 67, 70, 71). Programs for physical activity associated with community facilities that provide exercise for senior citizens may have some advantages over programs conducted at home, such as the ability to maintain a functional exercise regimen that is connected to social activities. However, these programs also have drawbacks, including higher costs, challenges with transportation and comfort, and user preferences (31, 17). Home-based programs are more accessible and potentially cost-effective, eliminating transportation barriers for many frail adults. This allows them to remain active, lead healthy lives, and stay longer in their local community, but they have the lowest adherence rate (73).

According to the Trondheim hip fracture RCT (71, 74), which compared the physical activity and mobility of older patients in the post-surgery period on a geriatric ward to that of a conventional post-surgery ward, early mobilization as early as 24 h after surgery can improve older people's physical performance (71). When gait was analyzed at 12 months, it was shown that a considerably higher number of patients in the same demographics had improved gait parameters, such as pace, rhythm, postural control, and decreased gait inequality. When compared to controls, a considerably higher percentage of individuals in the intervention group were able to walk independently, had improved outdoor mobility, and were more independent when utilizing public transit at 12 months. The intervention group's length of stay was marginally longer (12.6 vs. 11 days); yet, over the long term, the intervention was determined to have an 88% likelihood of being less expensive and more effective than orthopedic care.

The use of assistive technology and guidance of exercise protocols appears to be beneficial, as only the programs under supervision showed a discernible improvement in physical function (75). Evidence suggests that interventions could be delivered remotely via the telephone (69), mobile applications (70), or virtual reality and gaming technology (71). Though these treatments have not yet been evaluated in frail older persons, initial results seem promising with reported improvements in gait and balance (78).

The best physical exercise is the one that is actually carried out. Attendance to the exercise classes was reported in 17 studies and ranged from 42% over a 1-year period to 100% after 9 months. It is plausible that individuals who are frailer may be more likely to discontinue the exercise program, which could complicate the analysis of between-group differences during follow-up and perhaps introduce bias into the results when comparing the effectiveness of the exercise program. To increase adherence, greater focus should be placed on feasibility and enjoyment in more "real life" situations while developing workout regimens. Providing the patient with a comprehensive explanation of the program's potential dangers and advantages, together with a clear description, will help to enhance adherence. Extremely important is to follow the patients' preferences and acceptance (73), as well as to create a friendly and familiar training atmosphere and strict monitoring of the execution (74). There are also barriers to be considered from environmental factors like socioeconomic factors (lack of transportation, fixed incomes, or inclement weather) (75). In this case, an alternative could be home exercises or video games-based interventions (76).

Adverse events are uncommon, and administering exercise is safe and comparatively without any unpleasant side effects. It appears that the best way to minimize needless risks is to use a multidisciplinary team that includes geriatricians, nurses, and physiotherapists, especially when it comes to the most vulnerable participants (85).

Team-Based Approach

Implementing a multidisciplinary approach, with health care professionals collaborating to provide personalized care is an important issue This may include physiotherapists, occupational therapists, social workers, and geriatric specialists working together to assess the individual's needs and create tailored intervention plans. Furthermore, these programs often incorporate educational workshops and resources for both the frail individuals and their caregivers.

Due to the small number of studies, there are still several unanswered questions. First, the contribution of various monodomain interventions except exercise (e.g., hormonal intervention) and the impact of interventions on cognitive, social involvement, or certain functional outcomes are still unknown. Future studies should therefore concentrate on them.

VIRTUAL REALITY

As we discussed before, adherence and task-oriented exercises are very important confounders in the success of a function-based program; to this end, any type of modality that could positively impact enjoyment and outcomes, like virtual reality (VR), is a welcome adjunct. VR is the computer-generated simulation of a three-dimensional image or environment (VE) that can be interacted with in a seemingly real or physical way by a person, using special electronic equipment. It has two forms:

- The immersive virtual reality uses tools that are connected to the human body and designed to offer users a heightened sense of presence, and illusion of place and plausibility, jointly to an immersion in a digital environment in order to perform the task (77–82). A greater presence is felt when the VR device has a stereoscopic display, has a wider field of view, and has greater amounts of tracked user movement (81).
- In the non-immersive form of VR (exergames), the interaction might be achieved via a pointer, a joystick, or a representation of the user's body part (83). Exergames (exercises through video games), especially those utilizing existing, relatively inexpensive game technology (e.g., Nintendo Wii) (84), are gaining more and more interest in the rehabilitation of older people because they have several advantages in respect to conventional exercise such as: (1) motivation, (2) multiple training (both motor and cognitive skills) when users performing dual tasks (85), (3) focus, on the outcome of the movements in the game, not on the movements itself (85), and (4) facility of use, that is, they can also be used at home, either alone or within a small group, which may make the activity more accessible to many older adults (86).

In rehabilitation, the use of VR technology came out from the research in computational neuroscience involving motor learning mechanisms (87) and especially in the rehabilitation of older people is widely used (88, 89). It also has the advantage of a real real-time feedback (90), thereby increasing engagement in enjoyable rehabilitation tasks (91). It can be used to improve functional abilities by offering several features, such as goal-oriented tasks and repetition, increase exercise behavior (92), and influence cognitive abilities in this population segment (93). The controlled and safe environments that it creates as well as the personalised programs provide great flexibility of experiences and virtual tasks, which include the sequences of functional movements necessary to achieve autonomy in real-life activity. As a plus the exercises can be repeated over time, selecting the frequency and intensity, and that enables compliance with the principles of motor learning and neuroplasticity, as well as the generalization of this learning, and the achievement of goal not possible in real life setting due to restrictions whether environmental of physical (94).

As for the presence of adverse events in the immersive form, headache and dizziness were described by a few patients (95). The non-immersive systems have been shown to be highly accepted by older people, due to the lowest "cibersickness" symptomatology (95).

Analyzing the evidence, we see that most of the protocols are made with non-immersive systems due to the low cost and high availability. The impacts of this system in functional outcomes of older people have been studied and data shows significant improvements in at least one outcome measure (e.g., postural balance, human gait and mobility, muscle strength, self-efficacy for falls, or other physical/cognitive measures) (96) in the great majority of good quality and length studies. VR protocols, whether immersive or not, have also positive and significative effects on functional mobility (4 s reduction of TUG and 94 m more on 6MWT), after 18 sessions, as proved by a meta-analysis in a group of community-dwelling people (mainly women) aged 60–96 years (no measures of frailty) versus conventional physical activity or exercise (97). No conclusion can be made for the type of VR, and the strength of recommendation was low to very low due to the risk of bias, inconsistency, and imprecision of the analyzed data. Therefore, these studies should be treated with caution.

On pooling the data only for prefrail or frail older adults (98), although there was a variability in study designs, intensity, and duration of the intervention (ranging from 2 to 15 weeks), a clear positive trend emerged (99–103) for exergaming and mobility function and balance. Of note, positive effects and pleasure were reported only when older women practiced the exergame with moderate exercise intensity.

VR can also be used in combination with conventional training as in the work of Mirelman et al. (104) that adjunct VR in a treadmill training. They found a significant effect in reducing fall rates in the 6 months following the intervention by 42% compared with the treadmill training alone with a high retention rate (81%). The added value of the VR component is that by being a motor-cognitive intervention it requires attention plus an enhance executive function by training complex situations such as obstacle negotiation strategies in an enriched environment that requires focused attention and planning (105). Physical outcomes, like muscle strength, balance, and mobility functions that could enhance functioning, are also improved by adding an exergame in the treatment protocol (103).

The possibility to interact with and in a virtual environment makes the VR protocols very interesting in training ADLs and in the review of Corregidor-Sanchez (106) concerning the effectiveness of VRSs in improving the performance of ADLs, we can find out that the effect size of VRSs in improving the performance of BADLs is moderate but significant with a direct link between an increase in physical functional test scores (TUG, 6MWT) and the performance of basic activities. We assume, from these results, that intervention with VRSs could help older persons to live independently and safely in their own homes and could facilitate productive aging. However, the effectiveness of VRSs was smaller for the three studies which assessed IADLs as those activities are more complex and largely dependent on the environmental context (107, 108).

We can conclude that VR could contribute to maintaining independence in ADLs and delaying functional decline and the need for support. Unfortunately, we lack post-intervention follow-up to know the long-term and preventive effect of VRSs on the functional ability of older persons as well as a formal cost–benefit analysis.

NON-PHYSICAL TRAINING MOTOR IMAGERY

The non-physical training exercises, to improve balance and strength in older people, including those who are frail, comprise activities such as motor imagery (MI) and action observation (AO) (109, 110). MI stands for the mental simulation of an action without any corresponding motor output, but still causing neuronal activation (109). This simulated movement engages the frontoparietal cortical network, which overlaps with areas activated during motor execution, and can involve both kinesthetic and visual components (110). MI is known to facilitate movement and enhance motor learning (111), and training with MI is widely used in sports training and rehabilitation (112). Concurrently, AO has been presented as a tool to enhance neural activity by monitoring intentional behaviors and intending to mimic and carry out those actions. This tool is based on basic neurobiology and the mirror neuron system (113) and has been proposed to intensify neural activation induced by mental practice (109). These types of interventions are deemed safe and efficient and do not add additional neuromuscular fatigue when incorporated into any type of intervention, of value in frail people (109). Although the results are very promising, their effectiveness is still awaiting supporting evidence.

Older people have functional decrements in spatial cognition and the ability to mentally represent and effectively plan motor actions that has implications for movement efficiency and physical safety (114). For example, an imagined TUG or iTUG (115) version of the TUG test, in which patients imagine that they complete the task without actually performing it, demonstrated a lack of synchronization between executed (i.e., TUG) and imagined (i.e., iTUG) movements, caused by a decline in MI abilities in this age group. In a real-life setting, inappropriate planning can result in a fall. It is noteworthy that fear of falling (FoF) has been linked to numerous regions that are involved in MI abilities and gait control (e.g., hippocampus, primary motor cortex, cingulated motor area, etc.). Therefore, it is plausible that FoF influences MI performances. Age may increase the significance of sensory feedback for controlling movements, which could explain some of the reduction in MI ability. However, this tendency appears to be limited to unique or hard tasks and can be enhanced with practice (116). When comparing MI to rest, fMRI studies showed that activation of some part of the primary motor cortex, fronto-parietal areas, and cerebellum can be even greater in old than in young healthy adults and can become more bilaterally recruited with age. This over-activation has been attributed to a compensation for cognitive and attentional deficits associated with ageing, not related to task difficulty. Older people are able to mentally simulate motor action without producing any muscular EMG activity, emphasizing their ability to perform proper MI without unrequired motor output (117, 118).

These findings suggest that MI training can be a valuable tool for training or rehabilitation of motor performance, in terms of strength, locomotion, or simple daily tasks in healthy older adults and especially frail ones when movement must be restricted or there is an important compound of fear (109). One of the main advantages of using non-physical practices is their easy implementation in rehabilitation programs (111); they do not require specific materials and are particularly safe.

The strategies to improve mental imagery derive from training studies among brain and stroke patients (112). First-person internalizing, or concentrating on oneself as the mental image performance of the desired action, is one of these tactics. Comprehending one's physical limitations and possible outcomes is a component of this partnership. Second, the focus lies on the effector, which is the particular aspect of the self that is associated with the goal. The hand or arm when reaching for something, for instance. Third, concentrate on the target, object, or goal visual clues. The target is where the desired action is supposed to terminate. Fourth, reinforcement on the kinesthetically "feeling" way that movements are executed. Studies unambiguously show that "feeling" a movement rather than merely witnessing oneself perform it improves the mental model for movement through internalization. Fifth, combine physical with mental practice. Research also demonstrates that practicing both mentally and physically improves performance outcomes (112).

MI training can also benefit older people regarding more global functions such as gait, as it was shown to reduce postural oscillation (119), as proven by a pilot study on the benefits of MI training on older individuals' balance. The study involved twice weekly 12-week sessions of MI training, where participants were instructed to mentally cross a slackline or stand on one leg. The trained group's time to maintain an actual tandem standing position improved, while the control group's duration remained the same. Numerous psychological tests revealed no impacts from MI training, indicating that improvements in gait may be directly linked to the neurophysiological plasticity of the balance function.

MI can be combined with other types of training and be able to induce greater gains than physical training alone (111), sometimes being the only available option (109). The physical presence of a therapist is a guarantee of the success of non-physical protocols for older participants (111).

WELFARE TECHNOLOGY

Welfare technology (WT) is an umbrella term that covers different types of technologies that enable older people to remain in their homes for as long as safely possible (120), but also to improve their physical function. Safety equipment, robots, and sensors that communicate to the health care providers a probable harm, promote function. Workout watches, symptom diaries, self-management behavior change devices (addressing health issues such as sleep and exercise) (121) directly improve performance and, as a result, functional abilities.

Welfare technology enables frail older people in all levels of care and around the clock and simultaneously in a large geographical area to implement interventions with reduced costs (122), lessen the need for formal care, and as a result remain independent as long as possible. The field of assistive technology has undergone tremendous development, and today's assistive technologies and ICTs have the potential to improve quality of life by supporting age-related cognitive, sensory, and physical losses as well as help with everyday chores like mobility, hearing loss, memory loss, and wandering.

In the physical function domain, interventions focusing on outcomes such as self-care behavior and health-related quality of life appear to be effective (111). The technologies applied were mostly commercial solutions in the form of smartphones (122) and computers (123, 124), telemonitoring (125, 126), and robots (127, 128). The types of interventions varied and included balance training, physical exercise and function (122, 124, 128–130), CT (123, 131), monitoring of symptoms (132), activation of emergency medical systems (125), self-care (133), reduction of risk of death, and total length of stay at the hospital (126, 134). These studies included community-dwelling older people (no frailty criteria applied), usually both male and female, most of them enrolled small numbers of participants and had a huge variability of duration time from 8 weeks to 6 months. Although they were all RCT, there was a considerable risk of bias, but they showed notable effects on physical and psychological health status.

In the intervention of Bao et al. (122), a smartphone was used to provide for 8 weeks three 45-min exercise sessions using smartphone balance trainers, controlled by a PT, that provided written, graphic, and video guidance and monitored trunk sway. The experimental group received vibrotactile sensory augmentation via the smartphone balance trainers, while the control group performed exercises without sensory augmentation, with the experimental group outperforming in the testing. Hong et al. (124) provided a remote instructor tele-exercise program at home utilizing a 15-in. all-in-one PC and video conferencing software (Skype™), with broadband Internet connectivity, for 20–40 min a day and three times per week for 12 weeks, showing a significant impact on sarcopenia measures. In the work of Matz-Costa (129), 15 participants were randomized to the intervention group (technology-assisted self-monitoring of daily activity via pedometers and daily tablet-based surveys; psychoeducation + goal-setting via a 3-h workshop; and peer mentoring via phone 2×/week for 2.5 weeks) and 15 to the control (technology-assisted self-monitoring only). Although it was a feasibility study, there was a positive effect on the number of steps for the intervention group. Positive effects were also found with the ironHand system that could help people with hand problems (124, 131).

The work of Corbet et al. (123) used an online delivery of CT in older adults as a cost-effective measure. They enrolled an important number of participants, 2912 (no frailty evaluation), in a double-blind 6-month online randomized controlled trial in adults older than 50 randomized to General CT, Reasoning CT, or control. The primary outcome was IADL in adults older than 60. Secondary outcomes were reasoning, verbal short-term memory, spatial working memory, verbal learning (VL), and digit vigilance in adults older than 50. Secondary analyses were performed with a group defined as showing age-associated impairment in reasoning according to baseline scores in this domain. Both the general and the reasoning package demonstrated a clear benefit to IADL at 6 months comparable with in-person training. Also, positive results in the improvement of self-care and quality of life were found by just giving a tablet to 82 patients with heart failure and as a result a reduction in hospital readmissions (133).

Positive effects were also found from intervention of welfare technology in CT by a robot assisted training by measuring cortical thickness (129) or task performance (132), health monitoring by health survey forms (127), emergency system activation by the number of healthy days (134), and reduction of hospital readmission (128, 135).

The outcomes of those studies make WT promising, but there are still challenges with the use of WT in older people. Older people must be involved in the development of these technologies, and their physical constraints must be taken into account in order to increase acceptance of them. Another aspect that presents an ethical challenge of testing WT among frail older people is that the WT used in the home tends to be similar to that found in hospitals or care homes, making the home environment an extension of the institution. When technology becomes their only means of social engagement, people lose personal contact in addition to losing their privacy and having no idea where or how their data is kept or accessed. Additionally, older individuals may worry about who is in charge of their care and the technology – if their "care" is dependent on the technology or whether they have control over it. They may also feel objectified and powerless (120).

Acceptance of WT among older people depends on their previous experience of using technology and their familiarity with the device. Older persons prefer to interact with technology that uses both their native language and non-technical wording, through speech rather than visually.

From all the above, we can conclude that although there is a pressing demand for cutting-edge technology that encourages healthy aging and assists senior citizens in leading secure, comfortable, and independent lifestyles, there is still a lot of research needed on the use of this technology.

REFERENCES

1. Bland J. Defining function in the functional medicine model. Integr Med Encinitas Calif. 2017 Feb;16(1):22–5.
2. Rodriguez-Mañas L, Laosa O, Vellas B, Paolisso G, Topinkova E, Oliva-Moreno J, et al. Effectiveness of a multimodal intervention in functionally impaired older people with type 2 diabetes mellitus. J Cachexia Sarcopenia Muscle. 2019 Aug;10(4):721–33.
3. Stucki G. International Classification of Functioning, Disability, and Health (ICF): A promising framework and classification for rehabilitation medicine. Am J Phys Med Rehabil. 2005 Oct;84(10):733–40.
4. Bruce B, Fries JF, Ambrosini D, Lingala B, Gandek B, Rose M, et al. Better assessment of physical function: Item improvement is neglected but essential. Arthritis Res Ther. 2009 Dec;11(6):R191.
5. Dias JM. Physical Functioning (PF). In: Michalos AC, editor. Encyclopedia of Quality of Life and Well-Being Research [Internet]. Dordrecht: Springer Netherlands; 2014. pp. 4793–5. http://link.springer.com/10.1007/978-94-007-0753-5_2162
6. Hoogendijk EO, Afilalo J, Ensrud KE, Kowal P, Onder G, Fried LP. Frailty: Implications for clinical practice and public health. Lancet. 2019 Oct;394(10206):1365–75.
7. Ward DD, Wallace LMK, Rockwood K. Frailty and risk of dementia in mild cognitive impairment subtypes. Ann Neurol. 2021 Jun;89(6):1221–5.
8. Dong X, Milholland B, Vijg J. Evidence for a limit to human lifespan. Nature. 2016 Oct 13;538(7624):257–9.
9. World Health Organization. World Report on Aging and Health. WHO, Geneva, 2015
10. Robertson DA, Savva GM, Kenny RA. Frailty and cognitive impairment—A review of the evidence and causal mechanisms. Ageing Res Rev. 2013 Sep;12(4):840–51.
11. Roberts HC, Denison HJ, Martin HJ, Patel HP, Syddall H, Cooper C, et al. A review of the measurement of grip strength in clinical and epidemiological studies: Towards a standardised approach. Age Ageing. 2011 Jul;40(4):423–9.
12. Levinger I, Goodman C, Hare DL, Jerums G, Toia D, Selig S. The reliability of the 1RM strength test for untrained middle-aged individuals. J Sci Med Sport. 2009 Mar;12(2):310–6.
13. Mijnarends DM, Meijers JMM, Halfens RJG, Ter Borg S, Luiking YC, Verlaan S, et al. Validity and reliability of tools to measure muscle mass, strength, and physical performance in community-dwelling older people: A systematic review. J Am Med Dir Assoc. 2013 Mar;14(3):170–8.
14. Bohannon RW, Bubela DJ, Magasi SR, Wang YC, Gershon RC. Sit-to-stand test: Performance and determinants across the age-span. Isokinet Exerc Sci. 2010 Nov 5;18(4):235–40.
15. Izquierdo M, Ibañez J, Gorostiaga E, Garrues M, Zúñiga A, Antón A, et al. Maximal strength and power characteristics in isometric and dynamic actions of the upper and lower extremities in middle-aged and older men: Strength and power and ageing. Acta Physiol Scand. 1999 Sep;167(1):57–68.
16. Newman AB, Kupelian V, Visser M, Simonsick EM, Goodpaster BH, Kritchevsky SB, et al. Strength, but not muscle mass, is associated with mortality in the health, aging and body composition study cohort. J Gerontol A Biol Sci Med Sci. 2006 Jan 1;61(1):72–7.
17. Casas-Herrero Á, Sáez De Asteasu ML, Antón-Rodrigo I, Sánchez-Sánchez JL, Montero-Odasso M, Marín-Epelde I, et al. Effects of Vivifrail multicomponent intervention on functional capacity: A multicentre, randomized controlled trial. J Cachexia Sarcopenia Muscle. 2022 Apr;13(2):884–93.
18. Nnodim JO, Yung RL. Balance and its clinical assessment in older adults – A Review. J Geriatr Med Gerontol. 2015;1(1):003.
19. Abellan Van Kan G, Rolland Y, Andrieu S, Bauer J, Beauchet O, Bonnefoy M, et al. Gait speed at usual pace as a predictor of adverse outcomes in community-dwelling older people an International Academy on Nutrition and Aging (IANA) Task Force. J Nutr Health Aging. 2009 Dec;13(10):881–9.
20. Chou CH, Hwang CL, Wu YT. Effect of exercise on physical function, daily living activities, and quality of life in the frail older adults: A meta-analysis. Arch Phys Med Rehabil. 2012 Feb;93(2):237–44.
21. Guralnik JM, Simonsick EM, Ferrucci L, Glynn RJ, Berkman LF, Blazer DG, et al. A short physical performance battery assessing lower extremity function: Association with self-reported disability and prediction of mortality and nursing home admission. J Gerontol. 1994 Mar 1;49(2):M85–94.
22. Romero-García M, López-Rodríguez G, Henao-Morán S, González-Unzaga M, Galván M. Effect of a multicomponent exercise program (VIVIFRAIL) on functional capacity in elderly ambulatory: A non-randomized clinical trial in Mexican women with dynapenia. J Nutr Health Aging. 2021 Feb;25(2):148–54.
23. Seidel D, Brayne C, Jagger C. Limitations in physical functioning among older people as a predictor of subsequent disability in instrumental activities of daily living. Age Ageing. 2011 Jul 1;40(4):463–9.
24. Hopman-Rock M, Van Hirtum H, De Vreede P, Freiberger E. Activities of daily living in older community-dwelling persons: A systematic review of psychometric properties of instruments. Aging Clin Exp Res. 2019 Jul;31(7):917–25.

25. Giné-Garriga M, Roqué-Fíguls M, Coll-Planas L, Sitjà-Rabert M, Salvà A. physical exercise interventions for improving performance-based measures of physical function in community-dwelling, frail older adults: A systematic review and meta-analysis. Arch Phys Med Rehabil. 2014 Apr;95(4):753–769.e3.
26. Guralnik JM, Ferrucci L, Pieper CF, Leveille SG, Markides KS, Ostir GV, et al. Lower extremity function and subsequent disability: Consistency across studies, predictive models, and value of gait speed alone compared with the short physical performance battery. J Gerontol A Biol Sci Med Sci. 2000 Apr 1;55(4):M221–31.
27. Studenski S, Perera S, Wallace D, Chandler JM, Duncan PW, Rooney E, et al. Physical performance measures in the clinical setting. J Am Geriatr Soc. 2003 Mar;51(3):314–22.
28. Pijpers E, Ferreira I, Stehouwer CDA, Nieuwenhuijzen Kruseman AC. The frailty dilemma. Review of the predictive accuracy of major frailty scores. Eur J Intern Med. 2012 Mar;23(2):118–23.
29. Alley DE, Hicks GE, Shardell M, Hawkes W, Miller R, Craik RL, et al. Meaningful improvement in gait speed in hip fracture recovery. J Am Geriatr Soc. 2011 Sep;59(9):1650–7.
30. Perera S, Mody SH, Woodman RC, Studenski SA. Meaningful change and responsiveness in common physical performance measures in older adults. J Am Geriatr Soc. 2006 May;54(5):743–9.
31. Guralnik J. Assessing the building blocks of functionUtilizing measures of functional limitation. Am J Prev Med. 2003 Oct;25(3):112–21.
32. Brown GC. Living too long: The current focus of medical research on increasing the quantity, rather than the quality, of life is damaging our health and harming the economy. EMBO Rep. 2015 Feb;16(2):137–41.
33. Abdi S, Spann A, Borilovic J, De Witte L, Hawley M. Understanding the care and support needs of older people: A scoping review and categorisation using the WHO international classification of functioning, disability and health framework (ICF). BMC Geriatr. 2019 Dec;19(1):195.
34. Strout K, Ahmed F, Sporer K, Howard EP, Sassatelli E, Mcfadden K What are older adults wellness priorities? A qualitative analysis of priorities within multiple domains of wellness: Healthy Aging Res. 2018 Jun;7(2):e21.
35. Wu B. Social isolation and loneliness among older adults in the context of COVID-19: A global challenge. Glob Health Res Policy. 2020 Dec;5(1):27.
36. Beard JR, Officer A, De Carvalho IA, Sadana R, Pot AM, Michel JP, et al. The world report on ageing and health: A policy framework for healthy ageing. Lancet. 2016 May;387(10033):2145–54.
37. Dedeyne L, Deschodt M, Verschueren S, Tournoy J, Gielen E. Effects of multi-domain interventions in (pre)frail elderly on frailty, functional, and cognitive status: A systematic review. Clin Interv Aging. 2017 May;12:873–96.
38. Kenny AM, Boxer RS, Kleppinger A, Brindisi J, Feinn R, Burleson JA. Dehydroepiandrosterone combined with exercise improves muscle strength and physical function in frail older women: DEHYDROEPIANDROSTERONE, exercise, muscle strength, AND physical function. J Am Geriatr Soc. 2010 Sep;58(9):1707–14.
39. Tarazona-Santabalbina FJ, Gómez-Cabrera MC, Pérez-Ros P, Martínez-Arnau FM, Cabo H, Tsaparas K, et al. A multicomponent exercise intervention that reverses frailty and improves cognition, emotion, and social networking in the community-dwelling frail elderly: A randomized clinical trial. J Am Med Dir Assoc. 2016 May;17(5):426–33.
40. Tieland M, Dirks ML, Van Der Zwaluw N, Verdijk LB, Van De Rest O, De Groot LCPGM, et al. Protein supplementation increases muscle mass gain during prolonged resistance-type exercise training in frail elderly people: A randomized, double-blind, placebo-controlled trial. J Am Med Dir Assoc. 2012 Oct;13(8):713–9.
41. Puts MTE, Toubasi S, Andrew MK, Ashe MC, Ploeg J, Atkinson E, et al. Interventions to prevent or reduce the level of frailty in community-dwelling older adults: A scoping review of the literature and international policies. Age Ageing. 2017 May;46(3):383–92.
42. De Vries NM, Van Ravensberg CD, Hobbelen JSM, Olde Rikkert MGM, Staal JB, Nijhuis-van Der Sanden MWG. Effects of physical exercise therapy on mobility, physical functioning, physical activity and quality of life in community-dwelling older adults with impaired mobility, physical disability and/or multi-morbidity: A meta-analysis. Ageing Res Rev. 2012 Jan;11(1):136–49.
43. Vagetti GC, Barbosa Filho VC, Moreira NB, Oliveira VD, Mazzardo O, Campos WD. Association between physical activity and quality of life in the elderly: A systematic review, 2000-2012. Rev Bras Psiquiatr. 2014 Jan 17;36(1):76–88.
44. Daniels R, Metzelthin S, Van Rossum E, De Witte L, Van Den Heuvel W. Interventions to prevent disability in frail community-dwelling older persons: An overview. Eur J Ageing. 2010 Mar;7(1):37–55.
45. Bosaeus I, Rothenberg E. Nutrition and physical activity for the prevention and treatment of age-related sarcopenia. Proc Nutr Soc. 2016 May;75(2):174–80.

46. Malafarina V, Uriz-Otano F, Iniesta R, Gil-Guerrero L. Effectiveness of nutritional supplementation on muscle mass in treatment of sarcopenia in old age: A systematic review. J Am Med Dir Assoc. 2013 Jan;14(1):10–7.
47. Robinson S, Denison H, Cooper C, Aihie Sayer A. Prevention and optimal management of sarcopenia: A review of combined exercise and nutrition interventions to improve muscle outcomes in older people. Clin Interv Aging. 2015 May 11;10:859–69.
48. Ng TP, Feng L, Nyunt MSZ, Feng L, Niti M, Tan BY, et al. Nutritional, physical, cognitive, and combination interventions and frailty reversal among older adults: A randomized controlled trial. Am J Med. 2015 Nov;128(11):1225–1236.e1.
49. Brigola AG, Rossetti ES, Santos BRD, Neri AL, Zazzetta MS, Inouye K, et al. Relationship between cognition and frailty in elderly: A systematic review. Dement Neuropsychol. 2015 Jun;9(2):110–9.
50. Gill TM, Baker DI, Gottschalk M, Peduzzi PN, Allore H, Byers A. A program to prevent functional decline in physically frail, elderly persons who live at home. N Engl J Med. 2002 Oct 3;347(14):1068–74.
51. Littbrand H, Lundin-Olsson L, Gustafson Y, Rosendahl E. The effect of a high-intensity functional exercise program on activities of daily living: A randomized controlled trial in residential care facilities. J Am Geriatr Soc. 2009 Oct;57(10):1741–9.
52. Gates S, Fisher JD, Cooke MW, Carter YH, Lamb SE. Multifactorial assessment and targeted intervention for preventing falls and injuries among older people in community and emergency care settings: Systematic review and meta-analysis. BMJ. 2008 Jan 19;336(7636):130–3.
53. Chodzko-Zajko WJ, Proctor DN, Fiatarone Singh MA, Minson CT, Nigg CR, Salem GJ, et al. Exercise and physical activity for older adults. Med Sci Sports Exerc. 2009 Jul;41(7):1510–30.
54. Liu C, Latham NK. Progressive resistance strength training for improving physical function in older adults. Cochrane Database Syst Rev. 2009 Jul 8;2009(3):CD002759.
55. Peri K, Kerse N, Robinson E, Parsons M, Parsons J, Latham N. Does functionally based activity make a difference to health status and mobility? A randomised controlled trial in residential care facilities (The Promoting Independent Living Study; PILS). Age Ageing. 2007 Oct 11;37(1):57–63.
56. Shimada H, Uchiyama Y, Kakurai S. Specific effects of balance and gait exercises on physical function among the frail elderly. Clin Rehabil. 2003 Aug;17(5):472–9.
57. Chin APPSPMJM, Van Uffelen JGZ, Riphagen I, Van Mechelen W. The functional effects of physical exercise training in frail older people: A systematic review. Sports Med. 2008;38(9):781–93.
58. Judge JO, Schechtman K, Cress E. The relationship between physical performance measures and independence in instrumental activities of daily living. The FICSIT Group. Frailty and Injury: Cooperative Studies of Intervention Trials. J Am Geriatr Soc. 1996 Nov;44(11):1332–41.
59. Der Wiel AB, Gussekloo J, De Craen AJM, Van Exel E, Bloem BR, Westendorp RGJ. Common chronic diseases and general impairments as determinants of walking disability in the oldest-old population. J Am Geriatr Soc. 2002 Aug;50(8):1405–10.
60. Giné-Garriga M, Guerra M, Pagès E, Manini TM, Jiménez R, Unnithan VB. The effect of functional circuit training on physical frailty in frail older adults: A randomized controlled trial. J Aging Phys Act. 2010 Oct;18(4):401–24.
61. Skelton DA, Young A, Greig CA, Malbut KE. Effects of resistance training on strength, power, and selected functional abilities of women aged 75 and older. J Am Geriatr Soc. 1995 Oct;43(10):1081–7.
62. Binder EF, Schechtman KB, Ehsani AA, Steger-May K, Brown M, Sinacore DR, et al. Effects of exercise training on frailty in community-dwelling older adults: Results of a randomized, controlled trial. J Am Geriatr Soc. 2002 Dec;50(12):1921–8.
63. Sato D, Kaneda K, Wakabayashi H, Nomura T. The water exercise improves health-related quality of life of frail elderly people at day service facility. Qual Life Res. 2007 Nov 2;16(10):1577–85.
64. Brach JS, VanSwearingen JM. Physical impairment and disability: Relationship to performance of activities of daily living in community-dwelling older men. Phys Ther. 2002 Aug 1;82(8):752–61.
65. Cesari M, Vellas B, Hsu FC, Newman AB, Doss H, King AC, et al. A physical activity intervention to treat the frailty syndrome in older persons–Results from the LIFE-P Study. J Gerontol A Biol Sci Med Sci. 2015 Feb 1;70(2):216–22.
66. Frontera WR, editor. Rehabilitation of Sports Injuries: Scientific Basis [Internet]. 1st ed. Wiley; 2003. https://onlinelibrary.wiley.com/doi/book/10.1002/9780470757178
67. Garber CE, Blissmer B, Deschenes MR, Franklin BA, Lamonte MJ, Lee IM, et al. Quantity and quality of exercise for developing and maintaining cardiorespiratory, musculoskeletal, and neuromotor fitness in apparently healthy adults: Guidance for prescribing exercise. Med Sci Sports Exerc. 2011 Jul;43(7):1334–59.

68. Micheo W, Baerga L, Miranda G. Basic principles regarding strength, flexibility, and stability exercises. PM&R. 2012 Nov;4(11):805–11.
69. Kidd T, Mold F, Jones C, Ream E, Grosvenor W, Sund-Levander M, et al. What are the most effective interventions to improve physical performance in pre-frail and frail adults? A systematic review of randomised control trials. BMC Geriatr. 2019 Dec;19(1):184.
70. Opdenacker J, Delecluse C, Boen F. A 2-year follow-up of a lifestyle physical activity versus a structured exercise intervention in older adults. J Am Geriatr Soc. 2011 Sep;59(9):1602–11.
71. Silveira P, Reve EVH, Daniel F, Casati F, De Bruin ED. Motivating and assisting physical exercise in independently living older adults: A pilot study. Int J Med Inf. 2013 May;82(5):325–34.
72. Skjæret N, Nawaz A, Morat T, Schoene D, Helbostad JL, Vereijken B. Exercise and rehabilitation delivered through exergames in older adults: An integrative review of technologies, safety and efficacy. Int J Med Inf. 2016 Jan;85(1):1–16.
73. Rivera-Torres S, Fahey TD, Rivera MA. Adherence to exercise programs in older adults: Informative report. Gerontol Geriatr Med. 2019;5:233372141882360.
74. Cadore EL, Izquierdo M. muscle power training: A hallmark for muscle function retaining in frail clinical setting. J Am Med Dir Assoc. 2018 Mar;19(3):190–2.
75. O'Neill D, Forman DE. The importance of physical function as a clinical outcome: Assessment and enhancement. Clin Cardiol. 2020 Feb;43(2):108–17.
76. Buford TW, Anton SD, Clark DJ, Higgins TJ, Cooke MB. Optimizing the benefits of exercise on physical function in older adults. PM&R. 2014 Jun;6(6):528–43.
77. Henderson A, Korner-Bitensky N, Levin M. Virtual reality in stroke rehabilitation: A systematic review of its effectiveness for upper limb motor recovery. Top Stroke Rehabil. 2007 Mar;14(2):52–61.
78. Piron L, Turolla A, Agostini M, Zucconi CS, Ventura L, Tonin P, et al. Motor learning principles for rehabilitation: A pilot randomized controlled study in poststroke patients. Neurorehabil Neural Repair. 2010 Jul;24(6):501–8.
79. Dermody G, Whitehead L, Wilson G, Glass C. The role of virtual reality in improving health outcomes for community-dwelling older adults: Systematic review. J Med Internet Res. 2020 Jun 1;22(6): e17331.
80. Chesham RK, Malouff JM, Schutte NS. Meta-analysis of the efficacy of virtual reality exposure therapy for social anxiety. Behav Change. 2018 Sep;35(3):152–66.
81. Cummings JJ, Bailenson JN. How immersive is enough? A meta-analysis of the effect of immersive technology on user presence. Media Psychol. 2016 Apr 2;19(2):272–309.
82. Slater M. Place illusion and plausibility can lead to realistic behaviour in immersive virtual environments. Philos Trans R Soc B Biol Sci. 2009 Dec 12;364(1535):3549–57.
83. Sveistrup H. Motor rehabilitation using virtual reality. J NeuroEngineering Rehabil. 2004;1(1):10.
84. Choi SD, Guo L, Kang D, Xiong S. Exergame technology and interactive interventions for elderly fall prevention: A systematic literature review. Appl Ergon. 2017;65:570–81.
85. Proffitt R, Lange B. Considerations in the efficacy and effectiveness of virtual reality interventions for stroke rehabilitation: Moving the field forward. Phys Ther. 2015 Mar 1;95(3):441–8.
86. Chang AA, Kazemi E, Esmaeili V, Davies MS. The effectiveness of virtual reality training: A systematic review. J Organ Behav Manag. 2023 Aug 15;1–19.
87. Freeman D, Reeve S, Robinson A, Ehlers A, Clark D, Spanlang B, et al. Virtual reality in the assessment, understanding, and treatment of mental health disorders. Psychol Med. 2017 Oct;47(14):2393–400.
88. Ren Y, Lin C, Zhou Q, Yingyuan Z, Wang G, Lu A. Effectiveness of virtual reality games in improving physical function, balance and reducing falls in balance-impaired older adults: A systematic review and meta-analysis. Arch Gerontol Geriatr. 2023;108:104924.
89. Hall V, Conboy-Hill S, Taylor D. Using virtual reality to provide health care information to people with intellectual disabilities: Acceptability, usability, and potential utility. J Med Internet Res. 2011 Nov 14;13(4):e91.
90. Mohammadi R, Semnani AV, Mirmohammadkhani M, Grampurohit N. Effects of virtual reality compared to conventional therapy on balance poststroke: A systematic review and meta-analysis. J Stroke Cerebrovasc Dis. 2019 Jul;28(7):1787–98.
91. Ahn S, Hwang S. Virtual rehabilitation of upper extremity function and independence for stoke: A meta-analysis. J Exerc Rehabil. 2019 Jun 26;15(3):358–69.
92. Van Schaik P, Blake J, Pernet F, Spears I, Fencott C. Virtual augmented exercise gaming for older adults. Cyberpsychol Behav. 2008 Feb;11(1):103–6.
93. Zelinski EM, Reyes R. Cognitive benefits of computer games for older adults. Gerontechnology Int J Fundam Asp Technol Serve Ageing Soc. 2009;8(4):220–35.

94. Levin MF. Can virtual reality offer enriched environments for rehabilitation? Expert Rev Neurother. 2011 Feb;11(2):153–5.
95. An CM, Park YH. The effects of semi-immersive virtual reality therapy on standing balance and upright mobility function in individuals with chronic incomplete spinal cord injury: A preliminary study. J Spinal Cord Med. 2018 Mar 4;41(2):223–9.
96. Bevilacqua R, Maranesi E, Riccardi GR, Donna VD, Pelliccioni P, Luzi R, et al. Non-immersive virtual reality for rehabilitation of the older people: A systematic review into efficacy and effectiveness. J Clin Med. 2019 Nov 5;8(11):1882.
97. Corregidor-Sánchez AI, Segura-Fragoso A, Rodríguez-Hernández M, Jiménez-Rojas C, Polonio-López B, Criado-Álvarez JJ. Effectiveness of virtual reality technology on functional mobility of older adults: Systematic review and meta-analysis. Age Ageing. 2021 Feb 26;50(2):370–9.
98. Daniel K. Wii-hab for pre-frail older adults. Rehabil Nurs. 2012 Jul;37(4):195–201.
99. Zheng L, Li G, Wang X, Yin H, Jia Y, Leng M, et al. Effect of exergames on physical outcomes in frail elderly: A systematic review. Aging Clin Exp Res. 2020 Nov;32(11):2187–200.
100. Hagedorn DK, Holm E. Effects of traditional physical training and visual computer feedback training in frail elderly patients. A randomized intervention study. Eur J Phys Rehabil Med. 2010 Jun;46(2): 159–68.
101. Kubicki A, Bonnetblanc F, Petrement G, Mourey F. Motor-prediction improvements after virtual rehabilitation in geriatrics: Frail patients reveal different learning curves for movement and postural control. Neurophysiol Clin Neurophysiol. 2014 Jan;44(1):109–18.
102. Mugueta-Aguinaga I, Garcia-Zapirain B. FRED: Exergame to prevent dependence and functional deterioration associated with ageing. A pilot three-week randomized controlled clinical trial. Int J Environ Res Public Health. 2017 Nov 23;14(12):1439.
103. Santos GOR, Wolf R, Silva MM, Rodacki ALF, Pereira G. Does exercise intensity increment in exergame promote changes in strength, functional capacity and perceptual parameters in pre-frail older women? A randomized controlled trial. Exp Gerontol. 2019;116:25–30.
104. Mirelman A, Rochester L, Maidan I, Del Din S, Alcock L, Nieuwhof F, et al. Addition of a non-immersive virtual reality component to treadmill training to reduce fall risk in older adults (V-TIME): A randomised controlled trial. The Lancet. 2016 Sep;388(10050):1170–82.
105. Mirelman A, Maidan I, Herman T, Deutsch JE, Giladi N, Hausdorff JM. Virtual reality for gait training: Can it induce motor learning to enhance complex walking and reduce fall risk in patients with Parkinson's disease? J Gerontol A Biol Sci Med Sci. 2011 Feb 1;66A(2):234–40.
106. Corregidor-Sánchez AI, Segura-Fragoso A, Criado-Álvarez JJ, Rodríguez-Hernández M, Mohedano-Moriano A, Polonio-López B. Effectiveness of virtual reality systems to improve the activities of daily life in older people. Int J Environ Res Public Health. 2020 Aug 28;17(17):6283.
107. Farzin A, Ibrahim R, Madon Z, Basri H. The efficiency of a multicomponent training for prospective memory among healthy older adults: A single-blind, randomized controlled within-participants cross-over trial. Am J Phys Med Rehabil. 2018 Sep;97(9):628–35.
108. Hughes KJ, Salmon N, Galvin R, Casey B, Clifford AM. Interventions to improve adherence to exercise therapy for falls prevention in community-dwelling older adults: Systematic review and meta-analysis. Age Ageing. 2019 Mar 1;48(2):185–95.
109. Marusic U, Grosprêtre S. Non-physical approaches to counteract age-related functional deterioration: Applications for rehabilitation and neural mechanisms. Eur J Sport Sci. 2018 May 28;18(5):639–49.
110. Mouthon AA, Ruffieux J, Keller M, Taube W. Age-related differences in corticospinal excitability during observation and motor imagery of balance tasks. Front Aging Neurosci. 2016 Dec 23;8:317.
111. Malouin F, Jackson PL, Richards CL. Towards the integration of mental practice in rehabilitation programs. A critical review. Front Hum Neurosci. 2013 Sep 19;7:576.
112. Schuster C, Hilfiker R, Amft O, Scheidhauer A, Andrews B, Butler J, et al. Best Practice for motor imagery: A systematic literature review on motor imagery training elements in five different disciplines. BMC Med. 2011 Dec;9(1):75.
113. Shih TY, Wu CY, Lin KC, Cheng CH, Hsieh YW, Chen CL, et al. Effects of action observation therapy and mirror therapy after stroke on rehabilitation outcomes and neural mechanisms by MEG: Study protocol for a randomized controlled trial. Trials. 2017 Dec;18(1):459.
114. McAvinue LP, Robertson IH. Measuring motor imagery ability: A review. Eur J Cogn Psychol. 2008 Mar;20(2):232–51.
115. Grenier S, Richard-Devantoy S, Nadeau A, Payette MC, Benyebdri F, Duhaime MMB, et al. The association between fear of falling and motor imagery abilities in older community-dwelling individuals. Maturitas. 2018 Apr;110:18–20.

116. Saimpont A, Malouin F, Tousignant B, Jackson PL. Motor imagery and aging. J Mot Behav. 2013 Jan;45(1):21–8.
117. Gentili R, Papaxanthis C, Pozzo T. Improvement and generalization of arm motor performance through motor imagery practice. Neuroscience. 2006 Jan;137(3):761–72.
118. Personnier P, Ballay Y, Papaxanthis C. Mentally represented motor actions in normal aging: III. Electromyographic features of imagined arm movements. Behav Brain Res. 2010 Jan;206(2):184–91.
119. Kalicinski M, Kempe M, Bock O. Motor imagery: Effects of age, task complexity, and task setting. Exp Aging Res. 2015 Jan;41(1):25–38.
120. Pajalic Z, De Sousa DA, Strøm BS, Lausund H, Breievne G, Kisa S, et al. Welfare technology interventions among older people living at home—A systematic review of RCT studies. PLOS Digit Health. 2023 Jan 24;2(1):e0000184.
121. Unsworth H, Dillon B, Collinson L, Powell H, Salmon M, Oladapo T, et al. The NICE Evidence Standards Framework for digital health and care technologies – Developing and maintaining an innovative evidence framework with global impact. Digit Health. 2021 Jun 24:7:205520762110186.
122. Bao T, Carender WJ, Kinnaird C, Barone VJ, Peethambaran G, Whitney SL, et al. Effects of long-term balance training with vibrotactile sensory augmentation among community-dwelling healthy older adults: A randomized preliminary study. J NeuroEngineering Rehabil. 2018 Dec;15(1):5.
123. Corbett A, Owen A, Hampshire A, Grahn J, Stenton R, Dajani S, et al. The effect of an online cognitive training package in healthy older adults: An online randomized controlled trial. J Am Med Dir Assoc. 2015 Nov;16(11):990–7.
124. Hong J, Kim J, Kim SW, Kong HJ. Effects of home-based tele-exercise on sarcopenia among community-dwelling elderly adults: Body composition and functional fitness. Exp Gerontol. 2017;87:33–9.
125. Morgenstern LB, Adelman EE, Hughes R, Wing JJ, Lisabeth LD. The women independently living alone with a medical alert device (WILMA) trial. Transl Stroke Res. 2015 Oct;6(5):355–60.
126. Pedone C, Rossi FF, Cecere A, Costanzo L, Antonelli Incalzi R. Efficacy of a physician-led multiparametric telemonitoring system in very old adults with heart failure. J Am Geriatr Soc. 2015 Jun;63(6):1175–80.
127. Kim GH, Jeon S, Im K, Kwon H, Lee BH, Kim GY, et al. Structural brain changes after traditional and robot-assisted multi-domain cognitive training in community-dwelling healthy elderly. PLoS One. 2015 Apr 21;10(4):e0123251.
128. Radder B, Prange-Lasonder GB, Kottink AIR, Holmberg J, Sletta K, Van Dijk M, et al. Home rehabilitation supported by a wearable soft-robotic device for improving hand function in older adults: A pilot randomized controlled trial. PLoS One. 2019 Aug 6;14(8):e0220544.
129. Matz-Costa C, Lubben J, Lachman ME, Lee H, Choi YJ. A pilot randomized trial of an intervention to enhance the health-promoting effects of older adults' activity portfolios: The Engaged4Life program. J Gerontol Soc Work. 2018 Nov 17;61(8):792–816.
130. Simon SS, Tusch ES, Feng NC, Håkansson K, Mohammed AH, Daffner KR. Is computerized working memory training effective in healthy older adults? Evidence from a multi-site, randomized controlled trial. J Alzheimers Dis. 2018 Sep 11;65(3):931–49.
131. Kim H, Suzuki T, Kim M, Kojima N, Ota N, Shimotoyodome A, et al. Effects of exercise and milk fat globule membrane (MFGM) supplementation on body composition, physical function, and hematological parameters in community-dwelling frail Japanese women: A randomized double blind, placebo-controlled, follow-up trial. PLoS One. 2015 Feb 6;10(2):e0116256.
132. Mavandadi S, Benson A, DiFilippo S, Streim JE, Oslin D. A telephone-based program to provide symptom monitoring alone vs symptom monitoring plus care management for late-life depression and anxiety: A randomized clinical trial. JAMA Psychiatry. 2015 Dec 1;72(12):1211.
133. Melin M, Hägglund E, Ullman B, Persson H, Hagerman I. Effects of a tablet computer on self-care, quality of life, and knowledge: A randomized clinical trial. J Cardiovasc Nurs. 2018 Jul;33(4):336–43.
134. Ong NWR, Ho AFW, Chakraborty B, Fook-Chong S, Yogeswary P, Lian S, et al. Utility of a Medical Alert Protection System compared to telephone follow-up only for home-alone elderly presenting to the ED—A randomized controlled trial. Am J Emerg Med. 2018 Apr;36(4):594–601.
135. Rydwik E, Lammes E, Frändin K, Akner G. Effects of a physical and nutritional intervention program for frail elderly people over age 75. A randomized controlled pilot treatment trial. Aging Clin Exp Res. 2008 Apr;20(2):159–70.

8 Diabetes and Frailty-Interactions and Clinical Implications

Alan J. Sinclair and A. H. Abdelhafiz

INTRODUCTION

The global prevalence of diabetes mellitus is increasing due to increased life expectancy. The prevalence is particularly high in the older age group between 75 and 79 years reaching up to 24%.[1] With increasing age, frailty emerges as a new third category of complication of diabetes, in addition to the traditional diabetes-related micro and macrovascular disease.[2] Up to 48% of older people with diabetes are frail.[3] Although the prevalence of frailty increases with increasing age, frailty is not an inevitable part of growing old, nor is a synonymous with ageing.[4] Frailty is associated with an increased risk of adverse outcomes. For example, frail older people are at increased risk of injurious falls, disability, and mortality.[5–7] The adverse events associated with frailty increase the burden on the health care systems, such as increased frequency of emergency department visits, hospitalisation, and care home admissions.[8] Frailty, in older people with diabetes, acts as a clinical marker that can distinguish biological from chronological ageing and can be predictive of downstream adverse clinical outcomes. Therefore, it has now been recognised as a part of the routine clinical assessment in older people with diabetes to identify patients at higher risk of adverse outcomes.[9] In this chapter, we review the interrelationship between frailty and diabetes and the associated clinical implications.

DIABETES AND FRAILTY INTERACTIONS

The cause of diabetes-associated frailty is likely to be multifactorial. Diabetes is associated with other comorbidities such as hypertension and diabetes-related complications such as cardiac and renal disease. Frailty, therefore, is likely to be a result of these diabetes-associated morbidities and diabetes-related complications. Both morbidity and frailty follow the disablement process model (DPM) pathway, which incorporates four sequential stages from disease to impairment to functional decline then to disability.[10] The multimorbidity mediator effect to frailty appears to be synergistic when more than one morbidity coexists with diabetes. The Mexican health and nutrition survey of 7,164 older people, mean (SD) age 70.6 (8.1) years, showed diabetes to be independently associated with frailty (coefficient 0.28, $p < 0.001$) and an incremental association when hypertension (0.63, $p < 0.001$) or any diabetes-related complication was present (0.55, $p < 0.001$).[11] In a Japanese cross sectional study of 9,606 participants ≥65 years old, participants in the lowest quartile of estimated glomerular filtration rate (eGFR) <30.0 mL/min/1.73 m^2 showed an independent higher risk of frailty [odds ratio (OR) 1.83, 95% confidence interval (CI) 1.01 to 3.45] compared to those in the highest quartile (eGFR ≥60.0 mL/min/1.73 m^2). Subjects with a history of hypertension or diabetes mellitus showed a significantly increased risk of frailty in the group with the lowest eGFR (OR 2.53, 95% CI 1.45 to 5.12 and 2.76, 1.21 to 8.24, respectively) and the risk increased further when hypertension and diabetes coexisted (OR 3.67, 95% CI 1.13 to 14.05).[12] Comorbidities such as cerebrovascular disease, ischaemic heart disease, chronic

DOI: 10.1201/9781003398295-8

kidney disease, as well as hormonal factors and vitamin D deficiency, are more common in older people with diabetes and may increase the risk of frailty.[13–15] Geriatric syndromes such as dementia, depression, falls, and urinary incontinence are common in diabetes and are associated with increased risk of frailty and disability compared with other morbidities.[16] In addition to comorbidities, diabetes may exert a direct negative effect on skeletal muscle mass and muscle quality leading to sarcopenia and eventually frailty. For example, the health and body composition study showed accelerated loss of leg muscle strength and quality in older people with diabetes compared to those without diabetes.[17] In a Korean study of 414 participants ≥65 years old, type 2 diabetes was associated with a twofold to fourfold higher risk of low muscle mass compared to control subjects.[18] Other studies have also shown an accelerated decline in leg lean muscle mass, muscle strength, and functional capacity including gait speed in older people compared to those without diabetes.[19–21] Other factors such as poor nutrition, especially inadequate protein intake, reduced physical exercise, and decline in the neuromuscular junction may contribute to muscular weakness and the development of frailty.[22] When frailty develops, it increases the risk of falls, fractures, and leads to further deterioration in physical functions and the severity of frailty.[23] Due to reduced muscle mass and increased insulin resistance in frail older people, diabetes prevalence is higher in frail compared with non-frail subjects.[24] Muscle fat infiltration alters glucose metabolism setting off insulin resistance and promoting the development of sarcopenia and diabetes.[25] Muscle weakness has been shown to be associated with significantly increased odds of diabetes and metabolic syndrome.[26,27] Therefore, it appears that diabetes and frailty are interrelated and may have a common pathogenic pathway. Chronic low-grade inflammation is a link or a contributing factor to the occurrence of chronic diseases such as diabetes and frailty.[28] Furthermore, diabetes and frailty have been shown to share a common cardiovascular risk factors pathway as reported in the Whitehall II prospective cohort study (Figure 8.1).[29]

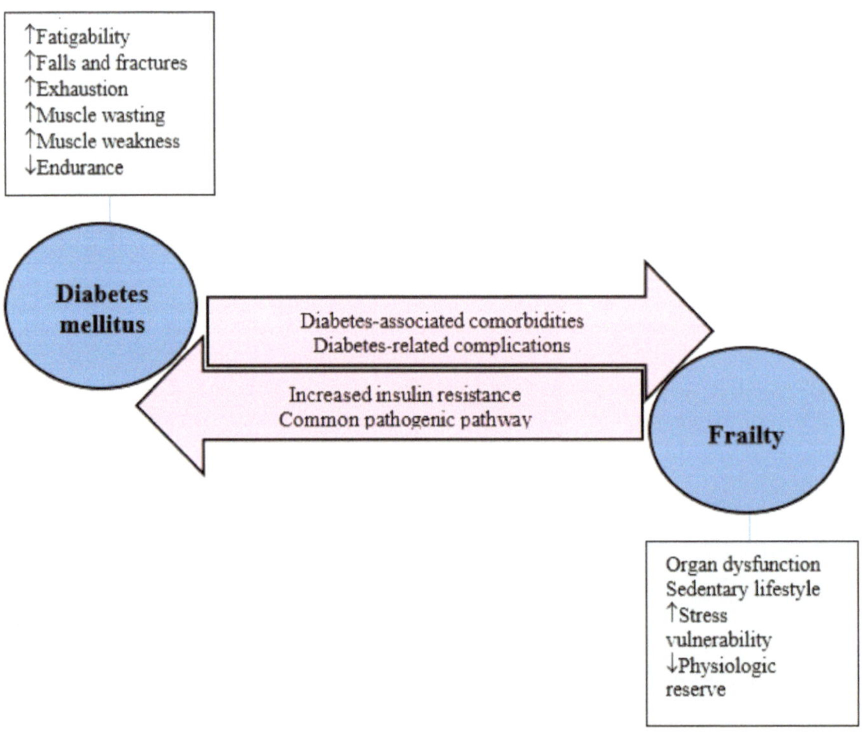

FIGURE 8.1 The reciprocal relationship between diabetes and frailty.

DIABETES AND FRAILTY CLINICAL IMPLICATIONS

The reciprocal relationship between frailty and diabetes will have significant clinical implications. Diabetes increases the risk of frailty and this appears to be related to both high and low blood glucose levels. On the other hand, frailty increases the risk of adverse outcomes and impacts on the glycaemic trajectory of diabetes.

Implications of Diabetes on Frailty

Dysglycaemia (both low and high blood glucose levels) increases the risk of frailty. MacKenzie et al. reported an inverse relationship between blood glucose levels and the severity of frailty as assessed by the clinical frailty scale (CFS).[30] Similarly, Chao et al. found that hypoglycaemia increased the risk of incident frailty by 60% using the FRAIL scale.[31] Yanagita et al. showed lower HbA1c to independently increase the risk of frailty and to have an inverse relationship with CFS.[32] Morita et al. reported that HbA1c <6.0% increased the risk of support or care-need certification compared with HbA1c ≥6.0%.[33] Similarly, McAlister et al. demonstrated that low HbA1c <6.0% was more common in moderately-severely frail (20.0%) and mildly frail (16.6%) compared with fit subjects (15.0%), $p < 0.0001$.[34] On the other hand, high blood glucose levels increased the risk of frailty. Lin et al. showed that HbA1c ≥8.0% is associated with frailty.[35] García-Esquinas et al. explained that poor glycaemic control and high HbA1c increased the risk of frailty in a dose-response relationship.[36] Nguyen et al., in their retrospective analysis of the Action in Diabetes and Vascular Disease: Preterax and Diamicron Modified Release Controlled Evaluation (ADVANCE) trial, demonstrated that frail subjects had a higher mean (SD) HbA1c of 7.85% (1.70) compared with non-frail 7.40% (1.48) subjects.[37] The U-shaped relationship between glycaemia and frailty was demonstrated by Zaslavsky et al. showing an increased risk of frailty with HbA1c above or below the 7.6% threshold.[38]

IMPACT OF FRAILTY ON DIABETES

Frailty impacts on the adverse outcomes and on the glycaemic trajectory of diabetes. For example, the UK Biobank data showed that frailty was associated with major adverse cardiovascular events (MACE) and mortality.[39] It has been shown that the increase in frailty incidence was associated with poor cognitive and physical functions.[40] In patients with diabetes and kidney disease, frailty accelerated the progression to end stage renal disease in a dose-response manner.[41] In addition to frailty, pre-frailty was associated with increased cardiovascular events, health care utilisation, hospitalisation, and mortality (Figure 8.2).[42]

Frailty impacts the glycaemic trajectory of diabetes due to its effects on insulin resistance. Frail older people with diabetes can be seen as a spectrum, which starts from a sarcopenic obese (SO) phenotype at one end, associated with increased insulin resistance and progressive glycaemic trajectory and an anorexic malnourished (AM) phenotype at the other end with reduced insulin resistance and regressive glycaemic trajectory.[43] The SO phenotype is characterised by lower oxidative capacity, increased fat storage, increased insulin resistance, hyperglycaemia, and dyslipidaemia.[44–46] Due to obesity and a sedentary lifestyle, the surplus of energy is stored as ectopic muscular adipose tissue, which further increases insulin resistance. Therefore, in this phenotype glycaemic trajectory is accelerated. The AM phenotype is characterised by anorexia, malnutrition, and significant weight loss. This phenotype is particularly common in institutionalised, compared to, patients living in the community.[47] The significant weight loss in this phenotype leads to reduction of glycaemia and, in some patients, normalisation of HbA1c.[48] Therefore, in this phenotype glycaemic trajectory is decelerated.

Effects of Hypoglycaemic Therapy on Frailty

Metformin is the most studied hypoglycaemic agent. Metformin has positive effects on frailty through the reduction of insulin resistance, hyperglycaemia, inflammation, myostatin levels, and

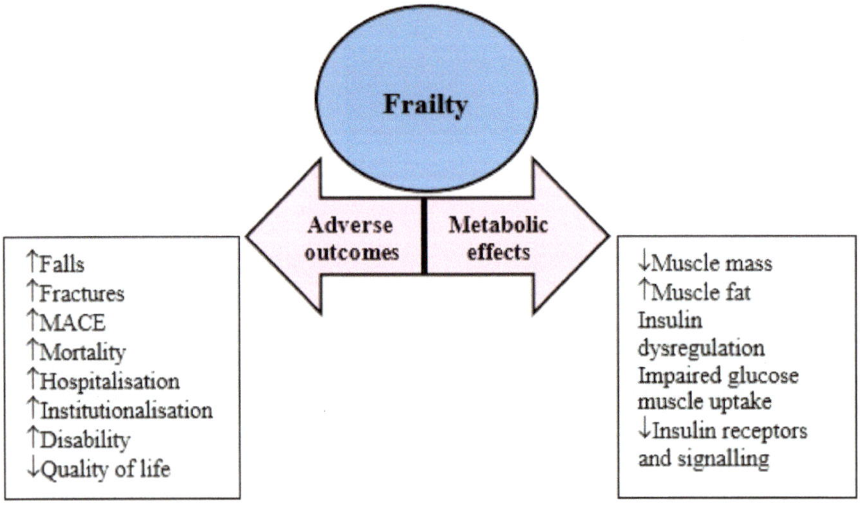

FIGURE 8.2 Impact of frailty on diabetes outcomes and metabolism.

activation of cellular metabolic adenosine monophosphate-activated protein kinase (AMPK).[49,50] The study by Huang et al. showed that metformin had a clinically meaningful improvement in the gait speed of pre-frail individuals.[51] Baskaran et al. showed that exposure to metformin therapy was associated with a lower risk of frailty.[52] The effects of metformin appear to be independent of glycaemic control, suggesting that this is likely to be an extra-glycaemic effect. Metformin has been shown to delay age-related morbidities including frailty in men with type 2 diabetes.[53] Metformin showed a positive effect on depression and cognitive function, which are related to physical frailty.[54] In newly diagnosed people with diabetes, the use of metformin may prevent the accumulation of morbidities and the development of frailty later in life, through its effects on cellular ageing process such as improving muscular protein synthesis, reducing oxidative stress, and the formation of advanced glycosylation end products (AGEs).[55,56] The effects of metformin appear more significant in non-frail compared to frail individuals, suggesting that the early use of metformin is important to achieve this benefit before the development of frailty.[57] This assumption has also been demonstrated by Liu et al. who found metformin to independently protect against frailty in non-frail subjects and this protective effect was attenuated once frailty was developed.[58] Mone et al. demonstrated the sodium glucose cotransporter-2 (SGLT-2) inhibitor empagliflozin to reduce frailty and improve cognitive function in older people with diabetes.[59] The multivariable regression analysis confirmed the marked amelioration of physical impairment, assessed by the 5-m gait speed test, in the empagliflozin and metformin groups but not in the insulin group.[60] This study also showed a significant correlation between the 5-m gait speed test and the Montreal Cognitive Assessment (MoCA) score in frail older people with diabetes and hypertension.[61] Authors concluded that empagliflozin effects are most likely due to decrease in the mitochondrial generation of reactive oxygen species in endothelial cells, thus reducing oxidative stress, which is involved in the pathogenesis of diabetes and frailty.[62–64] It has also been shown that microRNA may be involved in the regulation of endothelial function in frail subjects with diabetes and heart failure with preserved ejection fraction, which is improved by empagliflozin treatment.[65]

Choice of Hypoglycaemic Agents in Frailty

There is little data on the choice of hypoglycaemic agents in people with comorbid diabetes and frailty. The general recommendation is to avoid the risk of hypoglycaemia and the cautious use of the new

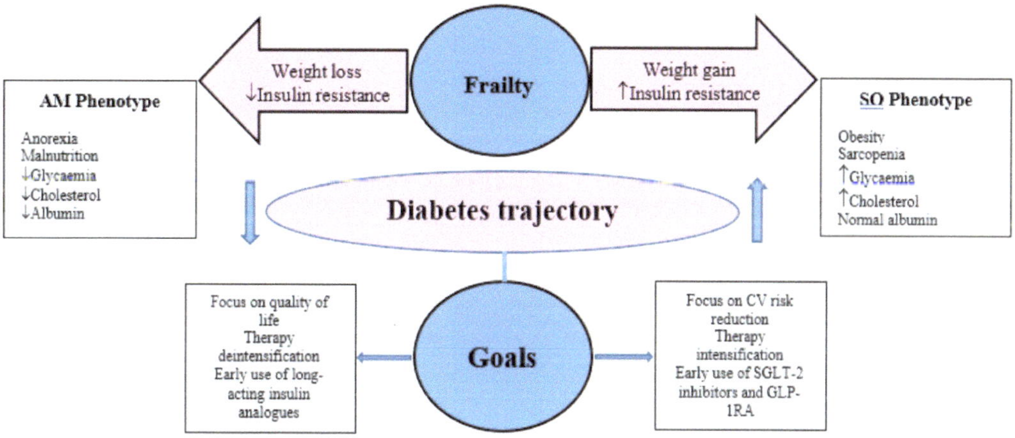

FIGURE 8.3 Impact of frailty on the diabetes trajectory and choice of hypoglycaemic therapy.

therapies of SGLT-2 inhibitors and glucagon-like peptide-1 receptor agonists (GLP-1RA) due to the high risk of dehydration, hypotension, falls, and weight loss in this vulnerable population.[66,67] However, due to the wide spectrum of frailty, these new therapies should be considered early in the SO frailty phenotype as they are likely to gain most from their positive effect on the metabolism and their weight-losing properties. Compared with obesity or sarcopenia alone, sarcopenic obesity is associated with an unfavourable metabolism such as increased insulin resistance, persistent hyperglycaemia, dyslipidaemia, and hypertension, which are associated with an increased risk of MACE including mortality.[68–71] Therefore, the absolute benefits in the SO phenotype are the highest due to their higher baseline risk.[72] The new therapies are associated with weight loss and osmotic diuresis reducing the blood pressure, which is required in this frailty phenotype.[73] They also improve liver pathology through reducing total body fat, lipogenesis, inflammatory markers, and oxidative stress and stimulating the oxidation of free fatty acids.[74,75] Other advantages of the new therapies include low risk of hypoglycaemia, early benefits after initiation of therapy, and the reduction of polypharmacy due to the multifaceted effects of such therapy. The new therapies, however, are not suitable in the AM frailty phenotype due to the undesirable weight loss induced by these agents. Insulin should be considered early in this phenotype due to its anabolic and weight gain properties.[76] Weight gain associated with insulin will be an advantage in this phenotype, but other insulin-associated side effects such as the inconvenience of frequent injections, the need for blood glucose monitoring, and the increased risk of hypoglycaemia should be considered. Therefore, a most convenient and simple regimen using long-acting basal insulin analogue at bedtime is suitable. Long-acting insulins (glargine or detemir) have been shown to reduce emergency department visits or hospitalisation due to hypoglycaemia compared with NPH insulin in older people with type 2 diabetes.[77] The ultra-long-acting basal insulin degludec demonstrates a flat and stable glucose lowering effect with once-daily administration.[78] Although many frail people will do well with this regimen, there will be a need to consider adding prandial or mealtime short-acting insulin to cover postprandial glucose excursions if they repeatedly occur. To reduce the risk of hypoglycaemia in patients with erratic eating patterns, short-acting insulin analogues should be administered after a meal is consumed (Figure 8.3).

Glycaemic Control

Uncontrolled diabetes with HbA1c ≥8.0% may increase the risk of frailty.[35] Uncontrolled diabetes, measured as less time blood glucose in range and more time above range, was associated with frailty while time spent below range was not related to frailty levels as reported by Fung et al.[79] In addition, HbA1c was higher in subjects with combined uncontrolled diabetes and frailty compared to those with controlled diabetes and less frailty (8.9% v 7.8% respectively, $p < 0.0001$).[79] Bilgin et al. reported that poor glycaemic control is associated with frailty and better glycaemic control may slow down the development of frailty.[80] Hyde et al. found similar findings with subjects having a higher HbA1c (≥6.5%) to have a higher prevalence of frailty (70.3%) compared to those with HbA1c <5.7%, and those with levels between 5.7% and 6.4% (frailty prevalence 47.6% and 51.8%, respectively).[81] There was a statistically significant trend for greater frailty by higher category of HbA1c ($p = 0.025$).[81] Aguayo et al. also demonstrated that higher HbA1c levels are significantly associated with higher frailty trajectories measured from the age ≥60 years.[82] Non-frail older people with diabetes and high HbA1c, at baseline, could become frail a decade earlier than their peers without diabetes.[82] Persistent hyperglycaemia causes microvascular disease, which may induce muscular mitochondrial dysfunction leading to poor muscular function and frailty.[83] Although low HbA1c was shown to be associated with frailty in other studies, low HbA1c was also associated with low body weight, low cholesterol, low serum albumin, and low Hb and suggesting that malnutrition could be an explanation of this relationship.[32] The study by Zaslavsky et al. suggested that HbA1c of 7.6% has the lowest risk of frailty, while levels around this cut-off point increase the risk in a U-shaped relationship.[38] This may suggest that a HbA1c target around 7.5% is reasonable in older people with diabetes and frailty.

Goals of Therapy

In the SO frailty phenotype, the goals should be intensification of therapy with the main aim of reducing cardiovascular risk. This phenotype has unfavourable metabolism and stands to benefit most from the new therapy due to the high baseline risk. In addition to pharmacologic therapy, improving muscle power may further reduce cardiovascular risk.[84] Exercise training combined with adequate nutrition may help restore muscle mass and function.[85] Muscle contraction during exercise releases protective myokines, which reduce the harmful effects of the pro-inflammatory adipokines associated with sarcopenic obesity.[86] The AM frailty phenotype is likely to be prevalent in the very frail group of patients, such as those living in care homes with significant anorexia and weight loss. Therefore, this group of patients tend to have lower HbA1c and are at high risk of hypoglycaemia. Relaxed targets are appropriate as aggressive lowering of HbA1c with agents that have a high risk of hypoglycaemia may be associated with increased mortality.[87] Therefore, the goals are deintensification of therapy and the focus is to maintain quality of life. Due to limited life expectancy in this phenotype, short-term glycaemic targets that avoid extreme blood glucose levels will maintain well-being and avoid hospitalisation rather than long-term targets of HbA1c.

Prevention

Screening for frailty should be part of the integrated patient care plan and annual review. Because no pharmacologic treatment to reverse frailty is available, the focus will be on prevention. Therefore, lifestyle interventions to maintain muscle mass and muscle power are the key to delay and prevent the development of frailty. In addition, early detection of pre-frailty and frailty will help to tailor interventions that may reduce the risk of progression of frailty to disability.[67] Frailty can be assessed using any of the available tools, as there is no specific tool to be used in older people with diabetes.[39] Comprehensive Geriatric Assessment (CGA) is a multidimensional assessment of interrelated domains such as physical, cognitive, psychological, functional, and social aspects and is central in any tool to assess frailty and can be used to tailor the care-needs of older people with diabetes. To maintain the muscle mass and delay or prevent frailty, adequate nutrition and exercise training are the main key interventions.[88] Nutritional management is mainly based on achieving ideal body weight and, at the same time, maintaining muscle mass. Insufficient dietary protein intake is associated with reduced lean body mass, which may lead to the development of frailty.[89] A higher daily protein intake is required in older people due to reduced protein absorption or anabolic resistance common in old age.[90] A daily protein intake of 1–1.2 g/kg may enhance muscle mass and function.[91] The inclusion of the essential amino acids, especially leucine, is important.[92] Vitamin D supplementation may increase muscle strength and reduce the frequency of falls, especially in older people with low vitamin D levels.[93] In addition to nutrition, resistance and aerobic exercise are useful, although a combination of both has been shown to be superior to either of them alone.[85] Resistance exercise improves muscle mass, while aerobic exercise reduces obesity. The combination of resistance training and protein supplementation has been shown to reduce fat mass and to improve lean muscle mass and strength in frail older people than exercise alone.[94] A structured multimodal intervention, which included nutrition, resistance exercise training, and optimal diabetes care, led to a clinically relevant and cost-effective improvement in the functional status of older frail and pre-frail participants with type 2 diabetes mellitus.[95]

Role of Technology

There is growing evidence of the use of technology in the management of diabetes, such as continuous glucose monitoring (CGM), insulin pumps, and closed loop systems, to reduce the incidence of hypoglycaemia, control hyperglycaemia, and reduce the variability of blood glucose levels in older people with diabetes in their sixties. Although there is no evidence yet in older age, but the reduction of diabetes-related complications by the use of technology may translate into a lower incidence of frailty at a later age.[96] A geriatric multimodal intervention including nutrition and exercise training improves functional decline in pre-frail and frail older people with diabetes, and this improvement is even better when supported by the CAPACITY technological ecosystem. The CAPACITY technological ecosystem is based on a low-cost home monitoring and remote intervention system aimed to prevent disability among the frail older population.[97] The FACET technological ecosystem was developed to gather data from patient's home to a medical team for early intervention and has been shown to help in the early identification of changes in functional status of pre-frail and frail older people and facilitating prompt clinical interventions and thereby to improve health outcomes in terms of frailty and functional status to potentially prevent disability and dependency.[98]

CONCLUSION

Frailty is a new and important complication of diabetes in older people, which significantly impacts diabetes-related outcomes. In addition, frailty influences the diabetes metabolic profile and changes its trajectory. In the SO phenotype, the diabetes trajectory is accelerated due to increased insulin resistance and intensification of therapy is required. In the AM phenotype, the diabetes trajectory is decelerated due to decreased insulin resistance and deintensification of therapy is required. The focus of therapy is cardiovascular risk reduction in the SO phenotype, while in the AM phenotype the focus is maintenance of quality of life. Adequate nutrition and exercise training are the key interventions to prevent the development of frailty and delay the progression of frailty to disability.

FUTURE PERSPECTIVES

Although frailty is increasingly recognised as an important complication of diabetes in older people, the metabolic impact of frailty on diabetes is not yet fully explored. Frailty is likely to span across a metabolic spectrum with an anorexic malnourished phenotype at one end and a sarcopenic obese phenotype at the other end. This metabolic variation will affect diabetes trajectory, choice of hypoglycaemic agents, and goals of therapy. Therefore, future studies on older people with diabetes should characterise frailty phenotypes rather than grouping them as one category. The cause of frailty is partly related to diabetes associated complications, especially renal and cardiovascular disease. The protective effects of the new therapies on renal and cardiovascular outcomes may lead to a reduction in the incidence of frailty. Therefore, frailty should be an important outcome in future studies in older people with diabetes. The new therapies have several extra glycaemic effects; therefore, novel therapies with extra glycaemic effects on frailty such as improving muscle function and muscle strength are required. The current evidence of the effect of glycaemic control on the incidence of frailty is largely indirect with no specific clinical trials directly investigating this as a cause and effect relationship. The anabolic properties of insulin and its effects on muscle function and its benefits in the AM frailty phenotype need future exploration.

KEY POINTS

- We have already recognised Frailty as an important complication of diabetes in older people.
- The relationship between frailty and diabetes appears to be bidirectional.
- Frailty is associated with increased risk of adverse events in older people with diabetes and affects the diabetes trajectory.
- In the sarcopenic obese frailty phenotype, the diabetes trajectory is accelerated, while in the anorexic malnourished, the trajectory is decelerated.
- Future studies are still required to precisely characterise frailty phenotypes, investigate the impact of glycaemic control on frailty, and explore the best hypoglycaemic therapy, which suits each phenotype.

REFERENCES

1. Sun H, Saeedi P, Karuranga S, Pinkepank M, Ogurtsova K, Duncan BB, Stein C, Basit A, Chan JCN, et al. IDF Diabetes Atlas: Global, regional and country-level diabetes prevalence estimates for 2021 and projections for 2045. Diabetes Res Clin Pract 2022; 183: 109119. DOI: 10.1016/j.diabres.2021.109119.
2. Sinclair AJ, Abdelhafiz AH, Rodríguez-Mañas L. Frailty and sarcopenia – Newly emerging and high impact complications of diabetes. J Diabetes Complications 2017; 31: 1465–73.
3. Collard RM, Boter H, Schoevers RA, Oude Voshaar RC. Prevalence of frailty in community-dwelling older persons: A systematic review. J Am Geriatr Soc 2012; 60: 1487–92.
4. Clegg A, Young J, Iliffe S, et al. Frailty in elderly people. Lancet 2013; 381: 752–62.
5. Kojima G. Frailty as a predictor of fractures among community-dwelling older people: A systematic review and meta-analysis. Bone 2016; 90: 116–22.
6. Kojima G. Frailty as a predictor of disabilities among community-dwelling older people: A systematic review and meta-analysis. Disabil Rehabil 2017; 39: 1897–908.
7. Kojima G, Iliffe S, Walters K. Frailty index as a predictor of mortality: A systematic review and meta-analysis. Age Ageing 2018; 47: 193–200.
8. Roe L, Normand C, Wren MA, et al. The impact of frailty on healthcare utilisation in Ireland: Evidence from the Irish longitudinal study on ageing. BMC Geriatr 2017; 17: 203. DOI: 10.1186/s12877-017-0579-0.
9. Turner G, Clegg A, British Geriatrics Society, Age UK, Royal College of General Practioners. Best practice guidelines for the management of frailty: A British Geriatrics Society, Age UK and Royal College of General Practitioners report. Age Ageing 2014; 43: 7447.
10. Nagi SZ. Disability concepts revisited: Implications for prevention In: Pope AM, Tarlov AR, editors. *Disability in America: Toward a National Agenda for Prevention*. Washington, DC: National Academy Press; 1991. p. 309–27.
11. Castrejón-Pérez RC, Gutiérrez-Robledo LM, Cesari M, Pérez-Zepeda MU. Diabetes mellitus, hypertension and frailty: A population-based, cross-sectional study of Mexican older adults. Geriatr Gerontol Int 2016. DOI: 10.1111/ggi.12805.
12. Lee S, Lee S, Harada K, et al. Relationship between chronic kidney disease with diabetes or hypertension and frailty in community-dwelling Japanese older adults. Geriatr Gerontol Int. 2016. DOI: 10.1111/ggi.12910.

13. Leod M, Tooke KM. Direct and indirect costs of cardiovascular and cerebrovascular complications of type II diabetes. Pharmacoeconomics 1995; 8: 46–51.
14. Kalyani RR, Corriere M, Ferrucci L. Age-related and disease-related muscle loss: The effect of diabetes, obesity, and other diseases. Lancet Diabetes Endocrinol 2014; 2: 819–29.
15. Ozfirat Z, Chowdhury TA. Vitamin D deficiency and type 2 diabetes. Postgrad Med J 2010; 86: 18–25.
16. Lu FP, Chang WC, Wu SC. Geriatric conditions, rather than multimorbidity, as predictors of disability and mortality among octogenarians: A population-based cohort study. Geriatr Gerontol Int 2016; 16: 345–51.
17. Park SW, Goodpaster BH, Strotmeyer ES, et al. Health, aging, and body composition study. Accelerated loss of skeletal muscle strength in older adults with type 2 diabetes: The health, aging, and body composition study. Diabetes Care 2007; 30: 1507–12.
18. Kim KS, Park KS, Kim MJ, et al. Type 2 diabetes is associated with low muscle mass in older adults. Geriatr Gerontol Int 2014; 14 (Suppl. 1): 115–21.
19. Leenders M, Verdijk LB, van der Hoeven L, et al. Patients with type 2 diabetes show a greater decline in muscle mass, muscle strength, and functional capacity with aging. J Am Med Dir Assoc 2013; 14: 585–92.
20. Volpato S, Bianchi L, Lauretani F, et al. Role of muscle mass and muscle quality in the association between diabetes and gait speed. Diabetes Care 2012; 35: 1672–9.
21. Kalyani RR, Tra Y, Yeh HC, et al. Quadriceps strength, quadriceps power, and gait speed in older U.S. adults with diabetes mellitus: Results from the National Health and Nutrition Examination Survey, 1999-2002. J Am Geriatr Soc 2013; 61: 769–75.
22. Gonzalez-Freire M, de Cabo R, Studenski SA, Ferrucci L. The neuromuscular junction: Aging at the crossroad between nerves and muscle. Front Aging Neurosci 2014; 6: 208. DOI: 10.3389/fnagi.2014.00208.
23. Xie WQ, He M, Yu DJ, et al. Mouse models of sarcopenia: Classification and evaluation. J Cachexia Sarcopenia Muscle 2021; 12: 538–54.
24. Aguilar-Navarro SG, Amieva H, Gutierrez-Robledo LM, et al. Frailty among Mexican community-dwelling elderly: A story told 11 years later. The Mexican Health and Aging Study. Salud Pública Mex 2015; 57: S62–9.
25. Sriram S, Subramanian S, Juvvuna PK, et al. Myostatin induces DNA damage in skeletal muscle of streptozotocininduced type 1 diabetic mice. J Biol Chem 2014; 289: 5784–98.
26. Peterson MD, Zhang P, Choksi P, et al. Muscle weakness thresholds for prediction of diabetes in adults. Sports Med 2016; 46: 619–28.
27. Peterson MD, McGrath R, Zhang P, et al. Muscle weakness is associated with diabetes in older Mexicans: The Mexican Health and Aging Study. J Am Med Dir Assoc 2016; 17: 933–8.
28. Franceschi C, Campisi J. Chronic inflammation (inflammaging) and its potential contribution to age-associated diseases. J Gerontol A Biol Sci Med Sci 2014; 69 (Suppl 1): S4–9.
29. Bouillon K, Kivimäki M, Hamer M, et al. Diabetes risk factors, diabetes risk algorithms, and the prediction of future frailty: The Whitehall II prospective cohort study. J Am Med Dir Assoc 2013; 851.e1-6. DOI: 10.1016/j.jamda.2013.08.016.
30. MacKenzie HT, Tugwell B, Rockwood K, et al. Frailty and diabetes in older hospitalized adults: The case for routine frailty assessment. Can J Diabetes 2020; 44: 241–45.
31. Chao CT, Wang J, Huang JW, Chan DC, Chien KL, COhort of GEriatric Nephrology in NTUH (COGENT) study group. Hypoglycemic episodes are associated with an increased risk of incident frailty among new onset diabetic patients. J Diabetes Complications 2020; 34: 107492. https://doi.org/10.1016/j.jdiacomp.2019.107492.
32. Yanagita I, Fujihara Y, Eda T, et al. Low glycated hemoglobin level is associated with severity of frailty in Japanese elderly diabetes patients. J Diabetes Investig 2018; 9: 419–25.
33. Morita T, Okuno T, Himeno T, et al. Glycemic control and disability-free survival in hypoglycemic agent-treated community-dwelling older patients with type 2 diabetes mellitus. Geriatr Gerontol Int 2017; 17: 1858–65.
34. McAlister FA, Lethebe BC, Lambe C, Williamson T, Lowerison M. Control of glycemia and blood pressure in British adults with diabetes mellitus and subsequent therapy choices: A comparison across health States. Cardiovasc Diabetol 2018; 17: 27. https://doi.org/10.1186/s12933-018-0673-4.
35. Lin CL, Yu NC, Wu HC, et al. Risk factors associated with frailty in older adults with type 2 diabetes: A cross-sectional study. J Clin Nurs 2022; 31: 967–74.
36. García-Esquinas E, Graciani A, Guallar-Castillón P, et al. Diabetes and risk of frailty and its potential mechanisms: A prospective cohort study of older adults. J Am Med Dir Assoc 2015; 16: 748–54.

37. Nguyen TN, Harris K, Woodward M, Chalmers J, Cooper M, Hamet P, Harrap S, Heller S, MacMahon S, Mancia G, et al. The impact of frailty on the effectiveness and safety of intensive glucose control and blood pressure-lowering therapy for people with type 2 diabetes: Results from the ADVANCE trial. Diabetes Care 2021; 44: 1622–9.
38. Zaslavsky O, Walker RL, Crane PK, et al. Glucose levels and risk of frailty. J Gerontol A Biol Sci Med Sci 2016; 71: 1223–9.
39. Hanlon P, Jani BD, Butterly E, et al. An analysis of frailty and multimorbidity in 20,566 UK Biobank participants with type 2 diabetes. Commun Med 2021; 28. https://doi.org/10.1038/s43856-021-00029-9.
40. Espeland MA, Justice JN, Bahnson J, et al. Eight year changes in multimorbidity and frailty in adults with type 2 diabetes mellitus: Associations with cognitive and physical function and mortality. J Gerontol A Biol Sci Med Sci 2022; 77: 1691–98. DOI: 10.1093/gerona/glab342.
41. Chao CT, Wang J, Huang JW, et al.·Frailty predicts an increased risk of end-stage renal disease with risk competition by mortality among 165,461 diabetic kidney disease patients. Aging Dis 2019; 10: 1270–81.
42. Chao CT, Wang J, Chien KL, COhort of GEriatric Nephrology in NTUH (COGENT) study group. Both pre-frailty and frailty increase healthcare utilization and adverse health outcomes in patients with type 2 diabetes mellitus. Cardiovasc Diabetol 2018; 17: 130. DOI: 10.1186/s12933-018-0772-2.
43. Sinclair AJ, Abdelhafiz AH. Metabolic impact of frailty changes diabetes trajectory. Metabolites 2023; 13: 295. DOI: 10.3390/metabo13020295.
44. Chitwood LF, Brown SP, Lundy MJ, et al. Metabolic propensity toward obesity in black vs white females: Responses during rest, exercise, and recovery. Int J Obes 1996; 20: 455–62.
45. Hickner RC, Privette J, McIver K, et al. Fatty acid oxidation in African-American and Caucasian women during physical activity. J Appl Physiol 2001; 90: 2319–24.
46. Brancati FL, Kao WH, Folson AR, et al. Incident type 2 diabetes mellitus in African American and white adults: The Atherosclerosis Risk in Communities Study. J Am Med Assoc 2000; 283: 2253–9.
47. Donini LM, Poggiogalle E, Piredda M, et al. Anorexia and eating patterns in the elderly. PLoS One 2013; 8: e63539. DOI: 10.137/journal.pone.00653539.
48. Abdelhafiz AH, Chakravorty P, Gupta S, Haque A, Sinclair AJ. Can hypoglycaemic medications be withdrawn in older people with type 2 diabetes? Int J Clin Pract 2014; 68: 790–2.
49. Graham GG, Punt J, Arora M, et al. Clinical pharmacokinetics of metformin. Clin Pharmacokinet 2011; 50: 81–98.
50. Salminen A, Hyttinen JM, Kaarniranta K. AMP-activated protein kinase inhibits NF-κB signaling and inflammation: Impact on healthspan and lifespan. J Mol Med (Berl) 2011; 89: 667–76.
51. Huang C-F, Shiao M-S, Mao T-Y. Effects of low-dose metformin on pre-frailty among middle-aged and elderly pre-diabetic people. JCSM Rapid Commun 2022; 5: 33–39.
52. Baskaran D, Aparicio-Ugarriza R, Ferri-Guerra J, et al. Is there an association between metformin exposure and frailty? Gerontol Geriatr Med 2020; 6: 2333721420924956. DOI: 10.1177/2333721420924956.
53. Wang CP, Lorenzo C, Habib SL, et al. Differential effects of metformin on age related comorbidities in older men with type 2 diabetes. J Diabetes Complications 2017; 31: 679–86.
54. Abdelhafiz AH, Davies PC, Sinclair AJ. Triad of impairment in older people with diabetes-reciprocal relations and clinical implications. Diabetes Res Clin Pract 2020; 161: 108065. DOI: 10.1016/j.diabres.2020.108065.
55. Gore DC, Wolf SE, Sanford A, et al. Influence of metformin on glucose intolerance and muscle catabolism following severe burn injury. Ann Surg 2005; 241: 334–42.
56. Esteghamati A, Eskandari D, Mirmiranpour H, et al. Effects of metformin on markers of oxidative stress and antioxidant reserve in patients with newly diagnosed type 2 diabetes: A randomized clinical trial. Clin Nutr 2013; 32: 179–85.
57. Wang CP, Lorenzo C, Espinoza SE. Frailty attenuates the impact of metformin on reducing mortality in older adults with type 2 diabetes. J Endocrinol Diabetes Obes 2014; 2: 1031.
58. Liu P, Pan Y, Song Y, et al. Association of metformin exposure with low risks of frailty and adverse outcomes in patients with diabetes. Eur J Med Res 2023; 28: 65. DOI: 10.1186/s40001-023-01017-6.
59. Mone P, Varzideh F, Jankauskas SS, et al. SGLT2 inhibition via empagliflozin improves endothelial function and reduces mitochondrial oxidative stress: Insights from frail hypertensive and diabetic patients. Hypertension 2022; 79: 1633–43.
60. Mone P, Lombardi A, Gambardella J, et al. Empagliflozin improves cognitive impairment in frail older adults with type 2 diabetes and heart failure with preserved ejection fraction. Diabetes Care 2022; 45: 1247–51.

61. Mone P, Gambardella J, Lombardi A, et al. Correlation of physical and cognitive impairment in diabetic and hypertensive frail older adults. Cardiovasc Diabetol. 2022; 21: 10. DOI: 10.1186/s12933-021-01442-z.
62. Vaidya AR, Wolska N, Vara D, et al. Diabetes and thrombosis: A central role for vascular oxidative stress. Antioxidants (Basel) 2021; 10: 706. DOI: 10.3390/antiox10050706.
63. Alabadi B, Civera M, De la Rosa A, et al. Frailty is associated with oxidative stress in older patients with type 2 diabetes. Nutrients 2021; 13: 3983. DOI: 10.3390/nu13113983.
64. Bernabeu-Wittel M, Gómez-Díaz R, González-Molina Á, et al. On behalf of the proteo researchers. Oxidative stress, telomere shortening, and apoptosis associated to sarcopenia and frailty in patients with multimorbidity. J Clin Med 2020; 9: E2669. DOI: 10.3390/jcm9082669.
65. Mone P, Lombardi A, Kansakar U, et al. Empagliflozin improves the microRNA signature of endothelial dysfunction in patients with heart failure with preserved ejection fraction and diabetes. J Pharmacol Exp Ther 2023; 384: 116–22.
66. Sinclair AJ, Abdelhafiz A, Dunning T, et al. An international position statement on the management of frailty in diabetes mellitus: Summary of recommendations 2017. J Frailty Aging 2018; 7: 10–20.
67. LeRoith D, Biessels GJ, Braithwaite SS, et al. Treatment of diabetes in older adults: An Endocrine Society Clinical Practice Guideline. J Clin Endocrinol Metab 2019; 104: 1520–74.
68. Xia M-F, Chen L-Y, Wu L, et al. Sarcopenia, sarcopenic overweight/obesity and risk of cardiovascular disease and cardiac arrhythmia: A cross-sectional study. Clin Nutr 2020; 40: 571–80.
69. Cho H-W, Chung W, Moon S, et al. Effect of sarcopenia and body shape on cardiovascular disease according to obesity phenotypes. Diabetes Metab J 2021; 45: 209–18.
70. Uchida S, Kamiya K, Hamazaki N, et al. Association between sarcopenia and atherosclerosis in elderly patients with ischemic heart disease. Heart Vessel 2020; 35: 769–75.
71. Yoo JH, Park SW, Jun JE, et al. Relationship between low skeletal muscle mass, sarcopenic obesity and left ventricular diastolic dysfunction in Korean adults. Diabetes Metabolism Res Rev 2020; 37: e3363. DOI: 10.1002/dmrr.3363.
72. Kutz A, Kim DH, Wexler DJ, et al. Comparative cardiovascular effectiveness and safety of SGLT-2 inhibitors, GLP-1 receptor agonists, and DPP-4 inhibitors according to frailty in type 2 diabetes. Diabetes Care 2023; 46: 2004–14.
73. Vasilakou D, Karagiannis T, Athanasiadou E, et al. Sodium–glucose cotransporter 2 inhibitors for type 2 diabetes: A systematic review and meta-analysis. Ann Intern Med 2013; 159: 262–74.
74. Sofogianni A, Filippidis A, Chrysavgis L, et al. Glucagon-like peptide-1 receptor agonists in non-alcoholic fatty liver disease: An update. World J Hepatol 2020; 12: 493–505.
75. Xing B, Zhao Y, Dong B, et al. Effects of sodium–glucose cotransporter 2 inhibitors on non-alcoholic fatty liver disease in patients with type 2 diabetes: A meta-analysis of randomized controlled trials. J Diabetes Investig 2020; 11: 1238–47.
76. Abdelhafiz A, Bisht S, Kovacevic I, et al. Insulin in frail, older people with type 2 diabetes—Low threshold for therapy. Diabetology 2022; 3: 369–83.
77. Bradley MC, Chillarige Y, Lee H, et al. Severe hypoglycemia risk with long-acting insulin analogs vs neutral protamine Hagedorn insulin. JAMA Intern Med 2021; 181: 598–607.
78. Heise T, Nosek L, Bottcher SG, et al. Ultra-long acting insulin degludec has a flat and stable glucose-lowering effect in type 2 diabetes. Diabetes Obes Metab 2012; 14: 944–50.
79. Fung E, Lui LT, Huang L, et al. Characterising frailty, metrics of continuous glucose monitoring, and mortality hazards in older adults with type 2 diabetes on insulin therapy (HARE): A prospective, observational cohort study. Lancet Healthy Longev. 2021; 2: e724–e35.
80. Bilgin S, Aktas G, Kurtkulagi O, et al. Edmonton frail score is associated with diabetic control in elderly type 2 diabetic subjects. J Diabetes Metab Disord 2020; 19: 511–4.
81. Hyde Z, Smith K, Flicker L, et al. HbA1c is associated with frailty in a group of aboriginal Australians. J Frailty Aging 2019; 8: 17–20.
82. Aguayo GA, Hulman A, Vaillant MT, et al. Prospective association among diabetes diagnosis, HbA1c, glycemia, and frailty trajectories in an elderly population. Diabetes Care 2019; 42: 1903–11.
83. Phielix E, Schrauwen-Hinderling VB, Mensink M, et al. Lower intrinsic ADP-stimulated mitochondrial respiration underlies in vivo mitochondrial dysfunction in muscle of male type 2 diabetic patients. Diabetes 2008; 57: 2943–9.
84. Liu HM, Zhang Q, Shen WD, et al. Sarcopenia-related traits and coronary artery disease: A bi-directional Mendelian randomization study. Aging (Albany NY) 2020; 12: 3340–53.
85. Villareal DT, Aguirre L, Gurney AB, et al. Aerobic or resistance exercise, or both, in dieting obese older adults. N Engl J Med 2017; 376: 1943–55.

86. Asano RY, Sales MM, Coelho JM, et al. Exercise, nitric oxide, and endothelial dysfunction: A brief review. J Exerc Physiol Online 2012; 15: 76–86.
87. Currie CJ, Holden SE, Jenkins-Jones S, et al. Impact of differing glucose-lowering regimens on the pattern of association between glucose control and survival. Diabetes Obes Metab 2018; 20: 821–30.
88. Rolland Y, Onder G, Morley JE, et al. Current and future pharmacologic treatment of sarcopenia. Clin Geriatr Med 2011; 27: 423–47.
89. Bartali B, Frongillo EA, Bandinelli S, et al. Low nutrient intake is an essential component of frailty in older persons. J Gerontol A Biol Sci Med Sci 2006; 61: 589–93.
90. Campbell WW, Trappe TA, Wolfe RR, et al. The recommended dietary allowance for protein may not be adequate for older people to maintain skeletal muscle. J Gerontol A Biol Sci Med Sci 2001; 56: M373–80.
91. Malafarina V, Uriz-Otano F, Iniesta R, et al. Effectiveness of nutritional supplementation on muscle mass intreatment of sarcopenia in old age: A systematic review. J Am Med Dir Assoc 2013; 14: 10–7.
92. Gryson C, Ratel S, Rance M, et al. Four-month course of soluble mil proteins interacts with exercise to improve muscle strength and delay fatigue in elderly participants. J Am Med Dir Assoc 2014; 15: 958.e1–9.
93. Beaudart C, Buckinx F, Rabenda V, et al. The effects of vitamin D on skeletal muscle strength, muscle mass, and muscle power: A systematic review and meta-analysis of randomized controlled trials. J Clin Endocrinol Metab 2014; 99: 4336–45.
94. Liao CD, Tsauo JY, Wu YT, et al. Effects of protein supplementation combined with resistance exercise on body composition and physical function in older adults: A systematic review and metaanalysis. Am J Clin Nutr 2017; 106: 1078–91.
95. Rodriguez-Mañas L, Laosa O, Vellas B, et al. European MID-Frail Consortium. Effectiveness of a multimodal intervention in functionally impaired older people with type 2 diabetes mellitus. J Cachexia Sarcopenia Muscle 2019; 10: 721–33.
96. Huang ES, Sinclair A, Conlin PR, et al. The growing role of technology in the care of older adults with diabetes. Diabetes Care 2023; 46: 1455–63.
97. Pérez-Rodríguez R, Guevara-Guevara T, Moreno-Sánchez PA, et al. Monitoring and intervention technologies to manage diabetic older persons: The CAPACITY case – A pilot study. Front Endocrinol (Lausanne) 2020 May; 11: 300. DOI: 10.3389/fendo.2020.00300.
98. Valdés-Aragonés M, Pérez-Rodríguez R, Carnicero JA, et al. Effect of monitoring frailty through a mobile application and a sensor kit to prevent functional decline in frail and prefrail older people: The FACET pilot randomized control trial. Journal of Medical Internet Research 2024. DOI: 10.2196/preprints.58312.

9 Complexity and Disentanglement of Frailty

The Biopsychosocial Model of Frailty and Resilience

M. Cristina Polidori

DETERMINANTS OF COMPLEXITY OF AGING AND FRAILTY

In the coming years, the number of patients suffering from age-related chronic non-communicable diseases (NCDs) and frailty, the multifactorial overarching syndrome of accelerated aging, is expected to increase significantly worldwide as a consequence of the aging population (Liotta et al., 2018; Kim and Rockwood, 2024). The impressive increase in *life expectancy* over the last 150 years – per se a triumph of development – has not been accompanied by a similar increase in *healthspan*. On the contrary, the demographic transition is marked by high rates of frailty, multimorbidity (two or more chronic conditions at the same time) (Skou et al., 2022), and disability (Liotta et al., 2018; Dent et al., 2019; Hoogendijk et al., 2019; Kim and Rockwood, 2024). In Europe, 80% of people over 65 years – one third of the population by 2050 – suffer from at least one NCD, with over one third of them suffering from multimorbidity and up to one fourth from frailty. The coexistence of two chronic conditions in older adults is being rapidly surpassed by the simultaneous presence of five to nine NCDs with increasing age, and evidence of cross-sectional and longitudinal studies showed that two-thirds of frail older adults had multimorbidity and a fifth of multimorbid adults presented with frailty, suggesting a plausible bidirectional relationship between multimorbidity and frailty (Vetrano et al., 2019). More recently, a nationwide population-based longitudinal study included 16,563 participants aged 65 years and older in the Chinese Longitudinal Healthy Longevity Survey who underwent the baseline visit in 2008 and were followed up until 2018 (She et al., 2024). The direct and indirect effects of multimorbidity and frailty as physical phenotypes on mortality were evaluated by means of a cross-lagged panel model and survival analysis. Of three multimorbidity patterns identified, the cognitive-sensory disease pattern showed a bidirectional relationship with frailty across waves, while cardiometabolic and arthritis-digestive-respiratory patterns predicted frailty change, not vice versa (She et al., 2024). In this investigation, frailty mediated 23–27% of the association between multimorbidity and mortality, suggesting it as a prominent pathway. Indeed, frailty, characterized by "invisible" accelerated aging of systems and organs, remains usually undiagnosed until it manifests with functional loss, poor quality of life, chronic disease including cognitive impairment, isolation, falls, depression, and caregiver stress, causing enormous cost to society. Once transitioned in chronic disability, frailty is not reversible anymore, and there is a wealth of literature showing that functional loss is a crucial factor for mortality risk over multimorbidity (Landi et al., 2010; Landi et al., 2016; Chen et al., 2023). However, functional loss with its main causes and clinical counterparts of geriatric syndromes largely escape traditional health care coding systems currently, with research, prevention, and treatment focusing on NCDs following a "one disease, one (or multiple) drug" approach. Clinical guidelines are based on evidence from randomized controlled trials (RCTs) of individual chronic conditions, largely excluding patients with multimorbidity and/or frailty (Buffel du Vaure et al., 2016; He et al., 2020). Indeed, if a person were to be graphically

DOI: 10.1201/9781003398295-9

FIGURE 9.1 *Traditional* approach to illness in Western health care systems. (Modified from Polidori and Ferrucci (2023) and Polidori (2024).)

seen as an iceberg, *traditional* clinical decision making still bases on its tip – chronological age and organ dysfunction (Figure 9.1).

On one end, the result of the traditional health care approach has enabled the above-mentioned doubling of life expectancy years in few decades; on the other end, multimorbidity and fragmented care, frequent medical visits, and high medication burden are increasing the risks of hospitalization due to drug-drug interactions, drug-disease interactions, and adverse drug events. The sustainability of this approach is therefore coming into question given the rapidly increasing number of older and chronically ill patients. The unmet clinical need of diagnosing frailty prior to overt and irreversible functional loss renders new approaches for maintaining health and robustness with aging a public health priority. Indeed, not only several actions with this aim are strongly supported internationally, but also resilience, intrinsic capacity (Carvalho et al., 2017; Cesari et al., 2018), and maintenance of health are increasingly dominating the public health scene.

In this sense, primary care is the ideal setting for systematic evaluation of intrinsic capacity with increasing age and also prompt detection of prefrailty/initial frailty. As the population of older adults grows, reliance on specialists for detection and diagnosis is unsustainable, highlighting the need for primary care-led frailty assessment. Initial successful frailty prevention strategies empower primary care to play a central role in early detection and intervention to enable the identification of reversible causes, support the management of medical conditions that are worsened by frailty, mitigate safety risks, and preserve quality of life.

However, together with the monodimensional, disease- and deficit-centered, intradisciplinary approach to cure, another crucial obstacle to functional loss control in the population is the lack of awareness and measurement tools for the established multidimensional nature of frailty. Frailty is a multidimensional and dynamic condition negatively impacting everyday life or coping with heterogeneous stressors due to the fact that multiple physiological systems are compromised. Frailty is largely accepted as an overarching geriatric syndrome, although the ICD code R54 refers to frailty as "age-related physical debility or senility", thereby limiting its definition to the physical nature of the condition. Indeed, a large body of research in the meantime recognizes not only the multidimensional and dynamic foundation of frailty but also its occurrence as a continuum, with a large portion of its development developing silently over the course of years (Figure 9.2).

In this sense, frailty tightly resembles advanced cognitive impairment, with the extreme expression of it – overt dementia – being incurable. Similar to cognitive impairment, whose early stages develop silently over time and are difficult to treat once they are overt, frailty manifests over time as increasingly accelerated aging in different domains of the organism until functional loss is irreversible and disability occurs. Thus, frailty as accelerated aging and neurodegeneration share both pathophysiological aspects (Hou et al., 2019) and the strong need of detection as soon as possible in order to prevent their irreversible consequences (Polidori, 2023). In particular, frailty as such can be prevented and even robustness can be restored (Figure 9.1).

Although the multidimensional and dynamic nature of frailty renders the early diagnosis outside of specialized settings challenging, a great advance was enabled by its description and

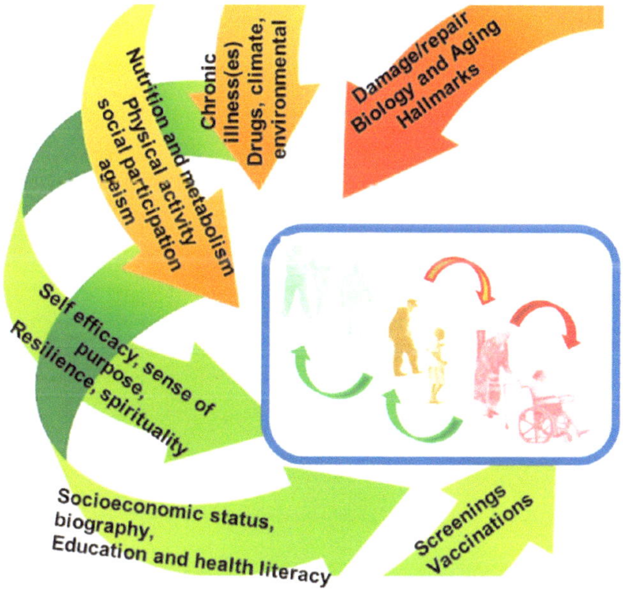

FIGURE 9.2 Biopsychosocial model of frailty, resilience, and well-being.

operationalization as physical phenotype (Fried et al., 2001) or accumulation deficit (Rockwood et al., 2001). This advance, which we termed *transitional* approach (Polidori and Ferrucci, 2023; Polidori, 2024a), enlarged the horizon of clinical decision making beyond chronological age and illness (Figure 9.3), but it also still largely confines frailty to the physical aspects of an inherently complex process.

In aging medicine, complexity is due to the evidence-based multifaceted nature of the aging process itself, biologically intended, as well as to the many diverse needs of the older population, which concern neuropsychological, functional, and socioeconomic aspects strongly influencing health trajectories, well-being, and recovery from disease (Polidori and Ferrucci, 2023; Polidori, 2024b) (Figure 9.2; Box 9.1). Therefore, in the graphical depiction of an (older or aging) adult as an iceberg, biological age measures, preclinical phenotypes and weakest links, multimorbidity, and polypharmacy must be included in a *revolutionary* paradigm urgently needed in the XXI century and beyond (Figure 9.4). Within this paradigm, the systematic evaluation of psychosocial, lifestyle, and functional factors is unavoidable.

Being aging medicine and geriatrics "young" metadisciplines, complexity derives also by the lack of studies on multidimensional frailty, of the consistent use of scientific methodology and

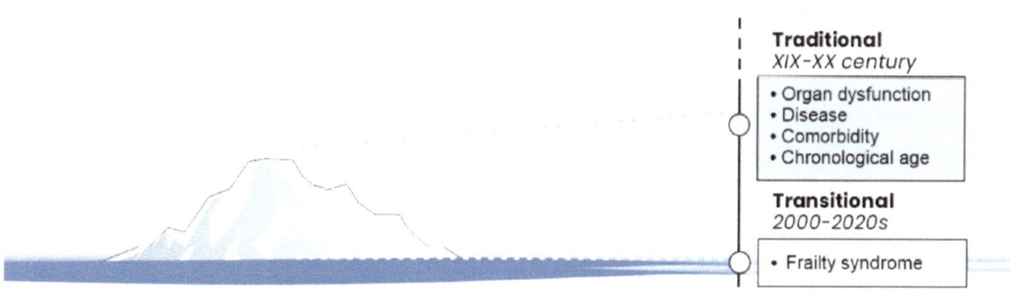

FIGURE 9.3 *Transitional* approach to older adults since the first descriptions of frailty in the medical literature. (Modified from Polidori and Ferrucci (2023) and Polidori (2024a).)

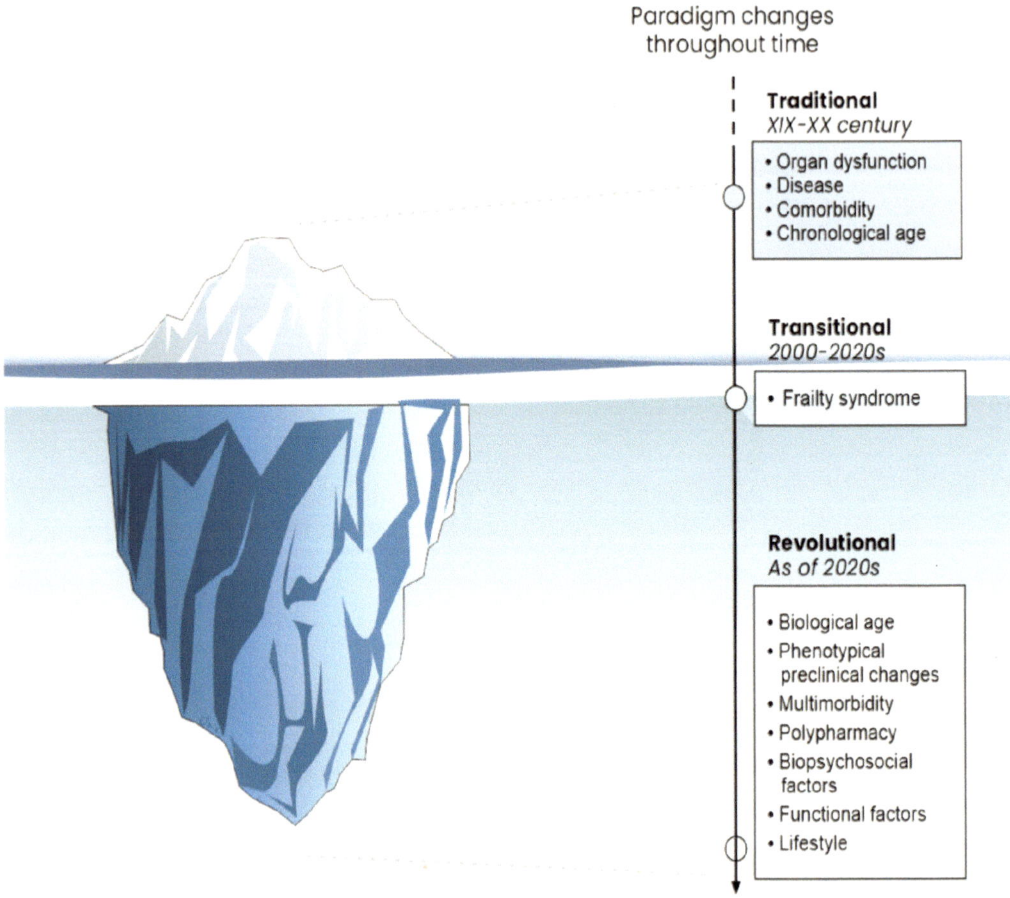

FIGURE 9.4 *Revolutional* paradigm change for the health- and well-being maintaining approach to older adults to be systematically implemented in the XXI century vs. older traditional approaches. (Modified from Polidori and Ferrucci (2023) and Polidori (2024b).)

terminology, of the adequate educational levers at under-, post, and no-graduate levels including self-empowerment of the civil society, as well as of familiarity of the interwoven relationship between biological and psychosocial determinants of aging and of its accelerated phenotype. For its multidimensional nature, it has been postulated that comprehensive geriatric assessment (CGA), the highly effective biopsychosocial cornerstone of geriatric medicine (Ellis et al., 2017), might best capture frailty and its prognosis (Veronese et al., 2022, 2024; Ho et al., 2024). In the meanwhile, a large body of evidence appears to support this postulate. However, even CGA-based tools do not allow yet the evaluation of all the determinants of (accelerated) aging described so far (Box 9.1).

Indeed, when not only comorbidities but also geriatric syndromes and geriatric resources were included in a complex analysis aimed at identifying prognostic clusters, five geriatric resources-dependent geriatric syndrome clusters were identified (fluid-dysbalance – *fluid and electrolyte imbalance* and *incoherence*; sensorimotor – *swallowing disorder, impoverishment, instability, sensorial impairment, and iatrogenic disease*; neurovegetative – *insomnia, chronic pain, and irritable colon*; psychosocial – *irritability, immobility, cognitive impairment, and social isolation*; and the iatrogenic – *incontinence, polypharmacy*, and *inanition*, each characterized by a specific risk for poor outcomes. Interestingly, the strongest prognostic signature was displayed in the iatrogenic and

BOX 9.1 DETERMINANTS OF COMPLEXITY OF AGING AND FRAILTY

- Biological and Physiological Multifactoriality. Aging involves biological changes that at least apparently involve parallel dysregulation of multiple biological pathways that affect various organ systems, increased vulnerability to age-related diseases, as well as deterioration of physical and cognitive function. Scientists have suggested that the overt limit of progression of these complex changes converges into the syndrome of frailty, intended as a state of accelerated aging when even the challenge of maintaining physiological homeostasis is not matched by adequate resilience and functional reserve. This conceptualization of frailty implies a nuanced, tailored approach to the study of aging/frailty, exceptional challenges when facing the diagnosis and treatment of frail older adults.
- Multimorbidity and Polypharmacy. The progressive imbalance between the accumulation of molecular and cellular damage and the resilience mechanisms that prevent, oppose, and repair such damage emerges as progressively higher susceptibility to chronic diseases with aging. Indeed, aging is the major risk factor for most chronic diseases, and the incidence and prevalence of chronic diseases increase with aging. Thus, multimorbidity, defined as the progressive accumulation of multiple diseases in the same individuals, is a clinical hallmark of frailty. The medical mantra of "each disease requires a specific treatment" complicates multimorbidity with polypharmacy. Prescriptions for older people are loaded with multiple drugs, in spite of the fact that their potential toxicity increases with aging due to changes in body composition and organ metabolic capacity, leading to difficult medication management. Due to age-related changes in pharmacokinetics and pharmacodynamics, decreased kidney excretory capacity and altered drug sensitivity, the presence of multiple chronic conditions and the use of numerous medications further increase the risk of adverse drug reactions and complicate treatment regimens. Intervening "crises" whose nature, acute disease events, or drug side effects may be difficult to clearly distinguish exert a profound impact on the management of the older patient and clinical outcomes.
- Psychosocial and Biographical Determinants. Social and behavioral factors such as low socioeconomic status, inadequate social support, adverse life events, or negative health behaviors can alter the aging process per se and act in a complex interaction with biological pathways. There is strong evidence in the literature that the social perception of aging and the stigma of being old may negatively affect several aspects of growing old and directly affect the pace of biological aging. However, the exact mechanisms are unclear and several hypotheses have been discussed, including stress, workplace discrimination, access to health care resources, and many others.
- Socioeconomic and Ethical Considerations. Poverty is highly prevalent in older patients and affects variables that negatively impact health, including access to healthy nutrition, and certain health care resources, for example, in the realm of oral health. Ethical considerations often contribute to the complexity of aging and frailty. For example, while older frail patients are often excluded from life-saving medical and surgical interventions, they are often offered extreme interventions that maintain them alive without any chance of improving their functional status or their overall prognosis. Early detection, prevention, and prompt treatment of age-related dysfunctions imply a proactive, patient-centered approach. However, prevention of frailty and related disability is not structured in modern health care systems. Longevity science, on the other hand, uses advanced biomedical technologies often emphasizing

anti-aging, ageistic solutions. Economic, medical, and ethical challenges are brought particularly in balancing the benefits of extended life expectancy with quality of life and health care costs.

- Reactive Nature of Disease-Centered Medicine. There is no doubt that medicine centered on the nosological classification of diagnosis and cure of disease has resulted in a substantial increase in life expectancy. However, the focus on prolonging life was not paralleled by attention to functional status. Without the substantial reduction in health expectancy, and given the aging of the population, the incidence and prevalence of major chronic diseases and a deterioration of quality of life are expanding and impose a great burden on the health and social support system of many countries around the world.

psychosocial cluster, with patients within this cluster showing significantly worse prognosis than patients outside the cluster ($p < 0.001$) (Stegemann et al., 2024).

In summary, the complexity of aging and frailty medicine arises from the interplay of biological changes, medication management challenges, multimorbidity, lifestyle, and the need for comprehensive, individualized care strategies. Within the demographic transition, the WHO definition of health as a "complete state of physical, mental and social wellbeing" reflects very well the biopsychosocial nature (Engel, 1977) of CGA (Rubenstein et al., 1981) and of the International Classification of Functioning, Disability and Health (WHO, 2001). In this sense, the "revolutional" paradigm change contemplated in the present and previous works (Polidori and Ferrucci, 2023; Polidori, 2024a) (Figure 9.4) is a "forward into yesterday" movement (Figure 9.5).

Despite all these (historical) efforts, the concept of health as absence of disease is still prevailing in our societies and much effort is still needed to implement the multidimensional, holistic concept

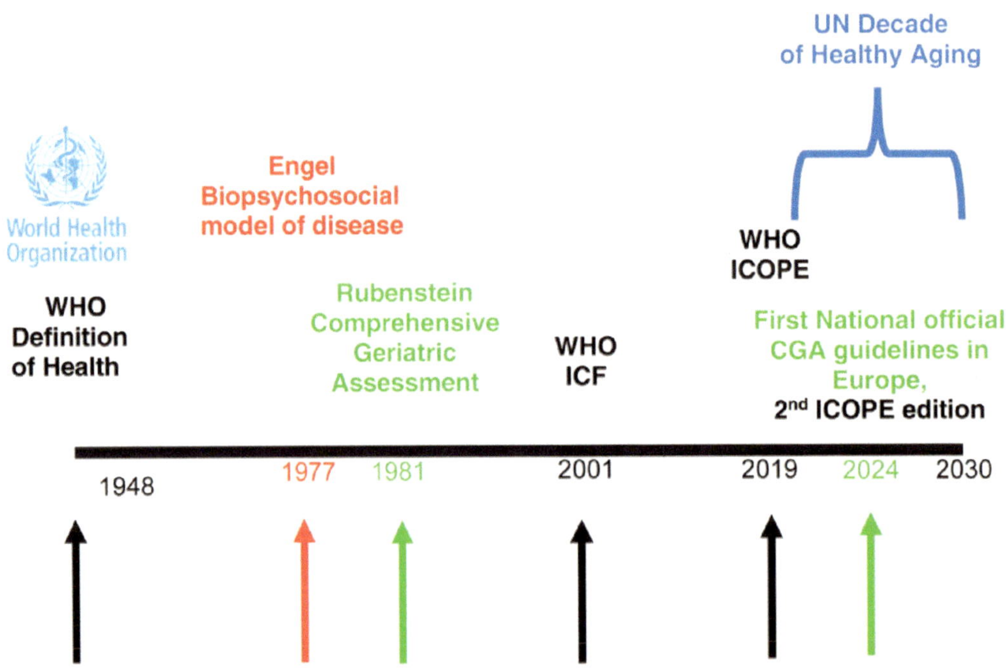

FIGURE 9.5 Timeline of the foundation of biophsychosocial operationalizations of health and disease from the WHO definition of health to the UN Decade of Healthy Aging.

of well-being needed for aging and frailty research, medicine (Rubin, 2023), and education. As described previously CGA is employed to prioritize interventions and tailor treatments to individual needs, improving outcomes and quality of life. The potential of comanagement and use of the CGA approach in both aging science and medicine of the aged person, however, has often been challenged by deterministic visions around aging. The human genome project has suggested that genes drive the aging process, but the understanding of the human genome does not appear to provide *insights into a process that is random and thermodynamically driven*. Besides the clear evidence of the absence, to date, of any intervention able to reverse the aging process in complex organisms, the practical observations distinguishing the aging process from age-associated disease (Hayflick, 2007) prove still true and are reported below.

> Unlike any disease, age changes: (1) occur in every multicellular animal that reaches a fixed size at reproductive maturity, (2) cross virtually all species barriers, (3) occur in all members of a species only after the age of reproductive maturation, (4) occur in all animals removed from the wild and protected by humans even when that species probably has not experienced aging for thousands or even millions of years, (5) occur in virtually all animate and inanimate matter, (6) have the same universal molecular etiology, that is, thermodynamic instability.
>
> Unlike aging, there is no disease or pathology that shares these six qualities.

THE BIOLOGICAL MODEL OF DISEASE AND THE END OF THE DISEASE ERA

Frailty, as a state of accelerated aging, is multifactorial, reversible, and measurable. It becomes more common as populations age and its prevalence among community-dwelling persons in 62 countries worldwide ranges from 11% among those who were 50 to 59 years of age to 51% among those who were 90 years of age or older (O'Caoimh et al., 2021). Interestingly enough, older persons in acute care hospitals and nursing homes, those in low- or middle-income countries, and those with a socially vulnerable status are all at increased risk for frailty (Hoogendijk and Dent, 2022), underscoring its multifactorial nature. Due to its multidimensional nature, frailty can be very well captured by CGA-based tools, which are able to address not only its main determinants but also the risk of poor outcomes associated with it. Most importantly, CGA is the very first example of *theranostics* and can accurately identify and activate the intervention targets. Unfortunately, instead of concentrating on the consistent, systematic use of very few agreed instruments depending on an indication, the medical community disagrees on tools and produces more of them, causing significant delay to the implementation of the patient-centered approach in clinical routine (Kim and Rockwood, 2024). For this reason, the deficit-centered, reactive biological disease model based on coded ICD diagnosis largely prevails and keeps displaying its flaws in the increasingly older population (Figure 9.1).

While medicine uses in principle the same roughly 40 diagnostic organ- and disease-centered biomarkers for decades, the expansion of the older population has clearly demonstrated the limits of such tools in predicting health and recovery course with increasing age. This fact not only reflects the role of prognosis-relevant factors in older patients beyond the biological ones but also supports the concept of accelerated aging as the individual, dynamic balance between damage and repair (Ferrucci et al., 2024). Even though mortality may not be the best outcome compared to quality of life, especially in advanced age, mortality prediction models support identifying older adults with short life expectancy for whom clinical care might need modifications. When external validations of mortality prediction models in older adults were systematically reviewed with up to 3 years of follow-up, an analysis of 36 studies reporting 74 validations of 64 unique models, risk of bias was generally found to be high. Illness in general and mostly cardiovascular diseases in particular were used as predictors in 70% of the models, suggesting a broad disease-centered, not person-centered, approach. However, models with more than 20 predictors were more likely to have acceptable discrimination, as were models including sex or predicting risk during CGA (Ho et al., 2024). The poor prediction power exerted in advanced age by tools based on single aspects of health, however, is well known, as shown by the earlier analysis of loss of predictive power of amyloid biomarkers in addressing prognosis of dementia course in advanced age (Mattsson et al., 2012). Of note, older patients admitted to intensive (ICU) or intermediate care units (IMC) enrolled to assess 90-day mortality using a CGA-based multidimensional tool and the Sequential Organ Failure Assessment (SOFA) score, showed that both multidimensional ($p = 0.039$) and organ-based scores ($p = 0.014$) successfully predicted mortality among older IMC and ICU patients in logistic regressions. Receiver operating characteristic (ROC) analyses demonstrated comparable areas under the curve, as well as a similar sensitivity and specificity. However, the CGA-based instrument at admission correlated significantly with quality of life ($p < 0.001$, $r = -0.631$ at discharge; $p = 0.005$, $r = -0.377$ at 30-day FU; $p = 0.004$, $r = -0.409$ at 90-day FU) and nursing needs (Mann-Whitney U-test, $p = 0.002$ at 30-day FU; $p = 0.011$ at 90-day FU) at FU, while the SOFA score did not show significant associations with respect to these parameters (Hochleitner et al., 2024). Similarly, a recent comparison of frailty and CGA-based tools with the Manchester Triage System in predicting mortality and rehospitalizations in older patients admitted to the emergency department uncovered large discrepancies in the predictive power of these instruments, with the highest clinimetric values displayed by the frailty- and CGA-tools (Dencker et al., 2024). Indeed, among 16,082 studies initially considered, 159 on almost 2.4 million patients with average age 78 years showed that several tools are used to predict poor prognosis in geriatric patients, but only those derived from a multidimensional evaluation have the characteristics of precision and accuracy needed to sustain adequate, non-futile clinical decision making (Veronese et al., 2024).

In 2004, i and Fried (Tinetti and Fried 2004) scientifically advocated for a medicine that recognizes the need to abandon disease as the focus of medical care. In the same year and in the same prestigious journal, Holman underlined "It is axiomatic that medical education should prepare students well for the clinical problems they will face in their future practice. However, that is not happening for the most prevalent problem in health care today: chronic disease" (Holman, 2004).

Two decades after these initial publications, and in line with the average time required to translate new knowledge into practice (Rubin, 2023), disease in general and acute symptomatic in particular are still object of all clinical algorithms – lack of evidence on multimorbidity and frailty dominates the health care scene and non-representative clinical guidelines.

THE END OF THE CHRONOLOGICAL AGE ERA: FRAILTY AND INTRINSIC CAPACITY AS THE CORE OF (AGING) MEDICINE AND GERIATRICS

Together with (acute) illness, chronological age plays a predominant role in clinical decision making (Figure 9.2). However, there is a wealth of literature showing that multidimensional, biopsychosocial frailty overcomes not only physical deficits but also chronological age for predictive risk stratification. The reason for this is that, with increasing chronological age, heterogeneity of phenotypes increases, and this heterogeneity of aging across individuals underscores the necessity of quantifying aging as a primary task in aging research (Ferrucci et al., 2020). Indeed, psychosocial dysfunction plays a major role in health trajectories in advanced age and can affect the hallmarks of aging (Picard, 2011; Mason et al., 2019; Diniz et al., 2024; Faria et al., 2024). The interaction of individual factors, behavioral factors, and psychosocial experiences across the life course influences health across the aging continuum. Chronic exposure to psychosocial stress and genetic-based mitochondrial dysfunction have similar biological consequences, predisposing individuals to adverse age-related health disorders and early mortality. Psychological factors are robust predictors of healthspan and longevity, and psychogenic aging research aims to integrate psychological influences into the Hallmarks of Aging framework. Social and behavioral factors, such as socioeconomic status and life events, can alter the aging process and interact with biological pathways. The relationship between social processes and biological aging was investigated in studies of psychoneuroimmunology on married couples to explore how social support is associated with biomarkers such as proinflammatory cytokines and insulin-like growth factor-1 (Wilson et al., 2021). Potential pathways from chronic marital distress to mental health problems such as depression were also described, able to increase the risk for obesity, metabolic syndrome, and cardiovascular and cardiometabolic diseases through behavioral (e.g., poor diet, sleep problems, low exercise levels) and biological factors (e.g., insulin, triglyceride) increasing the likelihood of gut dysbiosis and accelerated aging (e.g., inflammaging) (Kiecolt-Glaser et al., 2019). Biological hallmarks of aging including genomic instability, telomere attrition, epigenetic otinalteration, loss of proteostasis, deregulated nutrient sensing, mitochondrial dysfunction, cellular senescence, stem cell exhaustion, and altered intercellular communication are mechanisms, not mere causes of aging, and are increasingly recognized to be influenced by biographical, lifestyle, and non-physical pathophysiological mechanisms (Boccardi et al., 2024). Aging can be associated with mental health conditions, and chronic caregiving stress may accelerate biological aging. In summary, the interplay between psychosocial factors and biological aging is acknowledged, suggesting a potential influence of psychosocial dysfunction on the hallmarks of aging.

Unfortunately, however, a unified measure of biological aging still does not exist and aging quantification is an important goal for the accurate categorization not only of intrinsic human health status but also for the development of interventions aimed at slowing down aging (Ferrucci et al., 2024). Biological age reflects the overall health status of the individual, making aging possible to grasp and quantify. Although chronological age increases at the same rate, the rate of biological aging varies among individuals. Some people experience a slow pace of aging, while others experience a rapid increase. The latter is the (multifactorial) accelerated aging under which frailty is widely accepted. While indicators used for the measurement of biological age construction include molecular biomarkers (e.g., transcriptomics, epigenomics, proteomics, and metabolomics), clinical blood markers, physical measurements, and composite indicators (Polidori, 2023), their application in large populations is limited by intricate detection technologies and high costs (Rutledge et al., 2022). Routine clinical blood tests, non-invasive physical examinations, and patient self-reports or questionnaire assessments have the advantages of being convenient to operate, cost-effective, and easily interpretable for generalization. Previous studies have primarily focused on singular biological age indicators for aging quantification. However, composite indicators have gradually gained attention in recent years. This approach might be even more effective if applied to the concept of healthy aging proposed by the World Health Organization (Figure 9.5), emphasizing the functional ability and well-being among older populations. A healthy functional status, in fact, helps older adults maintain their daily living abilities and improves their quality of life. Consequently, functional indicators (e.g., cognitive, physical, mental function, and so on) appear excellent phenotypic composite proxies of biological age. CGA-tools, by offering a multidisciplinary approach to evaluate the older population's physical function, nutritional status, and psychosocial aspects, have been proven able to quantify aging in real life and serve as efficient monitors of interventions against frailty onset and progression (Meyer et al., 2019; Meyer et al., 2020;Pickert et al., 2021; Meyer et al., 2022; Zampino et al., 2022). However, intrinsic capacity (Carvalho et al., 2017; Cesari et al., 2018) and its resilience-boosting effect (Chhetri et al., 2021) appear to hold the highest level of promise for their multidimensional yet not deficit-oriented nature. Intrinsic capacity enables the focus on person-centered aspects that traditional approaches systematically ignore. For instance, cognitive and psycho-emotional domains of intrinsic capacity are determined not only by the individual characteristics of each person but also by the state of one's social environment and the social support of frail, pre-frail, and robust older adults varies accordingly. Though aging populations are often presented as a challenge, healthy aging can maintain or add quality and value to life and is associated with positive changes such as improvements in work performance, wisdom, emotion control, and happiness. Aging-associated gradual declines in physical, cognitive, and social performance associated with poorer well-being, lower quality of life, and a restriction of independence are heterogeneous, but engaging in physical activity at any age has numerous health and well-being advantages across psychological, physical, and social domains. Similarly, mental and social engagement may be beneficial for functional ability (Everard et al., 2000), reduced risk of mortality (Lennartsson and Silverstein, 2001) and decreased levels of loneliness and depression (Warr et al., 2004). A meta-analysis of the effects of real-world RCTs on interventions for cognitive aging in healthy older adults supports that real-world interventions, specifically physical activity interventions, may have specific benefits for cognitive function, with more consistent benefits reported for visuospatial abilities (Vaportzis et al., 2019), cardiovascular (Pinckard et al., 2019), central nervous system (Morgan et al., 2015), and immune system functions (Nieman & Wentz, 2019). Moderate-to-strong levels of sense of coherence- and self-efficacy are associated with lower odds of being frail and better response to complex geriatric interventions, respectively (Fang et al., 2022; Meyer et al., 2022). In addition, out-of-home mobility and social participation have been identified as resources to postpone frailty and data from the Survey of Health, Ageing, and Retirement in Europe (SHARE) waves 6-8 with 13,456 fit participants aged ≥50 years identified social network size, loneliness (UCLA 3-Item Loneliness Scale), and lack of motivation (EURO-D) as predictors and the number of performed social out-of-home activities as mediator variable on subsequent Edmonton Frailty Score (EFS) (Mümken et al., 2024) independent of age, education, gender, cohabitation, widowhood, urban environment, and country. In this study, direct effects of loneliness and lack of motivation on later EFS-scores were found to be partially mediated by social out-of-home activities (loneliness $\beta=0.005$; 95% CI = 0.003–0.008 and lack of motivation $\beta=0.014$; 95% CI = 0.009–0.019) and a significant effect of increasing social out-of-home activities on later reduction of EFS-scores ($\beta=-0.21$; 95% CI = −0.29–0.04; $p<0.001$) was observed, suggesting that social out-of-home activities may sustain frailty prevention (Mümken et al., 2024). Using latent cluster analysis, three patterns of participation restriction (low/moderate/high) in community-dwelling older adults ≥60 years with falls or risk of falls were identified, directly associated with impaired intrinsic capacity domains (Merchant et al., 2022) including nutrition. Due to recent pathophysiological and biomolecular advances, indeed, malnutrition and oxidative distress (Sies et al., 2024; Sørensen et al., 2024) appear to be a biomolecular correlate of frailty more than of chronological age and micronutrients with antioxidant properties are recognized as relevant biomarkers of frailty (Viña et al., 2018; Viña et al., 2019; Polidori and Mecocci, 2022), also in the frame of anorexia of aging and as indicators of inflammation and oxidative stress involvement. Due to their preferential uptake mechanisms, carotenoids in general and xantophylls in particular may constitute integral part of the success of nutritional interventions, along with physical exercise and other lifestyle strategies, in restoring robustness (Polidori et al., 2021). Once again, the rather disappointing results of studies attempting at ameliorating cognitive symptoms through carotenoids are due to the complexity of human (cognitive) impairment (Polidori, 2023) and problematic methodological applications, including confusing terms and definitions. Finally, a discussion on frailty cannot be complete without touching on inflammation. There is much evidence that inflammation is a common factor for all major chronic diseases, and the CANTOS trial used canakinumab therapy – blocking the interleukin-1 beta innate immunity – and showed a significantly reduced rate of recurrent cardiovascular events compared to placebo (Orkaby et al., 2024). Interestingly, a further analysis of the CANTOS data showed that the anti-inflammatory treatment also reduced the rates of lung cancer, severe diabetes, hip fracture, congestive heart failure, and anemia, although not frailty (Orkaby et al., 2024). While inflammation has been considered one of the hallmarks of aging, it is considered a resilience mechanism with dysfunction permissive of damage accumulation with aging, reflective of the burden of molecular damage. It is suggested that once inflammation is activated above a certain threshold, it initiates a vicious cycle that accelerates damage accumulation (Ferrucci et al., 2024). Thus, the regulation of inflammation to retain its beneficial functions while preventing the harmful effects of an exaggerated response might be a modulating strategy of biopsychosocial frailty.

CONCLUSIONS AND OUTLOOK

Within the current rapid demographic change, the quest for easily accessible, reliable, and feasible indicators of the intrinsic pace of aging beyond chronological age persists and in fact becomes more and more urgent. As aging is not only determined by biomolecular processes but also multifactorial and manifests at high and increasing heterogeneity with advancing chronological age, the operationalized concept of frailty as accelerated biopsychosocial aging indicator is proving adequate to measure and induce intrinsic aging acceleration and deceleration (Figure 9.3) (Polidori and Ferrucci, 2023; Polidori, 2023). In clinical routine, chronological age and morbidity are frequently not sufficient, in older adults, for successful clinical decision making. Fortunately, on the disease-management level, the concept of frailty is being increasingly recognized as a crucial diagnostic and prognostic tool. Within the conceptual bridge between aging science and aging medicine, frailty in its multidimensional nature is accepted as a reliable measure of accelerated aging beyond chronological age and illness. For instance, recently, a CGA-based frailty index obtained from older patients admitted to the emergency department undergoing blood withdrawal was used to explore biopsychosocial signatures of aging biomarkers and age-related processes. The well-established MetaboHealth score was shown to be strongly correlated with CGA-frailty. Lutein, suggested marker of frailty (Semba et al., 2007; Dent et al., 2019), among several lipophilic micronutrients, displayed a strong association with specific CGA subdomains and metabolomic profiles captured by computational analysis of differential expression of regulatory networks based on LASSO regression, stability selection, and GLay community cluster algorithms (unpublished data). Furthermore, transcriptome aging clocks created to analyze RNA sequencing data obtained from peripheral blood mononuclear cells of older emergency department patients and undergoing CGA evaluation were recently studied (unpublished data). Interestingly, clocks' error in predicting chronological age as well as difference between predicted and chronological age (age acceleration) applied to different CGA risk group patients showed that a transcriptomic clock with low scattering and precise prediction of chronological age is not as good to predict biological age phenotype like CGA tools do (unpublished data). Within this frame, the likely ability to discriminate the physiological relevance of aging mechanisms by means of multidimensional frailty tools was recently supported by an investigation on the age-related fasting response of the short-lived killifish *Nothobranchius furzeri*. A fasting-like transcriptional program in the adipose tissue of old fish that overrides the feeding response, setting the tissue in persistent metabolic quiescence, was uncovered. The fasting-refeeding regulatory cycle triggering an inverse oscillatory expression of genes encoding two AMP-activated protein kinase (AMPK) regulatory subunits [Prkag1 (γ1) and Prkag2 (γ2)] in young individuals was shown to be blunted by aging, resulting in reduced Prkag1 expression. Interestingly, the fasting-like transcriptional program was countered in transgenic fish with sustained $AMPK_{\gamma 1}$ activity, exhibiting a more youthful feeding and fasting response in older age (Ripa et al., 2023). While this transgenic model exhibited improved metabolic health and longevity, Prkag1 expression was shown to decline with age in human PBMCs, with γ1 expression anticorrelating, independent of chronological age, with multidimensional frailty as well as with its nutritional, functional, and multimorbidity subdomains. This observation suggests this gene as a possible hallmark of healthy aging and overall robustness in humans. Importantly, the evidence that γ2 expression does not correlate with multidimensional frailty highlights the relevance of the AMPK complex and could have therapeutic implications, as AMPK-activating compounds are often used without consideration of their subunits. This study, therefore, shows not only that selective activation of $AMPK_{\gamma 1}$ prevents metabolic quiescence and may preserve healthy aging in vertebrates, but also that multidimensional frailty as a biopsychosocial scaled measure of person-centered overall health and pace of aging is a useful tool to discriminate physiologically relevant features of aging in humans. In this sense, furthermore, the geroscientific conceptual basis of a common mechanism of aging is corroborated by recent evidence that multicomponent strategies against dementia development have been shown to decrease multimorbidity over time (Marengoni et al., 2018), and glucagon-like peptide-1 (GLP-1) receptor agonists, as well-established glucose-lowering medications, may impact not only type 2 diabetes but also cardiovascular risk, neurodegeneration, and overall mortality (Husain et al., 2019). In addition, vaccines appear to be powerful promoters of healthy aging, likely via their immunosenescence-control activity (Laupèze et al., 2021).

Due to the complexity and multifactorial nature, (accelerated) aging is not easy to operationalize beyond the counting of days after birth. The progressive decline of health with the passage of time is accompanied by great heterogeneity. This variability in phenotypes, particularly among older individuals, cannot be grasped by multimorbidity or physical frailty. If aging could be universally conceptualized, the geroscience hypothesis could be concretely tested and translated into community health and medicine.

For the clinical and real-life success of translational aging science, studies are needed in which gerontologists, including biologists, psychologists, sociologists, and geriatricians, use aging phenotypes and measures adequate to the complexity of human aging. Outcomes of aging and frailty studies should be integrated with a focus on maintenance of health, functioning, well-being, and quality of life.

REFERENCES

Boccardi V, Orr ME, Polidori MC, Ruggiero C, Mecocci P. Focus on senescence: Clinical significance and practical applications. J Intern Med. 2024;295(5):599–619. DOI: 10.1111/joim.13775.

Buffel du Vaure C, Dechartres A, Battin C, Ravaud P, Boutron I. Exclusion of patients with concomitant chronic conditions in ongoing randomised controlled trials targeting 10 common chronic conditions and registered at ClinicalTrials.Gov: A systematic review of registration details. BMJ Open. 2016 Sep 27;6(9):e012265. DOI: 10.1136/bmjopen-2016-012265.

Carvalho IA, Martin F, Cesari M, Sumi Y, Thiyagarajan JA, Beard J. Operationalising the concept of intrinsic capacity in clinical settings. In: Background Paper for the WHO Working Group on Metrics and Research Standards for Healthy Ageing. 2017.

Cesari M, Araujo de Carvalho I, Amuthavalli Thiyagarajan J, Cooper C, Martin FC, Reginster JY, et al. Evidence for the domains supporting the construct of intrinsic capacity. J Gerontol A, Biol Sci Med Sci. 2018;73:1653–60.

Chen H, Wang B, Lv R, Zhou T, Shen J, Song H, Xu X, Ma Y, Yuan C. Progression and trajectory network of age-related functional impairments and their combined associations with mortality. iScience. 2023 Oct 31;26(12):108368. DOI: 10.1016/j.isci.2023.108368.

Chhetri JK, Xue QL, Ma L, Chan P, Varadhan R. Intrinsic capacity as a determinant of physical resilience in older adults. J Nutr Health Aging. 2021;25:1006–11. 10.1007/s12603-021-1629-z

Dencker K, Becker I, Matthiessen U, Nouri B, Rhebaum V, Rupprecht L, Warmer N, Affeldt AM, Polidori MC. CGA- and physical phenotype-frailty-based vs. conventional risk tools in the Emergency Department. Annual EuGMS Congress, Valencia. 2024.

Dent, et al. Management of frailty: Opportunities, challenges, and future directions. Lancet. 2019;394(10206): 1376–86. DOI: 10.1016/S0140-6736(19)31785-4.

Diniz BS, Seitz-Holland J, Sehgal R, Kasamoto J, Higgins-Chen AT, Lenze E. Geroscience-centric perspective for geriatric psychiatry: Integrating aging biology with geriatric mental health research. Am J Geriatr Psychiatry. 2024;32(1):1–16. DOI: 10.1016/j.jagp.2023.09.014.

Ellis et al. Comprehensive geriatric assessment for older adults admitted to hospital. Cochrane Database Syst Rev. 2017;9(9):CD006211. DOI: 10.1002/14651858.CD006211.pub2.

Engel GL. The need for a new medical model: A challenge for biomedicine. Science. 1977;8:129–36. DOI: 10.1126/science.847460

Everard KM, Lach HW, Fisher EB, Baum MC. Relationship of activity and social support to the functional health of older adults. J Gerontol B Psychol Sci Soc Sci. 2000;55(4):S208–S212. https://doi.org/10.1093/geronb/55.4.S208

Fang J, Ren J, Ren L, Qiu X, Yuan S, Wang W, Wang J. Perceived social support and associated factors among community-dwelling older adults with frailty and pre-frailty in Hangzhou, China. Front Psychiatry. 2022 Jul 14;13:944293. DOI: 10.3389/fpsyt.2022.944293.

Faria M, Ganz A, Galkin F, Zhavoronkov A, Snyder M. Psychogenic aging: A novel prospect to integrate psychobiological hallmarks of aging. Transl Psychiatry. 2024 May 30;14(1):226. DOI: 10.1038/s41398-024-02919-7.

Ferrucci L, Gonzalez-Freire M, Fabbri E, Simonsick E, Tanaka T, Moore Z, Salimi S, Sierra F, de Cabo R. Measuring biological aging in humans: A quest. Aging Cell. 2020;19(2):e13080. DOI: 10.1111/acel.13080.

Ferrucci L, Tanaka T, Polidori MC. Can geroscience be translated into healthcare? Z Gerontol Geriatr. 2024;57(5):349–54. DOI: 10.1007/s00391-024-02326-z.

Fried LP, Tangen CM, Walston J, Newman AB, Hirsch C, Gottdiener J, Seeman T, Tracy R, Kop WJ, Burke G, McBurnie MA, Cardiovascular Health Study Collaborative Research Group. Frailty in older adults: Evidence for a phenotype. J Gerontol A Biol Sci Med Sci. 2001;56(3):M146–56. DOI: 10.1093/gerona/56.3.m146.

Hayflick L. Biological aging is no longer an unsolved problem. Ann N Y Acad Sci. 2007 Apr;1100:1–13. DOI: 10.1196/annals.1395.001.

Hayflick L. Entropy explains aging, genetic determinism explains longevity, and undefined terminology explains misunderstanding both. PLoS Genet. 2007;3(12):e220. DOI: 10.1371/journal.pgen.0030220.

Hayflick L. The greatest risk factor for the leading cause of death is ignored. Biogerontology. 2021;22(1):133–41. DOI: 10.1007/s10522-020-09901-y.

He J, Morales DR, Guthrie B. Exclusion rates in randomized controlled trials of treatments for physical conditions: A systematic review. Trials. 2020;21:228. https://doi.org/10.1186/s13063-020-4139-0

Ho L, Pugh C, Seth S, Arakelyan S, Lone NI, Lyall MJ, Anand A, Fleuriot JD, Galdi P, Guthrie B. Performance of models for predicting 1-year to 3-year mortality in older adults: A systematic review of externally validated models. Lancet Healthy Longev. 2024;5(3):e227–e235. DOI: 10.1016/S2666-7568(23)00264-7.

Hochleitner M, Pickert L, Nolting NA, Affeldt AM, Becker I, Benzing T, Kochanek M, Polidori MC. Patient- vs organ-based prognostic tools for older patients in critical care units: An observational study with a 3-month follow-up. Med Klin Intensivmed Notfmed. 2024. DOI: 10.1007/s00063-024-01179-z.

Holman H. Chronic disease—The need for a new clinical education. JAMA. 2004;292(9):1057–59. DOI: 10.1001/jama.292.9.1057.

Hoogendijk EO, et al. Frailty: Implications for clinical practice and public health. Lancet. 2019;394(10206): 1365–75. DOI: 10.1016/S0140-6736(19)31786-6.

Hoogendijk EO, Dent E. Trajectories, transitions, and trends in frailty among older adults: A review. Ann Geriatr Med Res. 2022;26:289–95.

Hou Y, Dan X, Babbar M, Wei Y, Hasselbalch SG, Croteau DL, Bohr VA. Ageing as a risk factor for neurodegenerative disease. Nat Rev Neurol. 2019;15(10):565–81. DOI: 10.1038/s41582-019-0244-7.

Husain M, Birkenfeld AL, Donsmark M, Dungan K, Eliaschewitz FG, Franco DR, Jeppesen OK, Lingvay I, Mosenzon O, Pedersen SD, Tack CJ, Thomsen M, Vilsbøll T, Warren ML, Bain SC; PIONEER 6 Investigators. Oral semaglutide and cardiovascular outcomes in patients with type 2 diabetes. N Engl J Med. 2019;381(9):841–51. DOI: 10.1056/NEJMoa1901118.

Kiecolt-Glaser JK, Wilson SJ, Madison A. Marriage and gut (microbiome) feelings: Tracing novel dyadic pathways to accelerated aging. Psychosom Med. 2019;81:704–10.

Kim DH, Rockwood K. Frailty in older adults. N Engl J Med. 2024;391(6):538–48. DOI: 10.1056/NEJMra2301292.

Landi F, Calvani R, Tosato M, Martone AM, Bernabei R, Onder G, Marzetti E. Impact of physical function impairment and multimorbidity on mortality among community-living older persons with sarcopaenia: Results from the ilSIRENTE prospective cohort study. BMJ Open. 2016 Jul 25;6(7):e008281. DOI: 10.1136/bmjopen-2015-008281.

Landi F, Liperoti R, Russo A, Capoluongo E, Barillaro C, Pahor M, Bernabei R, Onder G. Disability, more than multimorbidity, was predictive of mortality among older persons aged 80 years and older. J Clin Epidemiol. 2010;63(7):752–9. DOI: 10.1016/j.jclinepi.2009.09.007.

Laupèze B, Del Giudice G, Doherty MT, Van der Most R. Vaccination as a preventative measure contributing to immune fitness. NPJ Vaccines. 2021 Jul 27;6(1):93. DOI: 10.1038/s41541-021-00354-z.

Lennartsson C, Silverstein M. Does engagement with life enhance survival of elderly people in Sweden? The role of social and leisure activities. J Gerontol B Psychol Sci Soc Sci. 2001;56(6):S335–342. https://doi.org/10.1093/geronb/56.6.s335

Liotta G, et al. Frailty as the future core business of public health: Report of the activities of the A3 action group of the European innovation partnership on active and healthy ageing (EIP on AHA). Int J Environ Res Public Health. 2018;15(12):2843. DOI: 10.3390/ijerph15122843.

López-Otín C, Blasco MA, Partridge L, Serrano M, Kroemer G. Hallmarks of aging: An expanding universe. Cell. 2023 Jan 19;186(2):243–78. DOI: 10.1016/j.cell.2022.11.001.

López-Otín C, Blasco MA, Partridge L, Serrano M, Kroemer G. The hallmarks of aging. Cell. 2013 Jun 6;153(6):1194–217. DOI: 10.1016/j.cell.2013.05.039.

Marengoni A, Rizzuto D, Fratiglioni L, Antikainen R, Laatikainen T, Lehtisalo J, Peltonen M, Soininen H, Strandberg T, Tuomilehto J, Kivipelto M, Ngandu T. The effect of a 2-year intervention consisting of diet, physical exercise, cognitive training, and monitoring of vascular risk on chronic morbidity—the FINGER randomized controlled trial. J Am Med Dir Assoc. 2018;19(4):355–360.e1. DOI: 10.1016/j.jamda.2017.09.020.

Mason AE, Adler JM, Puterman E, Lakmazaheri A, Brucker M, Aschbacher K, Epel ES. Stress resilience: Narrative identity may buffer the longitudinal effects of chronic caregiving stress on mental health and telomere shortening. Brain Behav Immun. 2019 Mar;77:101–9. DOI: 10.1016/j.bbi.2018.12.010.

Mattsson N, Rosén E, Hansson O, Andreasen N, Parnetti L, Jonsson M, Herukka SK, van der Flier WM, Blankenstein MA, Ewers M, Rich K, Kaiser E, Verbeek MM, Olde Rikkert M, Tsolaki M, Mulugeta E, Aarsland D, Visser PJ, Schröder J, Marcusson J, de Leon M, Hampel H, Scheltens P, Wallin A, Eriksdotter-Jönhagen M, Minthon L, Winblad B, Blennow K, Zetterberg H. Age and diagnostic performance of Alzheimer disease CSF biomarkers. Neurology. 2012 Feb 14;78(7):468–76. DOI: 10.1212/WNL.0b013e3182477eed.

Merchant RA, Chan YH, Aprahamian I, Morley JE. Patterns of participation restriction among older adults at risk of falls and relationship with intrinsic capacity: A latent cluster analysis. Front Med (Lausanne). 2022 Nov 25;9:1023879. DOI: 10.3389/fmed.2022.1023879.

Meyer, et al. The multidimensional prognostic index in general practice: One-year follow-up study. Int J Clin Pract. 2019;13;e13403. DOI: 10.1111/ijcp.13403.

Meyer, et al. The prognostic significance of geriatric syndromes and resources. Aging Clin Exp Res. 2020;32(1):115–24. DOI: 10.1007/s00391-021-01978-5.

Meyer, et al. A tailored discharge program improves frailty and mood in patients undergoing usual rehabilitative care: A randomized controlled trial. J Am Med Dir Assoc. 2022;23(12):1962–e1-1962.e13. DOI: 10.1016/j.jamda.2022.09.003.

Mitnitski AB, Mogilner AJ, Rockwood K. Accumulation of deficits as a proxy measure of aging. Sci World J 2001; 8(1):323–36. DOI: 10.1100/tsw.2001.58.

Morgan JA, Corrigan F, Baune BT. Effects of physical exercise on central nervous system functions: A review of brain region specific adaptations. J Mol Psychiatry. 2015;3(1):3. https://doi.org/10.1186/s40303-015-0010-8

Mümken SA, Alonso-Perez E, Haeger C, et al. Prevention of frailty in relation with social out-of-home activities in older adults: Results from the survey of health, ageing, and retirement in Europe. Eur J Ageing. 2024;21;35. https://doi.org/10.1007/s10433-024-00829-7

Nieman DC, Wentz LM. The compelling link between physical activity and the body's defense system. J Sport Health Sci. 2019;8(3):201–17. https://doi.org/10.1016/j.jshs.2018.09.009

O'Caoimh R, Sezgin D, O'Donovan MR, et al. Prevalence of frailty in 62 countries across the world: A systematic review and meta-analysis of population-level studies. Age Ageing. 2021;50:96–104.

Orkaby AR, Thomson A, Macfadyen J, et al. Effect of canakinumab on frailty: A post hoc analysis of the CANTOS trial. Aging Cell. 2024;23:e14029.

Picard M. Pathways to aging: The mitochondrion at the intersection of biological and psychosocial sciences. J Aging Res. 2011;2011:814096. DOI: 10.4061/2011/814096.

Pickert, et al. Role of a multidimensional prognosis in-hospital monitoring for older patients with prolonged stay. Int J Clin Pract. 2021;75(5):e13989. DOI: 10.1111/ijcp.13989.

Pinckard K, Baskin KK, Stanford KI. Effects of exercise to improve cardiovascular health. Front Cardiovasc Med. 2019;6:69. https://doi.org/10.3389/fcvm.2019.00069.

Polidori MC. Embracing complexity of (brain) aging. FEBS Lett. 2024a;598(17):2067–73. DOI: 10.1002/1873-3468.14941.

Polidori MC. Aging hallmarks, biomarkers, and clocks for personalized medicine: (Re)positioning the limelight. Free Radic Biol Med. 2024b Mar;215:48–55. DOI: 10.1016/j.freeradbiomed.2024.02. 01 2.

Polidori MC, Ferrucci L. (2023). *Frailty from conceptualization to action: the biopsychosocial model of frailty and resilience.* Aging Clinical and Experimental Research, 35(4):725–727. doi:10.1007/s40520-022-02337-z

Polidori MC, Mecocci P. Modeling the dynamics of energy imbalance: The free radical theory of aging and frailty revisited. Free Radic Biol Med. 2022 Mar;181:235–40. DOI: 10.1016/j.freeradbiomed.2022.02.009.

Polidori MC, Stahl W, Griffiths HR. Nutritional cognitive neuroscience of aging: Focus on carotenoids and cognitive frailty. Redox Biol. 2021 Aug;44:101996. DOI: 10.1016/j.redox.2021.101996.

Ripa R, Ballhysa E, Steiner JD, Laboy R, Annibal A, Hochhard N, Latza C, Dolfi L, Calabrese C, Meyer AM, Polidori MC, Müller RU, Antebi A. Refeeding-associated $AMPK_{\gamma 1}$ complex activity is a hallmark of health and longevity. Nat Aging. 2023. DOI: 10.1038/s43587-023-00521-y.

Rockwood, K., & Mitnitski, A. (2001). Frailty in relation to the accumulation of deficits. *Journal of Gerontology: Series A, Biological Sciences and Medical Sciences*, 56(7), M418–M423. https://doi.org/10.1093/gerona/56.7.m418

Rubenstein LZ, Abrass IB, Kane RL. Improved care for patients on a new geriatric evaluation unit. J Am Geriatr Soc. 1981;29(11):531–6. DOI: 10.1111/j.1532-5415.1981.tb03357.x.

Rubin R It takes an average of 17 years for evidence to change practice—the burgeoning field of implementation science seeks to speed things up. JAMA. 2023;329(16):1333–6. DOI: 10.1001/jama.2023.4387.

Rutledge J, Oh H, Wyss-Coray T. Measuring biological age using omics data. Nat Rev Genet. 2022;23(12):715–27. https://doi.org/10.1038/s41576-022-00511-7.

Semba RD, Lauretani F, Ferrucci L. Carotenoids as protection against sarcopenia in older adults. Arch Biochem Biophys. 2007 Feb 15;458(2):141–5. DOI: 10.1016/j.abb.2006.11.025.

She R, Vetrano DL, Leung MKW, Jiang H, Qiu C. Differential interplay between multimorbidity patterns and frailty and their mutual mediation effect on mortality in old age. J Nutr Health Aging. 2024;28(8):100305. DOI: 10.1016/j.jnha.2024.100305.

Sies H, Mailloux RJ, Jakob U. Fundamentals of redox regulation in biology. Nat Rev Mol Cell Biol. 2024;25(9):701–19. DOI: 10.1038/s41580-024-00730-2.

Skou ST, et al. Multimorbidity. Nat Rev Dis Primers. 2022;8(1):48. DOI: 10.1038/s41572-022-00376-4.

Sørensen M, Pershagen G, Thacher JD, Lanki T, Wicki B, Röösli M, Vienneau D, Cantuaria ML, Schmidt JH, Aasvang GM, Al-Kindi S, Osborne MT, Wenzel P, Sastre J, Fleming I, Schulz R, Hahad O, Kuntic M, Zielonka J, Sies H, Grune T, Frenis K, Münzel T, Daiber A. Health position paper and redox perspectives – disease burden by transportation noise. Redox Biol. 2024 Feb;69:102995. DOI: 10.1016/j.redox.2023.102995.

Stegemann J, Affeldt AM, Mück L, Ferring A, Gerhards L, Pickert L, Bausch A, Benzing T, Antczak P, Polidori MC. Prognostic signature of multimorbidity, geriatric syndromes and resources cluster in older in- and outpatients: A pooled secondary analysis with A 6-month follow-up. BMJ Open. 2024 Dec 20;14(12):e086975. DOI: 10.1136/bmjopen-2024-086975.

Tinetti ME, Fried T The end of the disease era. Am J Med. 2004;116(3):179–85. DOI: 10.1016/j.amjmed.2003.09.031.

Vaportzis E, Niechcial MA, Gow AJ. A systematic literature review and meta-analysis of real-world interventions for cognitive ageing in healthy older adults. Ageing Res Rev. 2019;50:110–30. https://doi.org/10.1016/j.arr.2019.01.006

Veronese, et al. Comprehensive geriatric assessment in older people: An umbrella review of health outcomes. Age Ageing. 2022;51(5):afac104. DOI: 10.1093/ageing/afac104.

Veronese N, Fazzari A, Armata M, Parisi A, Parrinello A, Petralia V, Saccaro C, Vesco M, Tagliaferri F, Fittipaldo VA, Demurtas J, Smith L, Dominguez LJ, Pilotto A, Barbagallo M. Clinical prognostic factors for older people: A systematic review and meta-analysis. Ageing Res Rev. 2024 Jul;98:102345. DOI: 10.1016/j.arr.2024.102345.

Vetrano DL, Palmer K, Marengoni A, Marzetti E, Lattanzio F, Roller-Wirnsberger R, et al. Frailty and multimorbidity: A systematic review and meta-analysis. J Gerontol A Biol Sci Med Sci. 2019;74(5):659–66.

Viña J. The free radical theory of frailty: Mechanisms and opportunities for interventions to promote successful aging. Free Radic Biol Med. 2019 Apr;134:690–4. DOI: 10.1016/j.freeradbiomed.2019.01.045.

Viña J, Borras C, Gomez-Cabrera MC. A free radical theory of frailty. Free Radic Biol Med. 2018 Aug 20;124:358–63. DOI: 10.1016/j.freeradbiomed.2018.06.028.

Warr P, Butcher V, Robertson I. Activity and psychological well-being in older people. Aging Ment Health. 2004;8(2):172–83. https://doi.org/10.1080/13607860410001649662

WHO (World Health Organization) (2001). International classification of functioning, disability and health: ICF. World Health Organization. https://iris.who.int/handle/10665/42407

Wilson SJ, Bailey BE, Malarkey WB, Kiecolt-Glaser JK. Linking marital support to aging-related biomarkers: Both age and marital quality matter. J Gerontol B Psychol Sci Soc Sci. 2021;76:273–82.

Zampino, et al. Biomarkers of aging in real life: Three questions on aging and the comprehensive geriatric assessment. Geroscience. 2022;44(6):2611–22. DOI: 10.1007/s11357-022-00613-4.

10 The Impact of Polyphenol Intake on Cognitive Decline

Insights from Plasma and Urine Biomarker Analysis

Alba Tor-Roca, Cristina Andrés-Lacueva, Antonio Cherubini, and Mireia Urpí-Sardà

INTRODUCTION

Frailty in older subjects is defined as a clinical syndrome characterized by an accelerated decline of physiological reserves, with a high susceptibility to even minor stressors, which can induce adverse clinical outcomes, including the impairment of cognitive functions. At the same time, once frailty is established and progresses in its natural course, other geriatric syndromes like cognitive decline (CD) are more likely to emerge (1). Accumulating evidence from observational studies supports an association between frailty, lower cognitive performance, mild cognitive impairment (MCI), and dementia (2–4).

Reducing the prevalence and severity of frailty will likely benefit individuals, their families, and society. Given the bidirectional interplay between physical health and cognition, approaches targeting cognitive function across life might be of interest (5, 6).

A complex interplay of genetic and environmental factors, including early-life exposure to cognitive stimulation, socio-economic status, and educational background, influences the trajectory of cognitive function. Nutrition has garnered significant attention as a modifiable environmental factor, as it can potentially reduce the risk of age-related cognitive disorders through the diverse array of nutrients and food components that have been suggested to promote cognitive function (7).

Evidence from epidemiological and human intervention trials indicates that polyphenols may influence cognitive function, given their antioxidant properties (7–9). Polyphenols are a large family of bioactive compounds derived from secondary plant metabolism. They are widely distributed in plant-based foods and beverages such as fruits, vegetables, whole grains, nuts, seeds, wine, tea, coffee, and cocoa (10). These compounds can be classified into four main groups: flavonoids, phenolic acids, lignans, and stilbenes. They can further be subdivided into numerous subcategories depending on the number of phenol units within their molecular structure, substituent groups, and/or the type of linkage between phenol units (11). The daily consumption of polyphenols differs considerably across the globe – ranging from approximately 300 mg/d in Brazil to 1700 mg/d in Denmark – as do their major food sources (12, 13).

However, current epidemiological data on polyphenol intake mostly rely on self-reported questionnaires and food composition databases, which usually only consider polyphenols from the most representative foods (prespecified food lists). Portion size and frequency estimation are subject to participants' perception and memory, potentially leading to recall bias and measurement errors (14–16). In addition, the protective effects of polyphenols on cognitive function depend not only on the quantity consumed but also on their bioavailability, which largely differs among compounds and is influenced by large inter- and intra-subject variability (17, 18). Evaluating polyphenols and polyphenol metabolites in biological samples objectively estimates total dietary polyphenol consumption

DOI: 10.1201/9781003398295-10

(19, 20). This could be essential for providing a more accurate polyphenol exposure assessment and evaluating their protective effects on cognition (21).

In this chapter, several observational studies and randomized clinical trials (RCTs) are reviewed to understand better the effect of biomarkers of the intake of polyphenols on cognitive function in older adults. The research includes both observational studies and RCTs evaluating the association between urine or plasma biomarkers of polyphenol intake and cognitive function – assessed through cognitive tests – in older human subjects. Details of the design, subjects' characteristics, exposure or intervention, outcomes, and duration of the interventions or follow-up period of the RCTs and the observational studies included in this review are given in Tables 10.1–10.4.

OBSERVATIONAL STUDIES AND RANDOMIZED CLINICAL TRIALS OF BIOMARKERS OF POLYPHENOL INTAKE ON COGNITIVE FUNCTION

The results presented in this section lead to mixed conclusions regarding the potential benefits of polyphenol consumption for cognitive impairment (CI), which aligns with a previous systematic review (22). Flavonoids were tested in nine studies with cognitively healthy older subjects and in seven studies with older participants with some level of memory impairment, with inconsistent results. Heterogeneous findings for total polyphenols, which are examined exclusively in healthy older participants in this chapter, are also reported, as well as phenolic acids in individuals with MCI (23). Among the results included in this section, only executive function has consistently presented a non-improvement associated with total polyphenol intake. On the other hand, positive results have been shown on executive function when consuming phenolic acids. Even though some null associations have been described, 16 out of the 23 studies included here reported positive associations between the consumption of polyphenols and at least one cognitive domain or screening test. Only flavonoids significantly benefit the psychomotor domain, regardless of subjects' cognitive status (24–26). On the other hand, flavonoids have not shown a significant association in the attention domain (25, 27, 28), although positive correlations were observed.

FLAVONOIDS

The trials assessing flavonoid biomarkers reported mixed conclusions about the effects on cognitive performance.

Gleason et al. assessed isoflavones in patients with Alzheimer's disease (AD) and did not find any benefit in terms of cognition. Still, they did find a significant correlation between changes in plasma isoflavone and equol levels and alterations in speeded dexterity and verbal fluency score ($p \leq 0.05$) (25). Specifically, total plasma equol was significantly correlated with speeded dexterity ($p_{\text{non-dominant hand}} = 0.02$; and $p_{\text{dominant hand}} = 0.05$) as well as with verbal fluency ($p = 0.04$) in one study with AD patients (25). Vauzour et al. investigated the combined effect of eicosapentaenoic acid, docosahexaenoic acid, and cocoa flavanols on cognitive performance in older adults with subjective or MCI. The study found no significant impact on most cognitive outcomes, with reaction time variability and executive function being the exceptions, where a decline was observed in the intervention group compared to the control group (29). Changes to urinary flavanol metabolite concentrations did not correlate with changes in cognitive function.

However, in individuals with subjective cognitive complaints, reduced cognitive symptoms in everyday activities and improved recognition memory were found in the group consuming a daily dose equivalent to one cup of whole blueberry fruit (28). Such improvements were associated with the presence of urinary native anthocyanins, indicating recent blueberry consumption, but not with anthocyanin metabolites. Krikorian et al. also reported cognition performance improvements (semantic access [$p = 0.01$], visuospatial memory [$p = 0.05$], enhanced speed of processing [$p = 0.08$]) among patients with MCI who consumed blueberry supplementation. Moreover, urinary glycoside anthocyanin conjugates and total glycosidic anthocyanin metabolites were correlated with

TABLE 10.1
Summary of the Characteristics, Designs and Results of the Flavonoids-Based Studies Reviewed

Reference (Year)	Study Design/ Duration/N (% Women)	Subjects' Characteristics/ Age (SD)	Intervention	Cognitive Measurements	Polyphenol Measurements	Main Results

Gleason et al. (24)	Randomized, double-blind, placebo-controlled, pilot trial
	6 m.
	30 subjects
	Int: 15 (46.7%)
	Ctrl: 15 (53.3%)
		Older nondemented subjects
		73.7 y.
		Int: 73.0 (7.9) y.
		Ctrl: 74.3 (6.3) y.
			Int: 100 mg/d purified isoflavone glycosides (85% daidzein and genistein; Novasoy® brand isoflavones) capsules
			Ctrl: placebo tablets
				At baseline and at the 1-, 3-, 6-, and 8-m. visits
				• BSRT
				• Paragraph Recall Test
				• RCFT
				• VSLT
				• BNT
				• FAS word fluency test (letter fluency)
				• Animal fluency test (category fluency)
				• RCFT copy
				• GPT
				• SCWT
				• Mazes
				• TMT-B
					Plasma levels of total isoflavones, genistein, daidzein, and equol (*baseline and at the 1- and 6-m. visits)*
					Use of GC-MS (enzymatic hydrolysis of plasma)
					Total isoflavones
					M1: Int > Ctrl; $p < 0.001$
					M6: Int > Ctrl; $p < 0.001$
					Genistein
					M1: Int > Ctrl; $p < 0.001$
					M6: Int > Ctrl; $p < 0.001$
					Daidzein
					M1: Int > Ctrl; $p < 0.001$
					M6: Int > Ctrl; $p < 0.001$
						• Better performance in the following tests among isoflavone-treated group compared to placebo group:
						  • RCFT
						  • Category fluency-total (m. 3 and 6 only)
						  • GPT
						• Better performance in the following tests among placebo group compared to isoflavone-treated group:
						  • TMT-B
						  • SCWT

Gleason et al. (25)	Randomized, double-blind, parallel-group trial 6 m. 65 subjects **Int**: 32 (53.1%) **Ctrl**: 33 (51.5%)	Older subjects with AD 76.3 y. **Int**: 75.7 (7.7) y. **Ctrl**: 76.8 (6.8) y.	**Int**: 100 mg/d purified isoflavone glycosides (85% daidzein and genistein; Novasoy® brand isoflavones) capsules **Ctrl**: placebo capsules	*At baseline and at the 3- and 6-m. visits* • MMSEx • List Learning • Paragraph Recall Test • BVRT • Complex Figure Recall • Phonemic fluency test • Animal fluency test • Digit Symbol • Digit Span • SCWT • Mazes • TMT-A; TMT-B • Complex Figure Copy • GPT	Plasma levels of genistein, daidzein, and equol (at baseline and at the 3- and 6-m. visits) Use of GC-MS (enzymatic hydrolysis of plasma) **Genistein** M3: Int > Ctrl; $p < 0.001$ M6: Int > Ctrl; $p < 0.001$ **Daidzein** M3: Int > Ctrl; $p < 0.001$ M6: Int > Ctrl; $p < 0.001$ **Equol** M3: Int > Ctrl; $p = 0.03$ M6: Int > Ctrl; $p = 0.01$	• No differences in MMSE between groups. MMSE decline in both groups. • Significant correlation between total plasma equol, GPT, and Phonemic Fluency

John et al. (31)	Single-center, randomized, double-blind, placebo-controlled trial 2.5 y. 300 subjects **Int**: 150 (100%) **Ctrl**: 150 (100%)	Healthy older postmenopausal women 61.0 y. **Int**: 61.0 (7.0) y. **Ctrl**: 61.0 (7.0) y.	**Int**: 25 g/d isoflavone-rich soy protein (91 mg of aglycone isoflavones: 52 mg genistein, 36 mg daidzein, 3 mg glycitein) **Ctrl**: milk protein-matched placebo	*At baseline and at the 2.5-year visit* • WTAR • TMT-B • BNT • CVLT, immediate recall • CVLT, delayed recall • Faces I, immediate recall • Faces II, delayed recall • SDMT • Shipley Abstraction Test • Letter-Number Sequencing, from WAIS test • Block Design, from WAIS test • Judgment of Line Orientation • Category fluency • EBMT, immediate recall • EBMT, delayed recall	Urine excretion and plasma levels of isoflavonoids (at baseline and at the 6-m. visit) Use of LC-MS **Genistein, daidzein, glycitein, equol, dihydrogenistein, dihydrodaidzein, O-desmethylangolensin**	• No association between baseline overnight urine individual isoflavones, total isoflavones and their metabolites and baseline measures of global cognition or cognitive factor score • No significant association between change in overnight urine individual isoflavones, total isoflavones, and their metabolites and change in the global cognition score or changes in the verbal and visual episodic memory factors • Inverse association between change in overnight urine individual isoflavones, total isoflavones, and their metabolites and change in the general intelligence factor. Greater magnitude of association in women who had undergone surgical menopause than in those experiencing natural menopause • Similar but attenuated results in plasma
Igwe et al. (27)	Randomized, crossover, bioequivalence/ non-inferiority, pilot trial 3 hours on 2 different occasions with a 2-wk washout period 12 subjects (75%)	Healthy older subjects 77.0 (6.0) y.	**Int 1**: 1 × 300 mL QGP juice at baseline, 1 and 3 hours (369 mg anthocyanins) **Int 2**: 3 × 100 mL QGP juice at baseline, 1 and 3 hours (123 mg anthocyanins/serving)	*At baseline and at the 6-hour on each study day (2-wk washout period)* • TMT • RAVLT • Pattern and Letter Comparison task • Reaction task • Stroop task	Urine excretion of anthocyanins (at baseline and at the 0–2, 2–6, 6–12, and 12–24 hours after QGP juice intake on each study day) Use of HPLC-DAD/MS **Sum of cyanidin-3-glucoside, cyanidin-3-rutinoside, cyanidin-3-glucoside-G, cyanidin-G, peonidin-3-glucoside, peonidin-G, and pelargonidin-G** Int 1 = Int 2; $p \geq 0.05$	• No difference in cognitive tasks between dose-timing groups • No significant acute effect on cognition

(Continued)

TABLE 10.1 *(Continued)*

Summary of the Characteristics, Designs and Results of the Flavonoids-Based Studies Reviewed

Reference (Year)	Study Design/ Duration/N (% Women)	Subjects' Characteristics/ Age (SD)	Intervention	Cognitive Measurements	Polyphenol Measurements	Main Results
McNamara et al. (28)	Randomized, double-blind, placebo-controlled trial 76 subjects **Int 1**: 19 (57.9%) **Int 2:** 17 (41.2%) **Int 3:** 20 (65%) **Ctrl**: 20 (50%)	Older subjects with mild, self-perceived CD with aging 68.0 y. **Int 1**: 68 (3.9) y. **Int 2:** 69 (5.2) y. **Int 3:** 68 (4.7) y. **Ctrl**: 67 (4.9) y.	**Int 1**: 2 × 12.5 mg blueberry powder/d (417 GAEs, 269-mg C3G eq.) + corn oil capsules **Int 2:** 4 × fish oil capsule (400-mg EPA + 200-mg DHA/dose) + placebo powder **Int 3:** 2 × 12.5 mg blueberry powder/d (417 GAEs, 269-mg C3G eq.) + 4 × fish oil capsule (400-mg EPA + 200-mg DHA/dose) **Ctrl**: placebo powder and corn oil capsules	*At baseline and at the 6- and 12- m. visits* • DEX • TMT-A; TMT-B • Controlled Oral Word Production • HVLT	Urine excretion of anthocyanins (at baseline and at the 3-m. visit) Use of LC-MS/MS **Total anthocyanins (100%)** Int 1: wk 0 = wk 12 Int 2: wk 0 = wk 12 Int 3: wk 0 = wk 12 Ctrl: wk 0 = wk 12 **Deglycosylated anthocyanins-G (65%)** Int 1: wk 0 = wk 12 Int 2: wk 0 = wk 12 Int 3: wk 0 = wk 12 Ctrl: wk 0 = wk 12 **Glycoside anthocyanins (2%)** Int 1: wk 0 = wk 12 Int 2: wk 0 > wk 12; $p = 0.03$ Int 3: wk 0 = wk 12 Ctrl: wk 0 > wk 12; $p = 0.03$ **Native (food) form of anthocyanins (0.2%)** Int 1: wk 0 < wk 12 Int 2: wk 0 > wk 12; $p = 0.03$ Int 3: wk 0 < wk 12 Ctrl: wk 0 > wk 12; $p = 0.03$	• Fewer cognitive symptoms in the BB-treated group reported (DEX test) at 24-wk and 48-wk • Improved recognition memory (HVLT-discrimination test) in the BB-treated group at 24-wk but not at 48-wk • No effect in any cognitive domain for the combined fish oil and BB-treated group at 24-wk

Bensalem et al. (36)	Bicentric, randomized, double-blind, placebo-controlled trial 6 m. 190 subjects (71.1%) **Int**: 92 **Ctrl**: 98	Older subjects with lower level of memory performance 64.7 (2.9) y.	**Int**: 2 × 300 mg/d PEGB capsules (258 mg flavonoids) **Ctrl**: placebo capsules	*At baseline and at the 6- m. visit* • CANTAB-PALTEA test • VRMFR • VRMR • SSP test • Reverse SSP test	Urine excretion of phenolic metabolites (at baseline and at the 6-m. visit) Use of UHPLC-MS/MS **(+)-catechin, (-)-epicatechin, (epi) catechin-G, (epi)catechin-S, (epi) catechin-G-S, (epi)catechin methyl-G, (epi)catechin methyl-S, β-type procyanidins dimers, hydroxyphenyl-γ-valerolactone, dihydroxyphenyl-γ-valerolactone, 4-hydroxyphenylacetic acid, 3-hydroxybenzoic acid, gallic acid** Int: M0 < M6; $p < 0.001$ Ctrl: M0 = M6	• Almost significant association between changes in urinary concentration of biomarkers of PEGB consumption (flavan-3-ols) and changes in the performances at the VRMFR test • Association between variation of urinary concentration of native and conjugated derivatives of (epi)catechins and increase in the immediate VRMFR total correct responses • Association between variation of (epi) catechins and their conjugates and immediate VRMFR in decliners[a] • Association of change in urinary concentration of PEGB biomarkers (flavan-3-ols), (epi)catechins, and their conjugated metabolites with the VRMFR-delayed recognition total correct responses
Igwe et al. (33)	Randomized, crossover trial 4 m. with a 4-wk washout period 31 subjects **Int→Ctrl sequence**: 16 (63%) **Ctrl→Int sequence**: 15 (47%)	Healthy older subjects 70.0 y. **Int→Ctrl sequence**: 69.0 (10.0) y. **Ctrl→Int sequence**: 70.0 (4.8) y.	**Int**: 200 mL/d QGP nectar (3.7–5.3 mg C3G eq./100 mL) **Ctrl**: raspberry cordial beverage	*At baseline and at the 4-, 8-, 16- and 20-wk visits (4-wk washout period)* • RAVLT • Verbal fluency task • Digit-span backwards task • Stroop task • Counting span	Urine identification of anthocyanins and their common conjugated and methylated metabolites (at baseline and at the 4-, 8-, 16- and 20-wk visits) Use of UHPLC-PDA-ESI-MS/MS **Cyanidin-3-glucoside, cyanidin-G, cyanidin-3-rutinoside, peonidin-G, peonidin-3-rutinoside** Detected but below the limits of quantification	No significant effect of anthocyanin on a battery of cognitive tests

(Continued)

TABLE 10.1 *(Continued)*

Summary of the Characteristics, Designs and Results of the Flavonoids-Based Studies Reviewed

Reference (Year)	Study Design/ Duration/N (% Women)	Subjects' Characteristics/ Age (SD)	Intervention	Cognitive Measurements	Polyphenol Measurements	Main Results
Bergland et al. (30)	Open-label, pilot trial 4 m. 38 subjects **Int**: 27 (33%) **Ctrl**: 20 (55%)	Older subjects with MCI or mild dementia and/or stable non-obstructive coronary artery disease 59.7 y. **Int**: 61.0 (11.1) y.[b] **Ctrl**: 58.0 (5.2) y.[b]	**Int:** 4 × 80 mg/d anthocyanin capsules (320 mg anthocyanins) **Ctrl:** –	*At baseline and the 16-wk visit* • CERAD battery (learning, recall, and recognition) • TMT-A • TMT-B • Stroop Golden (word and color)	Plasma levels of 29 anthocyanin metabolites (at baseline and at the 16-wk visit) Use of LC-MS **Dihydroferulic acid-4-O-S** ΔM0-M4: Int < Ctrl; $p = 0.010$ Int: M0 > M4; $p < 0.05$ **o-Coumaric acid** ΔM0-M4: Int < Ctrl; $p = 0.019$ Int: M0 > M4; $p < 0.05$ Ctrl: M0 > M4; $p < 0.05$ **4-Hydroxybenzaldehyde** Int: M0 > M4; $p < 0.05$ Ctrl: M0 > M4; $p < 0.05$ **Pyrogallol-2-O-S, protocatechuic acid-3-O-S, pyrogallol-1-O-S, ferulic acid-4-O-β-D-G, isoferulic acid-3-O-β-D-G** Int: M0 < M4; $p < 0.05$ **3-Hydroxyhippuric acid, m-coumaric acid** Int: M0 > M4; $p < 0.05$ **Protocatechuic acid, 3,4-dihydroxyphenylacetic acid** Ctrl: M0 > M4; $p < 0.05$	• Improvements in the intervention group for CERAD learning, recall, and recognition and for STROOP test word and color

Krikorian et al. (26)	Randomized, double-blind, placebo-controlled trial 4 m. 37 subjects **Int**: 16 (57%) **Ctrl**: 21 (53%)	Older subjects with MCI 77.1 y. **Int**: 77.3 (6.3) y. **Ctrl**: 77.0 (5.4) y.	**Int:** 2 × 12 g/d whole freeze-dried blueberry fruit powder (401 GAEs, 258 mg C3G eq.) **Ctrl:** placebo powder	*At baseline and the 16-wk visit* • TMT-A • TMT-B • Controlled Oral Word Association task • HVLT • SPAL • DEX	Urine levels of anthocyanins (after overnight fast and before the morning dose at baseline and the 8-wk visit) Use of LC-MS/MS **Total anthocyanidin glycosides** $p_{group} \times_{time} = 0.002$ **Parent (food) form** $p_{group} \times_{time} < 0.001$	• Better SPAL and Controlled Oral Word Association task scores for the intervention group • Greater TMT-A trend for the intervention group • No differences in TMT-B, HVLT, and DEX • Correlation between effects on cognition (semantic access and visual-spatial memory) and urinary content of parent anthocyanins, glycoside anthocyanin conjugates, and total glycosidic anthocyanin forms
Sloan et al. (34)	Randomized, double-blind, controlled parallel-arm trial 3 m. with an 8-wk washout period 211 subjects **Int 1**: 53 (57%) **Int 2**: 53 (57%) **Int 3**: 53 (56%) **Ctrl**: 53 (59%)	Healthy, sedentary, normotensive older adults 62.0 y. **Int 1**: 61.9 (6.1) y. **Int 2**: 61.7 (6.7) y. **Int 3**: 62.1 (6.4) y. **Ctrl**: 62.4 (6.7) y.	**Int 1**: 260 mg/d cocoa flavonols capsules **Int 2**: 510 mg/d cocoa flavonols capsules **Int 3**: 770 mg/d cocoa flavonols capsules **Ctrl**: placebo capsules	*At baseline and the 1-, 3-, and 5-m. visits* • Modified Benton Recognition Task • Modified RAVLT • List Sorting Working Memory Test (NIH Toolbox Cognition Battery)	Plasma levels of gVLM (at baseline and at the 12- and 20-wk visit) Use of HPLC-MS/MS **Sum of 5-(3′,4′-dihydroxyphenyl)-γ-valerolactone-3′-S and 5-(3′,4′-dihydroxyphenyl)-γ-valerolactone-3′-O-G** M0: Int 1 = Int 2 = Int 3 = Ctrl; $p = 0.0617$ Blood tests after 12- and 20-wk revealed an intake amount-dependent increase and then a decrease in gVLM	• Modified Benton Recognition Task: no correlation with gVLM at baseline. No improvement after flavanol intake at 12-wk • Modified RAVLT: correlation with gVLM at baseline. Significant intake level dependent treatment effect compared to placebo after 12-wk of intervention. • List Sorting Working Memory Test: no correlation with gVLM at baseline. No improvement after flavanol intake at 12-wk

(Continued)

TABLE 10.1 *(Continued)*
Summary of the Characteristics, Designs and Results of the Flavonoids-Based Studies Reviewed

Reference (Year)	Study Design/ Duration/N (% Women)	Subjects' Characteristics/ Age (SD)	Intervention	Cognitive Measurements	Polyphenol Measurements	Main Results
Wood et al. (35)	Randomized, double-blind, controlled parallel trial 3 m. with a 4-wk washout period 61 subjects **Int**: 32 (63%) **Ctrl**: 29 (59%)	Healthy older individuals 70.1 y. **Int**: 69.4 (3.5) y. **Ctrl**: 70.8 (3.8) y.	**Int:** 26 g/d freeze-dried wild blueberry powder (302 mg anthocyanins) **Ctrl:** matched placebo (0 mg anthocyanins)	*At baseline and the 3- and 4-m. visits* • AVLT • Corsi Blocks • Serials Subtraction Tasks (3 s and 7 s) • TST	Plasma and urine levels of 87 polyphenol metabolites (at baseline and at the 3- and 4-m. visits) Use of LC-MS <u>Plasma</u> **Pyrogallol-O-S** M3: Int > Ctrl; $p = 0.017$ **2-Methylpyrogallol-O-S** M3: Int > Ctrl; $p = 0.042$ **4-Methylcatechol-O-S** M3: Int > Ctrl; $p = 0.028$ **4-Methylcatechol** M3: Int > Ctrl; $p = 0.011$ **Isoferulic acid** M3: Int > Ctrl; $p < 0.001$ <u>Urine</u> **Isovanillic acid 3-O-S** $p = 0.001$ **4-Feruloylquinic acid** $p = 0.001$ **Homovanillic acid-S sodium salt** $p = 0.002$ **(R)-(+)-2-(4-hydroxyphenoxy)-propionic acid** $p = 0.003$ **3-(2,3-Dihydroxyphenyl)propionic Acid** $p = 0.003$ **Isoferulic acid** $p = 0.02$ **3-Feruloylquinic acid** $p = 0.02$ **4-Methylcatechol** $p = 0.02$ **Ferulic acid 4-O-β-D-G** $p = 0.03$ **Catechol-O-1-G** $p = 0.03$	• Enhanced immediate recall on the AVLT after the intervention compared to placebo • Better delayed recall on the AVLT after placebo compared to treatment • Better accuracy on a TST after intervention compared with placebo • No differences in any other AVLT measure or cognitive outcome. • Changes for immediate recall positively correlated with plasma pyrogallol-S, 2,6-dihydroxybenzoic acid, isoferulic acid, 3-(2′,3′-dihydroxyphenyl) propanoic acid, benzoic acid, 2,5-dihydroxybenzoic acid, and 1-methylpyrogallol-O-S; and negatively correlated with plasma phenylacetic acid • Changes for delayed recall positively correlated with plasma dihydroisoferulic acid 3-G, dihydroisoferulic acid 3-O-S, (ferulic acid 4-G), and 2-hydroxyhippuric acid; and negatively correlated with plasma isovanillic acid 3-O-S, and quercetin-7-G • Changes for TST accuracy, positively correlated with plasma quercetin-7-G; and negatively correlated with plasma 2,4/2,6-dihydroxybenzoic acid

Curtis et al. (32)	Randomized, double-blind, controlled parallel trial 6 m. with a 3-wk washout period 115 subjects **Int 1**: 39 (28%) **Int 2**: 37 (35%) **Ctrl**: 39 (33%)	Subjects with overweight and obesity (BMI ≥25), and MetS (≥3 MetS components) 62.8 y. **Int 1**: 62.6 (7.2) y. **Int 2**: 63.0 (5.9) y. **Ctrl**: 62.9 (8.1) y.	**Int 1:** 13 g freeze-dried blueberries (182 mg anthocyanins) + 13 g placebo material **Int 2:** 26 g freeze-dried blueberries (364 mg anthocyanins) **Ctrl:** 26 g matched placebo powder	*At baseline and at the 6-m. visit preceding and 1 h and 4h after energy-dense drink* Single computerized test battery containing: • Simple Reaction Time • Digit Vigilance • Choice Reaction Time • Numeric Working Memory • Spatial Working Memory • Immediate Working Recall • Delayed Word Recall • Word Recognition • Picture Recognition • Executive Function Task	Serum and urine levels of 72 anthocyanin metabolites (serum: 4 h after energy-dense drink; urine: 24 h before/after energy-dense drink) Use of HPLC-MS	• No improvement in any cognitive function after anthocyanin intake at 6-m (both chronic and in postprandial substudy) • Attention negatively correlated with urine 3-hydroxybenzoic acid (chronic timeframe) and serum hydroxymethoxybenzoic acid-S (postprandial timeframe); and positively correlated with benzoic acid-4-S (postprandial timeframe) • Memory positively correlated with serum 3/4-methoxybenzoic acid-4/3-S (chronic timeframe) and 3-methoxyphenylacetic acid-4-S (postprandial timeframe) and negatively correlated with urinary 3,4-dihydroxyphenylacetic acid (postprandial timeframe) and chlorogenic acid (chronic timeframe) • Executive function positively correlated with urinary 4-hydroxybenzoic acid and benzoylglutamic acid (postprandial timeframe) • Picture recognition positively correlated with urinary 3-hydroxyhippuric acid and hippuric acid (chronic timeframe)
Vauzour et al. (29)	Randomized, double-blind, controlled parallel trial 12 m. 246 subjects **Int**: 125 (55%) **Ctrl**: 121 (59%)	Subjects with subjective cognitive impairment or MCI 65.5 y. **Int**: 66.0 (6.3) y. **Ctrl**: 65.0 (6.7) y.	**Int:** 3 × 1 g oil capsules (1.1 g/d DHA and 0.4 g/d EPA) and 33 g chocolate drops (500 mg/d flavan-3-ols) **Ctrl:** 3 × 1 g oil capsules (80% palm oil and 20% corn oil) and 33 g chocolate drops (isocaloric + 500 mg/d flavan-3-ols)	*At baseline and the 3- and 12-m. visits* • CDR test battery (Bracket Cognition) • MoCA • Delis–Kaplan Executive Function System • Verbal Fluency Test (0- and 12-m only) • iPosition task	Urine levels of 43 flavanol metabolites (at baseline and the 3- and 12-m. visits) Use of UHPLC-MS **5-(Phenyl)-γ-valerolactone-3′-S** Ctrl: M0 = M3 = M12 Int: M0 < M3, M12 $p_{group} \times {}_{time} = 0.073$ **Hydroxyphenyl-γ-valerolactone-sulfate** Ctrl: M0 = M3 = M12 Int: M0 < M3, M12 **5-(3′-Hydroxyphenyl)-γ-valerolactone-4′-O-G** Ctrl: M0 = M3 = M12 Int: M0 < M3, M12 $p_{group} \times {}_{time} = 0.523$	• No improvement in any cognitive outcome after combined intervention • Exception: decline in function for reaction time variability and executive function in the intervention group at baseline vs 12-m relative to the control • Changes to urinary flavonoid metabolite concentrations did not correlate with changes in cognitive function

(Continued)

TABLE 10.1 *(Continued)*

Summary of the Characteristics, Designs and Results of the Flavonoids-Based Studies Reviewed

Reference (Year)	Study Design/ Duration/N (% Women)	Subjects' Characteristics/ Age (SD)	Intervention	Cognitive Measurements	Polyphenol Measurements	Main Results
Yuan et al. (37)	Case-control 276 subjects **Cases:** 138 (64.5%) **Controls:** 138 (64.5%)	Older subjects with MCI (cases diagnostic criteria: >13/14 points for individuals with no formal education, 19/20 for individuals with 1-6 years of education and 24/25 for individuals with ≥7 years of education) 64.5 y. **Cases**: 65 (0.5) y. **Controls**: 64 (0.6) y.	*N.A.*	*At baseline* MoCA (MCI diagnosis)	*Method:* LC-MS *At baseline:* Plasma flavonoids (luteolin, quercetin, naringenin, apigenin, kaempferol, hesperetin) levels	No difference in plasma flavonoid levels between the MCI and control subjects
Igase et al. (38)	Cross-sectional, observational 152 subjects **Cases**[b]: 60 (74.0%) **Controls**[b]: 92 (66.0%)	Older healthy subjects (Case or equol-producer diagnostic criteria: urine equol level > 1000 nmol/L after intake of 2 × 50 mg isoflavones [soy food]) 69.2 y. **Cases**[b]: 69.1 (9.4) y. **Controls**[b]: 69.1 (9.0) y.	*N.A.*	*At baseline* • TDAS • MCI screening test	*Method:* immunochromatographic strip test *After isoflavones intake:* urine excretion of equol	• Higher score in TDAS cognitive test among equol producers compared to non-equol producers • Lower prevalence of MCI among equol producers compared to non-equol producers

Note: AD, Alzheimer's disease. AVLT, Auditory Verbal Learning Test. BMI, body mass index. BNT, Boston Naming Test. BSRT, Buschke Selective Reminding Test. BVRT, Benton Visual Retention Test. CANTAB, Cambridge Neuropsychological Test Automated Battery. CD, Cognitive Decline. Ctrl, Control. CVLT, California Verbal Learning Test. C3G eq., Cyanidin-3-Glucoside equivalents. DEX, Dysexecutive Questionnaire. EBMT, East Boston Memory Test. GAEs, Gallic Acid Equivalents. GC-MS, Gas Chromatography-Mass Spectrometry. GPT, Grooved Pegboard Test. gVLM, 5-(3′,4′-dihydroxyphenyl)-γ-Valerolactone Metabolites. HPLC-DAD/MS, High-Performance Liquid Chromatography with photodiode array-mass spectrometry. HPLC-MS/MS, High-Performance Liquid Chromatography-tandem mass spectrometry. HVLT, Hopkins Verbal Learning Test. IQR, Interquartile Range. Int, Intervention Group. LC-MS/MS, Liquid Chromatography-tandem Mass Spectrometry. m., month(s). MCI, Mild Cognitive Impairment. MetS, Metabolic Syndrome. MMSE, Mini-Mental Status Examination. MoCA, Montreal Cognitive Assessment. NA, Not Applicable. NIH, National Institutes of Health. PALTEA, Paired Associate Learning Total Errors Adjusted. PEGB, Polyphenol-rich Extract from Grape and Blueberry. QGP, Queen Garnet Plum. RAVLT, Rey Auditory Verbal Learning Test. RCFT, Rey Complex Figure Test. SCWT, Stroop Color and Word Test. SD, Standard Deviation. SDMT, Symbol Digit Modalities Test. SPAL, Spatial Paired Associate Learning Test. SSP, Spatial Span or Speech Sounds Pick. TDAS, Touch Panel-type Dementia Assessment Scale. TMT-A, Trail Making Test part A. TMT-B, Trail Making Test part B. TST, Task-Switching Test. UHPLC-MS/MS, Ultra-High-Performance Liquid Chromatography-tandem Mass Spectrometry VRMFR, Verbal Recognition Memory-Free Recall. VRMR, Verbal Recognition Memory-Recognition. VSLT, Visual Spatial Learning Test. WAIS, Wechsler Adult Intelligence Scale. wk, week. WTAR, Wechsler Test of Adult Reading. y., years. -G, Glucuronide. -S, sulfate

[a] Decliners: subgroup with advanced CD derived from the cohort stratification in quartiles based on PAL score at baseline.

[b] Cases were defined as equol producers, and controls were defined as equol non-producers.

TABLE 10.2
Summary of the Characteristics, Designs, and Results of the Phenolic Acids-Based Studies Reviewed

Reference (Year)	Study Design/ Duration/N (% Women)	Subjects' Characteristics/ Age (SD)	Intervention	Cognitive Measurements	Polyphenol Measurements	Main Results
Ochiai et al. (23)	Randomized, double-blind, placebo-controlled crossover trial 3 m. 34 subjects (56%) **Int-Ctrl sequence**: 17 **Ctrl-Int sequence**: 17	Older MCI patients 73.7 (6.0) y.	**Int**: 2 × 553.6 mg/d chlorogenic acids/100 mL water **Ctrl**: placebo beverage	*At baseline and the 12- and 28-wk visits (4-wk washout period)* • MMSE • ADAS-Cog • TMT-A • TMT-B	Plasma levels of chlorogenic acids (at baseline and at the 12- and 28-wk visits) Use of LC-ESI-MS/MS **3-caffeoylquinic acid** Int = Ctrl; $p = 0.691$ **4-caffeoylquinic acid** Int = Ctrl; $p = 0.386$ **5-caffeoylquinic acid** Int = Ctrl; $p = 0.150$ **3-feruloylquinic acid** Int = Ctrl; $p = 0.705$ **4-feruloylquinic acid** Int = Ctrl; $p = 0.827$ **5-feruloylquinic acid** Int = Ctrl; $p = 0.264$ **Dicaffeoylquinic acids** not detected	• Reduction in TMT-B errors in the active beverage condition compared to the placebo group after a 12-wk period • No difference between groups in test completion time • No significant change was between groups in the times taken to complete the MMSE, ADAS-Cog, and TMT-A • No significant correlation between change in blood chlorogenic acids levels and number of TMT-B errors before and after intake of the active beverage

(Continued)

TABLE 10.2 *(Continued)*
Summary of the Characteristics, Designs, and Results of the Phenolic Acids-Based Studies Reviewed

Reference (Year)	Study Design/ Duration/N (% Women)	Subjects' Characteristics/ Age (SD)	Intervention	Cognitive Measurements	Polyphenol Measurements	Main Results
Rutledge et al. (39)	Randomized, double-blind, placebo-controlled, parallel study 3 m. 38 subjects **Int**: 19 (72%) **Ctrl**: 19 (63%)	Older healthy subjects 67.6 y. **Int**: 67.8 (4.6) y. **Ctrl**: 67.3 (4.8) y.	**Int**: 24 g/d lyophilized, cultivated blueberries (36 mg total phenolics, 19.2 mg anthocyanins) **Ctrl**: placebo powder	*At baseline and the 45- and 90-day visits* • TST *At baseline and the 90-day visit* • CVLT-II	Plasma levels of anthocyanins and phenolic acid metabolites (after overnight fast and after breakfast + dose at baseline and the 45- and 90-day visits) Use of UHPLC-TQMS **Hippuric acid** $p_{group} \times_{time\ (fasting)} \leq 0.005$; $p_{group} \times_{time\ (postprandial)} \leq 0.005$ **Phloroglucinaldehyde, syringic acid; ferulic acid-G; cyanidin-3-O-galactoside; cyanidin-3-O-glucoside; malvidin-3-O-galactoside; malvidin-3-O-glucoside; peonidin-3-O-xyloside peonidin-G; petunidin-3-O-glucoside** $p_{group} \times_{time\ (postprandial)} \leq 0.005$	• Negative association between change in ferulic acid-G and change in task switching errors • Association between change in malvidin-3-O-galactoside and change in switching errors • No significant association between syringic acid and change in switching errors • Association between change in concentration of syringic acid and change in repetition errors • Association between change in concentration of ferulic acid-G and change in repetition errors • Negative association between change in hippuric acid and change in repetition errors

Tor-Roca et al. (40)	Nested case-control 12 years 418 subjects (discovery set) **Cases**: 209 (66.0%) **Controls**: 209 (66.0%) 422 subjects (validation set) **Cases**: 211 (63.0%) **Controls**: 211 (63.0%)	Older subjects with CD (cases diagnostic criteria: worst global cognition slope) 76.0 y. Discovery set: **Cases**: 75.9 (4.5) y. **Controls**: 75.7 (4.2) y. Validation set **Cases**: 76.5 (5.2) y. **Controls**: 76.2 (4.7) y.	*N.A.*	*At baseline and every 2–3 years of follow-up* Average z-score: • MMSE • TMT-A • TMT-B • BVRT • Isaac's Set Test	*Method:* UHPLC-QTRAP- MS *At baseline:* Serum levels of polyphenols and fatty acids biomarkers related to MD key food groups. Creation of an MD-based metabolomic score	• Inverse association between MD-based metabolomic score and CD in the discovery set • Comparable results in the validation set (trend toward significance)

Note: ADAS-Cog, Alzheimer's disease Assessment Scale-Cognitive Subscale. BVRT, Benton Visual Retention Test. CD, cognitive decline. Ctrl, control. CVLT, California Verbal Learning Test. HPLC-TQMS, High-performance liquid chromatography-triple quadrupole mass spectrometry. LC-ESI-MS/MS, Liquid chromatography-electrospray ionization-tandem mass spectrometry. m., month(s). MD, Mediterranean diet. MMSE, Mini-Mental Status Examination. N.A., Not applicable. SD, standard deviation. TMT-A, Trail Making Test part A. TMT-B, Trail Making Test part B. TST, Task-Switching Test. UHPLC-MS/MS, Ultra-high-performance liquid chromatography-tandem mass spectrometry. UHPLC-TQMS, Ultra-high-performance liquid chromatography-triple quadrupole mass spectrometry. wk, week. y., years. -G, glucuronide.

TABLE 10.3
Summary of the Characteristics, Designs, and Results of the Other Polyphenol-Based Studies Reviewed

Reference (Year)	Study Design/ Duration/ N (% Women)	Subjects' Characteristics/ Age (SD)	Intervention	Cognitive Measurements	Polyphenol Measurements	Main Results
Ringman et al. (42)	Randomized, double-blind, placebo-controlled trial 6 m. (with open-label extension to 12 m.) 30 subjects **Int 1:** 9 (67%) **Int 2:** 10 (70%) **Ctrl:** 11 (55%)	Older subjects with mild-to-moderate AD 73.9 y. **Int 1:** 76.7 (5.6) y. **Int 2:** 75.3 (6.9) y. **Ctrl:** 70.2 (12.4) y.	**Int 1:** 2 g/d Curcumin C3 Complex® capsules **Int 2:** 4 g/d Curcumin C3 Complex® capsules **Ctrl:** placebo capsules	*At baseline and the 6- and 12-m. visits* *Primary outcome:* • ADAS-Cog *Secondary outcomes:* • NPI • ADCS-ADL scale • MMSE	Plasma levels of curcumin, tetrahydrocurcumin, curcumin-G, tetrahydrocurcumin-G Use of LC-MS/MS **Curcumin** M0: not detected M6 pre-admin.: Int > Ctrl (not detected) M6 3-h post-admin: Int > Ctrl (not detected) **Tetrahydrocurcumin** M0: not detected M6 pre-admin.: Int > Ctrl (not detected) M6 3-h post-admin: Int > Ctrl (not detected) **Curcumin-G** M0: not detected M6 3-h post-admin: Int > Ctrl (not detected) **Tetrahydrocurcumin-G** M0: not detected M6 3-h post-admin: Int > Ctrl (not detected)	• No differences in ADAS-Cog, NPI, ADCS-ADL, or MMSE scores between groups

Bookheimer et al. (41)	Randomized, double-blind, placebo-controlled trial 1 m. 28 subjects **Int**: 15 (73.3%) **Ctrl**: 13 (76.9%)	Non-demented, older right-handed subjects with self-reported age-related memory complaints 62.7 y. **Int**: 63.1 (8.0) y. **Ctrl**: 62.0 (7.8) y.	**Int**: 236.59 mL PJ/d **Ctrl**: placebo drink	*At baseline and the 28-day visit* • Buschke-Fuld selective reminding test • Experimental unrelated word pair associates learning task, based on the Wechsler memory scale paired associates learning • Functional MRI with: • Verbal memory task • Visual memory task	Plasma levels of Uro-A-G (at baseline and at the 1-m. visit) Use of LC-MS Int: M0 < M1, $p < 0.001$ Ctrl: M0 < M1; $p \geq 0.05$	• Significantly better memory performance after intervention in PJ group but not in the placebo group. • Greater t2 vs t1 improvement in total recall measure in the PJ group compared to the placebo group • Greater consistent long-term retrieval score in the PJ group compared to the placebo group.

Note: AD, Alzheimer's disease. ADAS-Cog, Alzheimer's disease Assessment Scale-Cognitive Subscale. ADCS-ADL, Alzheimer's disease Cooperative Study-Activities of Daily Living. Ctrl, control. Int, intervention group. LC-MS/MS, Liquid chromatography-tandem mass spectrometry. m., month(s). MMSE, Mini-Mental Status Examination. MRI, magnetic resonance imaging. NPI, Neuropsychiatric Inventory. PJ, pomegranate juice. SD, standard deviation. Uro, urolithin. y., years. -G, glucuronide.

TABLE 10.4
Summary of the Characteristics, Designs, and Results of the Urinary Total Polyphenol-Based Studies Reviewed

Reference (Year)	Study Design/Duration/ N (% Women)	Subjects' Characteristics/ Age (SD)	Cognitive Measurements	Polyphenol Measurements	Main Results
Rabassa et al. (43)	Prospective cohort 3 years 652 subjects (55.4%)	Healthy older subjects (CD diagnostic criteria: Decrease in MMSE score of ≥3 points from baseline to 3-year follow-up and the worst 10% of the TMT-A and TMT-B distribution after subtracting baseline from 3-year follow-up scores [in seconds] or discontinuation of the test at follow-up) 73.4 (6.4) y.	*At baseline and after 3 years of follow-up* • MMSE • TMT-A; TMT-B	*Method:* Folin-Ciocalteu assay *At baseline:* Urine excretion of total polyphenols (expressed as mg GAEs/d)	• Association between TUP levels and lower risk of substantial CD on the MMSE and on the TMT-A, comparing extreme tertiles and in linear regression models • No association between TUP levels and risk of substantial CD on the TMT-B
Valls-Pedret et al. (45)	Cross-sectional in a longitudinal cohort 447 subjects (52%)	Older subjects at high cardiovascular risk 66.9 (6.38 [a]) y.	*At baseline* • MMSE • RAVLT (immediate recall and delayed recall) • Verbal paired associates test • SVF test • Digit span test, from WAIS test • CTT, part I and II	*Method:* Folin-Ciocalteu assay *At baseline:* Urine excretion of total polyphenols (expressed as mg GAEs/g of creatinine)	• Association between TUP excretion and immediate memory recall score (RAVLT immediate recall) – dose-relationship • Almost significant association between TUP excretion and delayed memory recall score (RAVLT delayed recall) – dose-relationship • No association between TUP excretion and other cognitive tests
Rabassa et al. (44)	Prospective cohort 3 years 119 subjects (47.9%)	Healthy older subjects (CD diagnostic criteria: decrease in MMSE score of ≥3 points from baseline to 3-year follow-up) 73.0 (5.9) y.	*At baseline and after 3 years of follow-up* MMSE	*Method:* HPLC-Q-ToF-MS *At baseline:* Urine metabolome	• Association between change in cognitive function (continuous variable) and urine excretion of Uro-A, Uro-A G, Uro-A-S, Uro-A-G-S, Uro-B, and Uro-B-G • Inverse association between cognitive decline (dichotomous variable) and urine excretion of Uro-A, Uro-A-S, Uro-B, Uro-B-G, and hydroxyphenylacetic acid

Note: CD, Cognitive decline. CTT, Color Trail Test. GAEs, Gallic Acid Equivalents. MMSE, Mini-Mental State Examination. RAVLT, Rey Auditory Verbal Learning Test. SD, standard deviation. SVF test, Semantic verbal fluency test. TMT-A, Trail Making Test, part A. TMT-B, Trail Making Test, part B. TUP, Total Urinary Polyphenols. Uro, urolithin. WAIS, Wechsler Adult Intelligence Scale.

[a] SD not reported and estimated as one quarter of the typical range of data values (119).

neurocognitive benefits (26). Anthocyanin supplementation also showed positive changes in cognitive test scores for Consortium to Establish a Registry for AD (CERAD) learning ($p = 0.016$), recall ($p < 0.001$), and recognition ($p = 0.047$), and for the STROOP word ($p < 0.001$) and color ($p = 0.044$) test, in people with increased risk of dementia followed for 16 weeks (30).

In healthy participants, John et al. reported that long-term changes in urinary isoflavonoids were not associated with global cognition or episodic memory (verbal episodic memory [i.e., list learning and logical memory] factors and visual episodic memory factors). However, greater isoflavonoid exposure was associated with decrements in the general intelligence factor (31). Interestingly, this inverse association was of smaller magnitude in women who had experienced natural menopause than in those who had undergone surgical menopause (31). The study of Igwe et al. also did not show a significant acute effect on various domains of cognitive function after a 300 mL serving of anthocyanin-rich Japanese plum juice, regardless of its two dose-timing consumption regimens (27). In adults with metabolic syndrome, cognitive function parameters were also unaffected after a postprandial and six-month blueberry intervention (32). However, an association was observed between microbial metabolites of anthocyanins and chlorogenic acid (i.e., hydroxycinnamic acids, benzoic acids, phenylalanine derivatives, and hippuric acids) and catechin with favorable chronic and postprandial memory, attention, and executive function (32). Likewise, no improvements were observed in verbal learning, short-term memory, or executive function following daily consumption of 200 mL of low-dose anthocyanin-rich plum nectar in healthy older adults (33). Sloan et al. showed that object-recognition and list-sorting performance did not improve after cocoa flavanol supplementation for 12 weeks. However, list-learning performance showed improvement in participants in the bottom tertile of baseline diet quality (34).

On the other hand, Gleason et al. showed that consuming 100 mg/d of soy isoflavones improved the performance in tests of visuospatial memory ($p < 0.01$) and construction ($p = 0.01$), verbal fluency ($p < 0.01$), and speeded dexterity ($p = 0.04$). In contrast, placebo-treated participants were significantly faster in executive function ($p < 0.05$) (24). Moreover, in healthy older individuals, enhanced immediate recall in the auditory verbal learning task, alongside better accuracy in a task-switch task, was also found following wild blueberry treatment (302 mg anthocyanins) compared with placebo ($p < 0.05$) (35).

Bensalem et al. also revealed in their study that urinary concentrations of native and conjugated derivatives of (epi)catechins from a polyphenol-rich extract from grape and blueberry (PEGB) were strongly associated with episodic verbal recall memory improvement [$p = 0.016$ and $p = 0.010$ for native and conjugate derivatives of (epi)catechins, respectively] (36). Similar associations in this outcome and in delayed verbal recognition memory (VRM) were found when focusing only on cognitive decliners [$p = 0.036$ and $p = 0.053$ in episodic verbal recall memory for native and conjugate derivatives of (epi)catechins, respectively; $p = 0.032$, $p = 0.055$, and $p = 0.032$ in VRM-delayed for flavan-3-ol biomarkers of PEGB, and native and conjugate derivatives of (epi)catechins, respectively] (36).

With regard to observational studies, no significant differences in plasma flavonoid concentrations were detected between adults with MCI and control subjects ($p > 0.05$) (37, 38), although compared to controls, lower plasma total antioxidant capacity was detected in patients with MCI ($p < 0.05$) (37). Conversely, a cross-sectional study evaluating cognitive function according to equol-production status showed that, compared with equol non-producers, equol producers were less susceptible to suffering from MCI (38). In fact, regression analysis to assess the presence of MCI showed that being an equol non-producer was an independent risk factor for MCI (OR = 3.961, 95% CI = 1.048–14.969, $p = 0.042$) (38).

Phenolic Acids

The study of Ochiai et al. examined the effect of beverages containing chlorogenic acids on older subjects with MCI (23). In the Trail Making Test – part B (TMT-B; attention and executive memory), participants produced significantly fewer errors when consuming the active beverage as compared to the placebo beverage ($p < 0.05$). There was still no significant correlation between a change in blood chlorogenic acid levels and the number of TMT-B errors before and after intake of the active beverage ($r = 0.011$, $p = 0.571$) (23). The authors thus concluded that the effect of chlorogenic acids on cognitive function was probably not related to chlorogenic acid levels in the blood. Scores in the Mini-Mental State Examination (MMSE), AD assessment scale-cognitive (ADAS-Cog), and Trail Making Test – part A (TMT-A) did not significantly differ between groups (23).

Phenolic acids from a blueberry powder intervention showed conflicting results as a change in plasma concentrations of ferulic acid glucuronide was negatively associated with a change in switching errors ($ß = -0.150$, $p < 0.01$) but positively associated with a change in California Verbal Learning Test II (CVLT-II) repetition errors ($ß = 0.044$, $p < 0.001$). Furthermore, a change in the plasma concentrations of syringic acid was positively associated with a change in CVLT-II repetition errors ($ß = 0.070$, $p < 0.001$). In contrast, a change in hippuric acid was negatively associated with this ($ß < -0.001$, $p < 0.001$) (39).

However, in a nested case-control study over 12 years, a Mediterranean diet-based metabolomic score was associated with lower odds of CD in older adults (40). Interestingly, some of the phytochemicals composing the metabolomic score were phenolic acids derived from the gut microbiota, such as hippuric acid, 3,4-dihydroxybenzoic acid, and 3,4-dihydroxyphenylacetic acid (40).

Other Polyphenols

Experimental studies of other polyphenol classes and MCI obtained mixed results. Bookheimer et al. evaluated the effects of drinking 236.59 mL of pomegranate juice in older adults with pre-existing mild memory complaints (41). As a result, the group that received the pomegranate juice intervention performed significantly better in the Buschke Selective Reminding Test of verbal memory after four weeks (two-tailed $t = 2.8$, $p = 0.017$) in contrast to the placebo group, which did not report any difference (two-tailed $t = 1.1$, $p > 0.3$). Between groups, the improvement in memory after the intervention was significantly greater in the pomegranate juice group than in the placebo group, specifically in the total recall measure (two-tailed $t = 2.3$, $p = 0.029$) as well as in consistent long-term retrieval (two-tailed $t = 2.4$, $p = 0.022$). After the intervention, a significant increase in plasma trolox-equivalent antioxidant capacity and urolithin A-glucuronide was observed only in the pomegranate group (41). Elsewhere, in another intervention study evaluating curcumin and the efficacy of its metabolites (tetrahydrocurcumin, glucuronidated curcumin, and glucuronidated tetrahydrocurcumin) in improving cognitive tasks in individuals with AD, investigators were not able to evidence differences between treatment groups (2 and 4 g/d of curcumin) and placebo-treated subjects in the severity of cognitive and non-cognitive symptoms of dementia (ADAS-Cog, Neuropsychiatric Inventory, AD Cooperative Study Activities of Daily Living [ADCS-ADL] and MMSE tests) (42). When combining the 2 and 4 g/d curcumin groups, there was a trend for placebo-treated participants to perform better than the intervention group in the MMSE ($p = 0.08$).

Urinary Total Polyphenols

Studies evaluating urinary total polyphenol (UTP) concentration showed a beneficial relationship between CI and polyphenol excretion in older individuals. In fact, Rabassa et al. showed in their three-year study that higher UTP levels were associated with a lower risk of CD, according to the MMSE ($OR_{\text{extreme tertiles}} = 0.53$, 95% CI = 0.34–0.85; *p-trend* = 0.008), and with enhanced performance in visual attention and processing speed ($OR_{\text{TMT-A}} = 0.52$, 95% CI = 0.28–0.96, *p-trend* = 0.03), but not in executive abilities (TMT-B) (43). Urinary polyphenols derived from nut intake were also associated with better cognitive function (OR = 0.006, 95% CI = 0.001–0.011; $p = 0.029$) and a lower risk of CD (OR = 0.995, 95% CI = 0.991–0.999; $p = 0.016$) (44). Similarly, a cross-sectional trial in a longitudinal cohort at high cardiovascular risk concluded that UTPs were associated with better scores in immediate verbal memory ($\beta = 1.208$, 95% CI = 0.236–2.180; $p = 0.015$) but not in the MMSE test (45).

INFLUENCING FACTORS OF BIOMARKERS OF POLYPHENOL INTAKE ON COGNITIVE FUNCTION

Polyphenols, Bioavailability, and Cognition

The concentration of urinary polyphenols depends on their bioavailability. It may be affected by biological traits, absorption, and metabolism, which differ greatly among compounds and have a large intra- and inter-individual variability (17, 46, 47). In fact, aging is known to impair gastrointestinal physiology and function, which may compromise the capacity for absorption and tolerance of polyphenols (48–51). For instance, one study concluded that the urinary phenolic acid profile varied between younger and older adults after a polyphenol-rich meal (52). However, these findings differed from those of Igwe et al., who concluded that urinary anthocyanin/metabolite profiles in younger and older adults were similar (27).

Such gastrointestinal disruptions can lead, in turn, to an increase in intestinal permeability (53), boosting the mucosal inflammation and the low-grade systemic inflammation detected in older adults (i.e., the inflammaging process) (54). It has been suggested that increased intestinal permeability plays a part in multiple neurodegenerative disorders, such as Parkinson's disease and AD (55, 56). There is also evidence indicating the potential of polyphenols in regulating the gut barrier function and preventing the increase of intestinal permeability (57).

Polyphenols, Gut Microbiota, and Cognition

One of the most common causes behind the increase in intestinal permeability is the changes in the gut microbiota composition (58). During aging, the composition of gut microbiota changes (59) and is reduced in terms of species richness, while instability and inter-individual variability increase (60–62). Moreover, recent findings allow us to hypothesize that the reduction in the diversity of the microbiome (63–65) and the presence of a distinctive gut microbial signature in patients with dementia, CI, or AD are characterized by dysbiosis with differences in beta diversity, the reduction of short-chain fatty-acid-producing bacteria, and overrepresentation of Gram-negative bacteria (56, 66). It is important to note that inconclusive results regarding participants' ages were observed in the included studies in this chapter.

In fact, the specific role of the commensal microbial population in regulating immunity and inflammation beyond the gastrointestinal tract may represent a relevant influence on the neurodegeneration process: changes in the proteolytic processing of the amyloid precursor protein may result in the aggregation of the amyloid beta peptide, which might prompt an immune response that leads to neuroinflammation and neurodegeneration in AD, specifically interfering with neuroplasticity and memory consolidation (67).

This is particularly relevant since there is growing evidence concerning the two-way interaction between (poly)phenols and gut microbiota (68), i.e., that the gut microbiota plays a key role in the metabolism and absorption of polyphenols, converting them into bioavailable and bioactive metabolites and contributing to the beneficial effect attributed to these compounds (69–71). At the same time, phenolic compounds may improve immune response modulation, inhibit the growth of detrimental gut species, and suppress the adhesion of gut pathogens to the human gut tract (72–74). Therefore, the colon microbiome is relevant for the inter-individual variability in response to polyphenol consumption, leading to unequal polyphenol effectiveness in all subjects. Results from an intervention study with polyphenol-rich foods supported the idea that the low bioavailability of polyphenols in patients with increased intestinal permeability might be attributed to alterations in the gut microbial metabolism. Moreover, they suggested that microbiota-derived metabolites could be largely responsible for the biological activity of dietary polyphenols against age-related disrupted IP (75). In this chapter, only Igwe et al. and Wood et al. examined the effect of polyphenol intervention on microbial composition (33, 35). They concluded that, contrary to the extensive evidence, there was no difference in gut microbiota between intervention arms or before and after treatment.

Polyphenols in the Gut–Brain Axis through Metabolomic Studies

Despite the increasing scientific literature on the bioavailability of polyphenols in the digestive system, it remains unclear whether they can reach the brain (47). In animal research, several brain-available polyphenol metabolites have been identified, such as quercetin-3-O-glucuronide, malvidin-3-O-glucoside, 3-hydroxybenzoic acid, 3-(3′-hydroxyphenyl) propionic acid, 3′-O-methyl-epicatechin-5-O-β-glucuronide, and urolithins (76–80). It has been suggested that these metabolites have potential benefits in relation to the onset and progression of AD-type pathophysiology. Nevertheless, special attention should be given to polyphenol interventions, as it is still unclear whether chronic consumption compared to acute consumption could influence the detected levels (81).

Thus, a nutritional biomarker of total dietary polyphenols and polyphenol metabolites is essential for accurately evaluating polyphenol exposure to study their specific cognitive effects (19, 54). Indeed, polyphenol metabolome and endogenous metabolites modulated by polyphenol intake are being increasingly considered, given their beneficial health effects. One exploratory study identified more than 20 metabolites associated with age-related CD, six of which were food-derived (82). Although no polyphenol was determined, hydroxycinnamate intake was significantly correlated with atractyligenin glucuronide and cyclo(leucyl-prolyl), two metabolites from the serum signature. Hydroxycinnamic acid intake was also correlated with an unidentified ion, which was not in the signature but had a univariate odds ratio for CD of 0.90. Atractyligenin glucuronide and cyclo(leucyl-prolyl) were also significantly correlated with coffee consumption, one of the primary sources of polyphenols in European countries and Mexico (13, 83). In 16 out of the 23 studies included in this chapter, polyphenol metabolites, such as valerolactones, equol, dihydrogenistein, dihydrodaidzein, enterolactones, O-desmethylangolensin, peonidin-glycosides and peonidin-glucuronides, urolithin A-glucuronide, and glucuronidated tetrahydrocurcumin, among others, were identified and/or quantified. This is particularly relevant since various studies have indicated that some of the beneficial effects of some polyphenols might be due to their microbiota-derived metabolites (i.e., equol, urolithins, 3,4-DHPPA, and 3/4-HPPA) (80), especially considering the usual low bioavailability of the parent compounds (70), such as anthocyanins (84, 85). However, current evidence is ambiguous concerning whether the proportion of some metabolite producers changes with age (86–94).

Long-Term Consumption of Polyphenols and Cognition

The length of time needed to perceive the effect of polyphenols on CI has not been fully established. The studies used a wide range of time periods, regardless of their design. This inconsistency is due to the difficulty in recognizing whether the effect is led by parent polyphenols or their metabolites (95). For instance, McNamara et al. and Bergland et al. discussed the influence of the very long-term retention of anthocyanins in the discrimination between their native and metabolized forms, as well as the effect of having a substantial reservoir of anthocyanins from habitual consumption on absorption (28, 30). This could explain the lack of favorable results in the Igwe et al. trial, which conducted a one-time intervention (27). The mechanism of anthocyanins influencing cognition is still unclear, but it seems to depend on the exposure period. Prior studies did not find a significant acute effect of fruit anthocyanins on cognitive processes (96).

Moreover, only four RCTs included an open-label extension for two and six months to evaluate whether any cognitive effects of the treatment remained without consumption, given the secondary absorption of flavonoid-derived metabolites in the gut (24, 28, 34, 35). This is noteworthy as recent investigations demonstrated that non-extractable polyphenols metabolized by the gut microbiome led to a sustained release of bioactive metabolites with antioxidant and anti-inflammatory effects in preclinical models (97).

Matrix Effect on the Impact of Polyphenols on Cognition

The vehicle for providing polyphenols also warrants consideration since the presence or absence of a food matrix influences polyphenol kinetics and absorption. Thus, the same dose of the same phenolic compound can result in misleading conclusions when ingested in isolation or mixed with other polyphenols, nutrients, or phytochemicals. For example, anthocyanins have been shown to have a greater bioavailability when consumed together with other polyphenols (98–101). Six studies tested flavonoids in an intervention administered in capsules (24, 25, 29, 30, 34, 36), while another six offered them as a powder (26, 28, 31, 32, 35, 39). The remaining three provided participants with juices or soy (27, 33, 38). Phenolic acids were administered in both beverage and powder forms, whereas total polyphenols were estimated from the participants' baseline diet. Interestingly, studies that included participants with cognitive dysfunction and assessed polyphenols from fruits showed promising results (26, 28, 30, 36, 40, 41).

The Role of Background Diet in Polyphenols and Cognition Trials

Another important point to consider is the baseline diet, which was taken into account as an exclusion criterion in trials conducted by Bookheimer et al., Bensalem et al., and Vauzour et al. (29, 36, 41). Ten RCTs restricted polyphenol-rich foods and supplements before and/or during the intervention (24, 26–29, 31, 32, 36, 39, 41). Most of the investigations that included subjects with some memory impairment and controlled the background diet showed positive findings (26, 28, 36, 41). Disregarding the background diet control, those investigations performed in Mediterranean countries (specifically in France, Italy, and Spain) and Japan had positive results (23, 36, 38, 40, 43–45). This is consistent with previous systematic reviews and meta-analyses (102–104), where the Mediterranean diet was the most investigated dietary pattern, with evidence supporting its protection against CD among older adults. A positive effect of greater adherence to the Mediterranean diet on neurodegenerative diseases has been found, particularly on AD and dementia (102). A two-point increase in adherence to the Mediterranean diet, characterized by high consumption of plant-based food and poly- and mono-unsaturated fatty acids and low consumption of processed foods, was associated with a significant 13% reduction in neurodegenerative diseases (103). Tor-Roca et al. also showed that greater adherence to the Mediterranean diet, assessed by a serum metabolomic score, was associated with lower odds of CD in older adults (40). Characterized by a high intake of plant foods (soybean products, seaweed, vegetables, and fruits) and fish, and a decreased intake of refined carbohydrates and animal fats, the traditional Japanese diet could also evidence protective effects against CI (105, 106).

Inter-Individual Variability in the Metabolism of Polyphenols

However, the above-mentioned inconclusive outcomes could be partly explained by the sizeable inter-individual variability in the host metabolism, absorption, and excretion of polyphenols (18). For instance, previous research has suggested that genetic polymorphisms of the drug transporters MRP2 and OATP1B1 may contribute to the high variability in the pharmacokinetic parameters of catechins. This can, in turn, influence the real exposure to polyphenols and partly explain the inter-individual variation in the beneficial effects of polyphenols (107). Miller et al. evidenced differences in small vessel tone according to the catechol-O-methyltransferase (COMT) genotype after acute green tea extract consumption (108), while in a population-based case-control study, only those tea drinkers with the low-activity COMT enzyme showed a decreased risk of developing breast cancer compared to those who did not drink tea (109). In addition, Xing et al. reported a synergistic effect between this polymorphism and apolipoprotein E ε4 (APOE ε4) status on the susceptibility to AD (110). However, the effect of polyphenol genetic polymorphism on CD has not been investigated to date, although there have been studies on the neurological effects of genetic variants related to caffeine metabolism (111). In a recent clinical trial, caffeine enhanced CYP1A2 fast metabolizers' cognitive performance during exercise more than slow metabolizers (112), although evidence is still inconclusive and limited (113). On the other hand, the susceptibility to AD might be modulated by NAT2, an enzyme responsible for caffeine metabolism and elimination. Some research seems to link the expression of NAT2 rapid acetylating phenotypes to the potential pathogenesis of AD (114).

In this chapter, among the 23 studies we examined, only one publication considered an inter-individual variability factor in response to polyphenol intake, particularly the equol production capacity (38). This limitation makes it harder to understand the role that genetic factors may play in individual responses to polyphenol-rich food intake. Nonetheless, some investigations used plasma and urine polyphenols as a dependent variable (23, 39, 40, 43–45), which accounts for the real bioavailability and may, therefore, be a more precise measure of true exposure as it considers polyphenol absorption and metabolism variability, besides the fact that it is an objective biomarker of intake that is not susceptible to the likelihood of bias or imprecise reporting (115).

Strengths and Limitations

To the best of our knowledge, this chapter represents the first comprehensive examination of the effect of total polyphenols and polyphenol classes detected in plasma or urine on memory impairment. By encompassing both interventional and observational studies, we aimed to provide a thorough understanding of the existing evidence. Notably, 17 of the 23 articles included in this chapter were RCTs (23–36, 39, 41, 42), considered the gold standard for providing internally valid and compelling evidence. Moreover, the investigation into the relationship between biomarkers of dietary polyphenols and CD includes both cognitively healthy subjects (24, 27, 31–35, 38, 39, 43–45) and those experiencing memory difficulties (23, 25, 26, 28–30, 36, 37, 40–42). This chapter has also identified the limitations of the current literature regarding biomarkers of polyphenol intake and their association with CD. First, the conclusions are based on the limited number of studies (n = 23), and only a few have examined the association between plasma or urine polyphenol levels and cognition. Furthermore, only six studies evaluated CI through biomarkers of AD (25, 28, 33, 41, 42, 45), which are generally considered more objective than cognitive tests (116). Thirdly, due to a lack of consensus regarding the suitability of cognitive tests, several tests were used to evaluate the same cognitive domain, resulting in more than 60 different cognitive tests or tasks. This diversity in exposure and outcome definition and measurements has compromised the comparability of studies, making it challenging to interpret and generalize the results. Finally, the article search was conducted across three databases, excluding non-English articles. Thus, not all the evidence may be represented, potentially introducing language bias (117).

Polyphenol Intake Recommendations

Given the overall scarcity of available evidence, it is not possible to make strong recommendations regarding polyphenol consumption in relation to AD and cognitive impairment until more robust and higher-quality evidence becomes available. Future research should focus on accurate measurement of all polyphenol-derived metabolites in blood and urine to better understand the specific components responsible for the potential cognitive protective effects of polyphenols. In addition, further evidence from RCTs is required to elucidate the role of polyphenols at various stages of cognitive performance to validate the observed results in older adults. In this regard, nutritional trials involving sustained but controlled polyphenol consumption could also provide a more reliable approach to exploring the intricate relationship between polyphenols and cognitive function (118, 119). This, in turn, will enable healthcare professionals and public health authorities to make informed and prudent recommendations.

CONCLUSIONS

In this chapter, current evidence regarding the association between biomarkers of polyphenol intake and cognitive function in old age has been evaluated in all of its stages and varieties. The relationship between polyphenol intake and cognitive function is complex and likely influenced by factors such as the type of polyphenol, the background diet, and the subject's metabotype, among others. Studies investigating cognitive performance have yielded contradictory findings. Standardization of tools used to assess cognitive function is essential for evaluating the effectiveness of dietary factors in preventing CI. Moreover, it remains unclear whether the absence of demonstrable benefits can be attributed to the bioavailability of a specific formulation of certain polyphenol classes or to the inefficacy of these polyphenol classes as interventions for AD or cognitive dysfunction. Further research is needed in the field of metabolomics as there is a notable gap in the current knowledge.

REFERENCES

1. Xue QL. The frailty syndrome: Definition and natural history. Clin Geriatr Med. 2011;27:1–15.
2. Fabrício DM, Chagas MHN, Diniz BS. Frailty and Cognitive Decline. Transl Res. 2020;221:58–64.
3. Robinson TL, Gogniat MA, Miller LS. Frailty and cognitive function in older adults: A systematic review and meta-analysis of cross-sectional studies. Neuropsychol Rev. 2022;32:274–93.
4. Kojima G, Taniguchi Y, Iliffe S, Walters K. Frailty as a predictor of Alzheimer disease, vascular dementia, and all dementia among community-dwelling older people: A systematic review and meta-analysis. J Am Med Dir Assoc. 2016;17:881–8.
5. Calderón-Larrañaga A, Vetrano DL, Ferrucci L, Mercer SW, Marengoni A, Onder G, et al. Multimorbidity and functional impairment–bidirectional interplay, synergistic effects and common pathways. J Intern Med. 2019;285: 255–71.
6. Grande G, Haaksma ML, Rizzuto D, Melis RJF, Marengoni A, Onder G, et al. Co-occurrence of cognitive impairment and physical frailty, and incidence of dementia: Systematic review and meta-analysis. Neurosci Biobehav Rev. 2019;107:96–103.
7. Scarmeas N, Anastasiou CA, Yannakoulia M. Nutrition and prevention of cognitive impairment. Lancet Neurol. 2018;17(11):1006–15. http://dx.doi.org/10.1016/S1474-4422(18)30338-7
8. Caruso G, Godos J, Privitera A, Lanza G, Castellano S, Chillemi A, et al. Phenolic acids and prevention of cognitive decline: Polyphenols with a neuroprotective role in cognitive disorders and Alzheimer's disease. Nutrients. 2022;14(4):819.
9. Cheng N, Bell L, Lamport DJ, Williams CM. Dietary flavonoids and human cognition: A meta-analysis. Mol Nutr Food Res. 2022 Nov 1;66(21): e2100976.
10. Pérez-Jiménez J, Neveu V, Vos F, Scalbert A. Identification of the 100 richest dietary sources of polyphenols: An application of the Phenol-Explorer database. Eur J Clin Nutr. 2010;64:S112–20.
11. Singla RK, Dubey AK, Garg A, Sharma RK, Fiorino M, Ameen SM, et al. Natural polyphenols: Chemical classification, definition of classes, subcategories, and structures. J AOAC Int. 2019;102(5):1397–400.
12. Miranda AM, Steluti J, Fisberg RM, Marchioni DM. Dietary intake and food contributors of polyphenols in adults and elderly adults of Sao Paulo: A population-based study. Br J Nutr. 2016 Jan 26;115(6):1061–70.
13. Zamora-Ros R, Knaze V, Rothwell JA, Hémon B, Moskal A, Overvad K, et al. Dietary polyphenol intake in Europe: The European prospective investigation into cancer and nutrition (EPIC) study. Eur J Nutr. 2016;55(4):1359–75.
14. Emmett PM, Jones LR, Northstone K, Pounis G, Taylor CM. Collection and Management of Dietary Data. Analysis in Nutrition Research: Principles of Statistical Methodology and Interpretation of the Results. Elsevier Inc.; 2018. pp. 43–73.
15. Subar AF, Freedman LS, Tooze JA, Kirkpatrick SI, Boushey C, Neuhouser ML, et al. Addressing current criticism regarding the value of self-report dietary data. J Nutr. 2015;145(12):2639–45.
16. Poslusna K, Ruprich J, De Vries JHM, Jakubikova M, Van'T Veer P. Misreporting of energy and micronutrient intake estimated by food records and 24 hour recalls, control and adjustment methods in practice. Br J Nutr. 2009;101(SUPPL. 2): S73–S85.
17. Bento-Silva A, Koistinen VM, Mena P, Bronze MR, Hanhineva K, Sahlstrøm S, et al. Factors affecting intake, metabolism and health benefits of phenolic acids: Do we understand individual variability? Eur J Nutr. 2020;59(4):1275–93. https://doi.org/10.1007/s00394-019-01987-6
18. Manach C, Milenkovic D, Van de Wiele T, Rodriguez Mateos A, de Roos B, Garcia-Conesa MT, et al. Addressing the inter-individual variation in response to consumption of plant food bioactives: Towards a better understanding of their role in healthy aging and cardiometabolic risk reduction. Mol Nutr Food Res. 2017;61(6):1–16.
19. Zamora-Ros R, Rabassa M, Llorach R, González CA, Andres-Lacueva C Application of dietary phenolic biomarkers in epidemiology: Past, present, and future. J Agric Food Chem. 2012;60(27):6648–57.
20. Zamora-Ros R, Rabassa M, Cherubini A, Urpi-Sarda M, Llorach R, Bandinelli S, et al. Comparison of 24-h volume and creatinine-corrected total urinary polyphenol as a biomarker of total dietary polyphenols in the Invecchiare InCHIANTI study. Anal Chim Acta. 2011;704(1–2):110–5.
21. Landberg R, Karra P, Hoobler R, Loftfield E, Huybrechts I, Rattner JI, et al. Dietary biomarkers—An update on their validity and applicability in epidemiological studies. Nutr Rev. 2023 Oct 3; 82(9):1260–1280.

22. Colizzi C. The protective effects of polyphenols on Alzheimer's disease: A systematic review. Alzheimer's Dement. 2018;5:184–96.
23. Ochiai R, Saitou K, Suzukamo C, Osaki N, Asada T. Effect of chlorogenic acids on cognitive function in mild cognitive impairment: A randomized controlled crossover trial. J Alzheimer's Dis. 2019;72(4): 1209–16.
24. Gleason CE, Carlsson CM, Barnet JH, Meade SA, Setchell KDR, Atwood CS, et al. A preliminary study of the safety, feasibility and cognitive efficacy of soy isoflavone supplements in older men and women. Age Ageing. 2009;38(1):86–93.
25. Gleason CE, Fischer BL, Dowling NM, Setchell KDR, Atwood CS, Carlsson CM, et al. Cognitive effects of soy isoflavones in patients with Alzheimer's disease. J Alzheimer's Dis. 2015 Aug;47(4);1009–19.
26. Krikorian R, Kalt W, McDonald JE, Shidler MD, Summer SS, Stein AL. Cognitive performance in relation to urinary anthocyanins and their flavonoid-based products following blueberry supplementation in older adults at risk for dementia. J Funct Foods. 2020;64:1–10.
27. Igwe EO, Charlton KE, Roodenrys S, Kent K, Fanning K, Netzel ME. Anthocyanin-rich plum juice reduces ambulatory blood pressure but not acute cognitive function in younger and older adults: A pilot crossover dose-timing study. Nutr Res. 2017;47:28–43.
28. McNamara RK, Kalt W, Shidler MD, McDonald J, Summer SS, Stein AL, et al. Cognitive response to fish oil, blueberry, and combined supplementation in older adults with subjective cognitive impairment. Neurobiol Aging. 2018;64:147–56.
29. Vauzour D, Scholey A, White DJ, Cohen NJ, Cassidy A, Gillings R, et al. A combined DHA-rich fish oil and cocoa flavanols intervention does not improve cognition or brain structure in older adults with memory complaints: Results from the CANN randomized, controlled parallel-design study. Am J Clin Nutr. 2023 Aug 1;118(2):369–81.
30. Bergland AK, Soennesyn H, Dalen I, Rodriguez-Mateos A, Berge RK, Giil LM, et al. Effects of anthocyanin supplementation on serum lipids, glucose, markers of inflammation and cognition in adults with increased risk of dementia -- A pilot study. Front Genet. 2019;10(JUN):1–11.
31. John JAS, Henderson VW, Hodis HN, Kono N, McCleary CA, Franke AA, et al. Associations between urine excretion of isoflavonoids and cognition in postmenopausal women in the Women's Isoflavone Soy Health clinical trial. J Am Geriatr Soc. 2014;62(4):629–35.
32. Curtis PJ, van der Velpen V, Berends L, Jennings A, Haag L, Minihane AM, et al. Chronic and postprandial effect of blueberries on cognitive function, alertness, and mood in participants with metabolic syndrome – Results from a six-month, double-blind, randomized controlled trial. Am J Clin Nutr. 2024 Mar 1;119(3):658–68.
33. Igwe EO, Roodenrys S, Probst YC, do Rosario V, Netzel ME, Hong HT, et al. Low anthocyanin plum nectar does not impact cognition, blood pressure and gut microbiota in healthy older adults: A randomized crossover trial. Nutr Res. 2020;82:74–87.
34. Sloan RP, Wall M, Yeung LK, Feng T, Feng X, Provenzano F, et al. Insights into the role of diet and dietary flavanols in cognitive aging: Results of a randomized controlled trial. Sci Rep. 2021 Dec 1;11(1):3837.
35. Wood E, Hein S, Mesnage R, Fernandes F, Abhayaratne N, Xu Y, et al. Wild blueberry (poly)phenols can improve vascular function and cognitive performance in healthy older individuals: A double-blind randomized controlled trial. Am J Clin Nutr. 2023 Jun 1;117(6):1306–19.
36. Bensalem J, Dudonné S, Etchamendy N, Pellay H, Amadieu C, Gaudout D, et al. Polyphenols from grape and blueberry improve episodic memory in healthy elderly with lower level of memory performance: A bicentric double-blind, randomized, placebo-controlled clinical study. J Gerontol A Biol Sci Med Sci. 2019;74(7):996–1007.
37. Yuan L, Liu J, Ma W, Dong L, Wang W, Che R, et al. Dietary pattern and antioxidants in plasma and erythrocyte in patients with mild cognitive impairment from China. Nutrition. 2016;32(2):193–8.
38. Igase M, Igase K, Tabara Y, Ohyagi Y, Kohara K. Cross-sectional study of equol producer status and cognitive impairment in older adults. Geriatr Gerontol Int. 2017;17(11):2103–8.
39. Rutledge GA, Sandhu AK, Miller MG, Edirisinghe I, Burton-Freeman BB, Shukitt-Hale B Blueberry phenolics are associated with cognitive enhancement in supplemented healthy older adults. Food Funct. 2021;12(1):107–18.
40. Tor-Roca A, Sánchez-Pla A, Korosi A, Pallàs M, Lucassen PJ, Castellano-Escuder P, et al. A Mediterranean diet-based metabolomic score and cognitive decline in older adults: A case–control analysis nested within the three-city cohort study. Mol Nutr Food Res. 2023 Oct 24; 68(13):e2300271.
41. Bookheimer SY, Renner BA, Ekstrom A, Li Z, Henning SM, Brown JA, et al. Pomegranate juice augments memory and fMRI activity in middle-aged and older adults with mild memory complaints. Evid Based Complementary Alternat Med. 2013;2013:946298.

42. Ringman JM, Frautschy SA, Teng E, Begum AN, Bardens J, Beigi M, et al. Oral curcumin for Alzheimer's disease: Tolerability and efficacy in a 24-week randomized, double blind, placebo-controlled study. Alzheimers Res Ther. 2012;4(5):43.
43. Rabassa M, Cherubini A, Zamora-Ros R, Urpi-Sarda M, Bandinelli S, Ferrucci L, et al. Low levels of a urinary biomarker of dietary polyphenol are associated with substantial cognitive decline over a 3-year period in older adults: The invecchiare in chianti study. J Am Geriatr Soc. 2015;63(5): 938–46.
44. Rabassa M, Zamora-Ros R, Palau-Rodriguez M, Tulipani S, Miñarro A, Bandinelli S, et al. Habitual nut exposure, assessed by dietary and multiple urinary metabolomic markers, and cognitive decline in older adults: The InCHIANTI Study. Mol Nutr Food Res. 2020;64(2):1–11.
45. Valls-Pedret C, Lamuela-Raventós RM, Medina-Remón A, Quintana M, Corella D, Pintó X, et al. Polyphenol-rich foods in the mediterranean diet are associated with better cognitive function in elderly subjects at high cardiovascular risk. J Alzheimer's Dis. 2012;29(4):773–82.
46. Scalbert A, Brennan L, Manach C, Andres-Lacueva C, Dragsted LO, Draper J, et al. The food metabolome: A window over dietary exposure. Am J Clin Nutr. 2014;99(6):1286–308.
47. Rodriguez-Mateos A, Vauzour D, Krueger CG, Shanmuganayagam D, Reed J, Calani L, et al. Bioavailability, bioactivity and impact on health of dietary flavonoids and related compounds: An update. Arch Toxicol. 2014;88(10):1803–53.
48. Rodriguez-Mateos A, Cifuentes-Gomez T, Gonzalez-Salvador I, Ottaviani JI, Schroeter H, Kelm M, et al. Influence of age on the absorption, metabolism, and excretion of cocoa flavanols in healthy subjects. Mol Nutr Food Res. 2015;59(8):1504–12.
49. Soenen S, Rayner CK, Jones KL, Horowitz M. The ageing gastrointestinal tract. Curr Opin Clin Nutr Metab Care. 2016;19(1):12–8.
50. Milan AM, Cameron-Smith D. Digestion and postprandial metabolism in the elderly. Adv Food Nutr Res. 2015;76:79–124.
51. Domínguez-Avila JA, Wall-Medrano A, Velderrain-Rodríguez GR, Chen CYO, Salazar-López NJ, Robles-Sánchez M, et al. Gastrointestinal interactions, absorption, splanchnic metabolism and pharmacokinetics of orally ingested phenolic compounds. Food Funct. 2017;8(1):15–38.
52. Alkhaldy A, Edwards CA, Combet E. The urinary phenolic acid profile varies between younger and older adults after a polyphenol-rich meal despite limited differences in in vitro colonic catabolism. Eur J Nutr. 2019;58(3):1095–111.
53. Thevaranjan N, Puchta A, Schulz C, Naidoo A, Szamosi JC, Verschoor CP, et al. Age-associated microbial dysbiosis promotes intestinal permeability, systemic inflammation, and macrophage dysfunction. Cell Host Microbe. 2017;21(4):455–466.e4.
54. Ulaszewska MM, Weinert CH, Trimigno A, Portmann R, Andres Lacueva C, Badertscher R, et al. Nutrimetabolomics: An integrative action for metabolomic analyses in human nutritional studies. Mol Nutr Food Res. 2019;63(1):1–38.
55. Seguella L, Capuano R, Sarnelli G, Esposito G. Play in advance against neurodegeneration: Exploring enteric glial cells in gut-brain axis during neurodegenerative diseases. Expert Rev Clin Pharmacol. 2019;12(6):555–64.
56. Stadlbauer V, Engertsberger L, Komarova I, Feldbacher N, Leber B, Pichler G, et al. Dysbiosis, gut barrier dysfunction and inflammation in dementia: A pilot study. BMC Geriatr. 2020;20(1):1–13.
57. Suzuki T. Regulation of intestinal epithelial permeability by tight junctions. Cell Mol Life Sci. 2013; 70(4):631–59.
58. Nagpal R, Mainali R, Ahmadi S, Wang S, Singh R, Kavanagh K, et al. Gut microbiome and aging: Physiological and mechanistic insights. Nutr Healthy Aging. 2018;4(4):267–85.
59. Mariat D, Firmesse O, Levenez F, Guimarães VD, Sokol H, Doré J, et al. The firmicutes/bacteroidetes ratio of the human microbiota changes with age. BMC Microbiol. 2009;9:1–6.
60. Gavini F, Cayuela C, Antoine JM, Lecoq C, Lefebvre B, Membré JM, et al. Differences in the distribution of bifidobacterial and enterobacterial species in human faecal microflora of three different (children, adults, elderly) age groups. Microb Ecol Health Dis. 2001;13(1):40–5.
61. Vaiserman AM, Koliada AK, Marotta F. Gut microbiota: A player in aging and a target for anti-aging intervention. Ageing Res Rev. 2017;35:36–45.
62. Claesson MJ, Cusack S, O'Sullivan O, Greene-Diniz R, De Weerd H, Flannery E, et al. Composition, variability, and temporal stability of the intestinal microbiota of the elderly. Proc Natl Acad Sci U S A. 2011;108(SUPPL. 1):4586–91.
63. Vogt NM, Kerby RL, Dill-McFarland KA, Harding SJ, Merluzzi AP, Johnson SC, et al. Gut microbiome alterations in Alzheimer's disease. Sci Rep. 2017;7(1):1–11.

64. Saji N, Niida S, Murotani K, Hisada T, Tsuduki T, Sugimoto T, et al. Analysis of the relationship between the gut microbiome and dementia: A cross-sectional study conducted in Japan. Sci Rep. 2019;9(1):1–9.
65. Araos R, Andreatos N, Ugalde J, Mitchell S, Mylonakis E, D'Agata EMC. Fecal microbiome among nursing home residents with advanced dementia and Clostridium difficile. Dig Dis Sci. 2018;63(6):1525–31.
66. Ticinesi A, Tana C, Nouvenne A, Prati B, Lauretani F, Meschi T. Gut microbiota, cognitive frailty and dementia in older individuals: A systematic review. Clin Interv Aging. 2018;13:1497–511.
67. Köhler C, Maes M, Slyepchenko A, Berk M, Solmi M, Lanctôt K, et al. The gut-brain axis, including the microbiome, leaky gut and bacterial translocation: Mechanisms and pathophysiological role in Alzheimer's disease. Curr Pharm Des. 2016 Dec 14;22(40):6152–66.
68. Cortés-Martín A, Selma MV, Tomás-Barberán FA, González-Sarrías A, Espín JC. Where to look into the puzzle of polyphenols and health? The postbiotics and gut Microbiota associated with human metabotypes. Mol Nutr Food Res. 2020;64(9):1–17.
69. Tomás-Barberán FA, Selma MV, Espín JC. Interactions of gut microbiota with dietary polyphenols and consequences to human health. Curr Opin Clin Nutr Metab Care. 2016;19(6):471–6.
70. Luca SV, Macovei I, Bujor A, Miron A, Skalicka-Woźniak K, Aprotosoaie AC, et al. Bioactivity of dietary polyphenols: The role of metabolites. Crit Rev Food Sci Nutr. 2020;60(4):626–59.
71. Domínguez-López I, López-Yerena A, Vallverdú-Queralt A, Pallàs M, Lamuela-Raventós RM, Pérez M. From the gut to the brain: The long journey of phenolic compounds with neurocognitive effects. Nutr Rev. 2024 Apr 30; 4(5):43.
72. Bub A, Watzl B, Blockhaus M, Briviba K, Liegibel U, Müller H, et al. Fruit juice consumption modulates antioxidative status, immune status and DNA damage. J Nutr Biochem. 2003;14(2):90–8.
73. Bialonska D, Kasimsetty SG, Schrader KK, Ferreira D. The effect of pomegranate (Punica granatum L.) byproducts and ellagitannins on the growth of human gut bacteria. J Agric Food Chem. 2009;57(18):8344–9.
74. Nohynek LJ, Alakomi HL, Kähkönen MP, Heinonen M, Helander IM, Oksman-Caldentey KM, et al. Berry phenolics: Antimicrobial properties and mechanisms of action against severe human pathogens. Nutr Cancer. 2006;54(1):18–32.
75. Hidalgo-Liberona N, González-Domínguez R, Vegas E, Riso P, Del Bo C, Bernardi S, et al. Increased intestinal permeability in older subjects impacts the beneficial effects of dietary polyphenols by modulating their bioavailability. J Agric Food Chem. 2020;68(44):12476–84.
76. Ho L, Ferruzzi MG, Janle EM, Wang J, Gong B, Chen TY, et al. Identification of brain-targeted bioactive dietary quercetin-3-O-glucuronide as a novel intervention for Alzheimer's disease. FASEB J. 2013;27(2):769–81.
77. Zhao W, Wang J, Bi W, Ferruzzi M, Yemul S, Freire D, et al. Novel application of brain-targeting polyphenol compounds in sleep deprivation-induced cognitive dysfunction. Neurochem Int. 2015;89:191–7.
78. Wang D, Ho L, Faith J, Ono K, Janle EM, Lachcik PJ, et al. Role of intestinal microbiota in the generation of polyphenol-derived phenolic acid mediated attenuation of Alzheimer's disease β-amyloid oligomerization. Mol Nutr Food Res. 2015;59(6):1025–40.
79. Wang J, Ferruzzi MG, Ho L, Blount J, Janle EM, Gong B, et al. Brain-targeted proanthocyanidin metabolites for Alzheimer's disease treatment. J Neurosci. 2012;32(15):5144–50.
80. Yuan T, Ma H, Liu W, Niesen DB, Shah N, Crews R, et al. Pomegranate's neuroprotective effects against Alzheimer's disease are mediated by urolithins, its ellagitannin-gut microbial derived metabolites. ACS Chem Neurosci. 2016;7(1):26–33.
81. Carecho R, Carregosa D, dos Santos CN. Low molecular weight (poly)phenol metabolites across the blood-brain barrier: The underexplored journey. Brain Plast. 2020;6(2):193–214.
82. Low DY, Lefèvre-Arbogast S, González-Domínguez R, Urpi-Sarda M, Micheau P, Petera M, et al. Diet-related metabolites associated with cognitive decline revealed by untargeted metabolomics in a prospective cohort. Mol Nutr Food Res. 2019;63(18):1–10.
83. Zamora-Ros R, Biessy C, Rothwell JA, Monge A, Lajous M, Scalbert A, et al. Dietary polyphenol intake and their major food sources in the Mexican Teachers' Cohort. Br J Nutr. 2018;120(3):353–60.
84. Feliciano RP, Istas G, Heiss C, Rodriguez-Mateos A. Plasma and urinary phenolic profiles after acute and repetitive intake of wild blueberry. Molecules. 2016;21(9):14–6.
85. Zhong S, Sandhu A, Edirisinghe I, Burton-Freeman B. Characterization of wild blueberry polyphenols bioavailability and kinetic profile in plasma over 24-h period in human subjects. Mol Nutr Food Res. 2017;61(12):1–13.
86. Frankenfeld CL, Atkinson C, Thomas WK, Goode EL, Gonzalez A, Jokela T, et al. Familial correlations, segregation analysis, and nongenetic correlates of soy isoflavone-metabolizing phenotypes. Exp Biol Med. 2004;229(9):902–13.

87. Ko TF, Tsai HS, Lin SM, Liu CD, Learn SP, Chiou RY. GC-MS determined distribution of urinary equol producers as affected by age, gender, and repeated ingestions of soymilk. J Food Sci. 2010; 75(9);H306–10.
88. Liu W, Tanabe M, Harada KH, Koizumi A. Levels of urinary isoflavones and lignan polyphenols in Japanese women. Environ Health Prev Med. 2013;18(5):394–400.
89. Takeda T, Ueno T, Uchiyama S, Hiramatsu K, Shiina M. Relation between premenstrual syndrome and equol-production status. J Obstet Gynaecol Res. 2016;42(11):1575–80.
90. Iino C, Shimoyama T, Iino K, Yokoyama Y, Chinda D, Sakuraba H, et al. Daidzein intake is associated with equol producing status through an increase in the intestinal bacteria responsible for equol production. Nutrients. 2019;11(2);433.
91. Haudum C, Lindheim L, Ascani A, Trummer C, Horvath A, Münzker J, et al. Impact of short-term isoflavone intervention in polycystic ovary syndrome (PCOS) patients on microbiota composition and metagenomics. Nutrients. 2020;12(6):1–22.
92. Tomás-Barberán FA, García-Villalba R, González-Sarrías A, Selma MV, Espín JC. Ellagic acid metabolism by human gut microbiota: Consistent observation of three urolithin phenotypes in intervention trials, independent of food source, age, and health status. J Agric Food Chem. 2014;62(28):6535–8.
93. Cortés-Martín A, García-Villalba R, González-Sarrías A, Romo-Vaquero M, Loria-Kohen V, Ramírez-De-Molina A, et al. The gut microbiota urolithin metabotypes revisited: The human metabolism of ellagic acid is mainly determined by aging. Food Funct. 2018;9(8):4100–6.
94. Cortés-Martín A, Selma MV, Espín JC, García-Villalba R. The human metabolism of nuts proanthocyanidins does not reveal urinary metabolites consistent with distinctive gut microbiota metabotypes. Mol Nutr Food Res. 2019;63(2):1–10.
95. Jenner AM The importance of bioavailability and metabolism on the health benefits of dietary flavonoids – Advances in biomarkers of intake. In: Plenary Session 4: Revisiting Nutritional Biomarkers. 2012. pp. 682–98.
96. Caldwell K, Charlton KE, Roodenrys S, Jenner A. Anthocyanin-rich cherry juice does not improve acute cognitive performance on RAVLT. Nutr Neurosci. 2016;19(9):423–4.
97. González-Sarrías A, Espín JC, Tomás-Barberán FA. Non-extractable polyphenols produce gut microbiota metabolites that persist in circulation and show anti-inflammatory and free radical-scavenging effects. Trends Food Sci Technol. 2017;69:281–8.
98. Liu RH. Health benefits of fruit and vegetables are from additive and synergistic combinations of phytochemicals. Am J Clin Nutr. 2003;78(3 SUPPL.):3–6.
99. Manach C, Williamson G, Morand C, Scalbert A, Rémésy C. Bioavailability and bioefficacy of polyphenols in humans. I. Review of 97 bioavailability studies. Am J Clin Nutr. 2005;81(1 Suppl):230–42.
100. Zhang H, Yu D, Sun J, Liu X, Jiang L, Guo H, et al. Interaction of plant phenols with food macronutrients: Characterisation and nutritional-physiological consequences. Nutr Res Rev. 2014;27(1):1–15.
101. Manach C, Williamson G. Bioavailability and bioefficacy of polyphenols in humans. II. Review of 97 bioavailability studies. Am J Clin Nutr. 2005;81(1 Suppl):230S–242S.
102. Dinu M, Pagliai G, Casini A, Sofi F. Mediterranean diet and multiple health outcomes: An umbrella review of meta-analyses of observational studies and randomised trials. Eur J Clin Nutr. 2018;72(1):30–43.
103. Sofi F, Abbate R, Gensini GF, Casini A. Accruing evidence on benefits of adherence to the Mediterranean diet on health: An updated systematic review and meta-analysis. Am J Clin Nutr. 2010;92(5):1189–96.
104. Chen X, Maguire B, Brodaty H, O'Leary F. Dietary patterns and cognitive health in older adults: A systematic review. J Alzheimer's Dis. 2019;67(2):583–619.
105. Dohrmann DD, Putnik P, Bursać Kovačević D, Simal-Gandara J, Lorenzo JM, Barba FJ. Japanese, Mediterranean and Argentinean diets and their potential roles in neurodegenerative diseases. Food Res Int. 2019;120:464–77.
106. Ozawa M, Ninomiya T, Ohara T, Doi Y, Uchida K, Shirota T, et al. Dietary patterns and risk of dementia in an elderly Japanese population: The Hisayama study. Am J Clin Nutr. 2013;97(5):1076–82.
107. Scholl C, Lepper A, Lehr T, Hanke N, Schneider KL, Brockmöller J, et al. Population nutrikinetics of green tea extract. PLoS One. 2018;13(2):1–22.
108. Miller RJ, Jackson KG, Dadd T, Mayes AE, Brown AL, Lovegrove JA, et al. The impact of the catechol-O-methyltransferase genotype on vascular function and blood pressure after acute green tea ingestion. Mol Nutr Food Res. 2012;56(6):966–75.
109. Wu AH, Tseng CC, Van Den Berg D, Yu MC. Tea intake, COMT genotype, and breast cancer in Asian-American women. Cancer Res. 2003;63(21):7526–9.
110. Xing Y, Jia JP, Ji XJ, Tian T. Estrogen associated gene polymorphisms and their interactions in the progress of Alzheimer's disease. Prog Neurobiol. 2013;111:53–74.

111. Kapellou A, King A, Graham CAM, Pilic L, Mavrommatis Y. Genetics of caffeine and brain-related outcomes – A systematic review of observational studies and randomized trials. Nutr Rev. 2023; 81(12):1571–1598.
112. Carswell AT, Howland K, Martinez-Gonzalez B, Baron P, Davison G. The effect of caffeine on cognitive performance is influenced by CYP1A2 but not ADORA2A genotype, yet neither genotype affects exercise performance in healthy adults. Eur J Appl Physiol. 2020;120(7):1495–508.
113. Salinero JJ, Lara B, Ruiz-Vicente D, Areces F, Puente-Torres C, Gallo-Salazar C, et al. CYP1A2 genotype variations do not modify the benefits and drawbacks of caffeine during exercise: A pilot study. Nutrients. 2017;9(3):1–12.
114. Nehlig A. Interindividual differences in caffeine metabolism and factors driving caffeine consumption. Pharmacol Rev. 2018;70(2):384–411.
115. Cuparencu C, Bulmuş-Tüccar T, Stanstrup J, La Barbera G, Roager HM, Dragsted LO. Towards nutrition with precision: Unlocking biomarkers as dietary assessment tools. Nat Metabol. 2024;6(8):1438–1453.
116. Huynh RA, Mohan C. Alzheimer's disease: Biomarkers in the genome, blood, and cerebrospinal fluid. Front Neurol. 2017;8(MAR).
117. Morrison A, Polisena J, Husereau D, Moulton K, Clark M, Fiander M, et al. The effect of English-language restriction on systematic review-based meta-analyses: A systematic review of empirical studies. Int J Technol Assess Health Care. 2012;28(2):138–44.
118. Mena P, Del Rio D. Gold standards for realistic (poly)phenol research. J Agric Food Chem. 2018;66(31):8221–3.
119. Higgins JP, Li T, Deeks JJ. Chapter 6: Choosing Effect Measures and Computing Estimates of Effect. In: Higgins J, Thomas J, Chandler J, Cumpston M, Li T, Page M, et al., editors. Cochrane Handbook for Systematic Reviews of Interventions. 2022.

11 The Role of the Microbiota in Aging and Frailty

Mónica De la Fuente and Eduardo Ortega

INTRODUCTION

Our body is colonized by trillions of microorganisms, primarily bacteria, but also including archaea, viruses, and eukaryotes (such as fungi, protozoa, and even worms). Together, these microorganisms form what we refer to as the Human Microbiota. It is estimated that microbial cells are present in the human body at a ratio of approximately 1:1 with our cells. Yet the genes of these microorganisms are significantly more diverse than those of human cells.

From the study of the interactions between the colonizing microbiota and the host, three key corollaries can be established:

a. There is a strong tendency for the microbiota present in each habitat within the same individual to persist. Even when altered by external factors, there is a tendency to recover a similar microbiota after these disturbances cease. This persistence is a result of the balance established among all microorganisms in the community, where each member's activities influence others' behavior.
b. There is less variability among the microbiota that colonize a specific biotope across different individuals than there is among the microbiota residing in different locations within the same individual.
c. The microbiota functions remain relatively constant in each body location. This stability is due to the functional redundancy of the microbiota; if one component of the microbial community is eliminated, other members will take over its functions.

Among all autochthonous microbiota locations, the digestive tract has the most extensive colonization, particularly in the oral cavity and gut, where microbiota complexity reaches its highest levels, especially in the colon. This chapter will focus solely on gut microbiota, and more concretely on large intestine microbiota, although the term gut microbiota will be used. However, it is important to note that microbiota also exist in other niches, such as the skin, the genitourinary tract, and the mammary gland (1) (see Figure 11.1).

Several basic concepts are given in Box 11.1.

GUT MICROBIOTA

General Concepts and Functions

The gut microbiota constitutes the most abundant community of microorganisms within the human body, accounting for over 92% of our microbial population. This community engages in a mutualistic symbiosis with the host, wherein we provide a suitable habitat for these microorganisms, and in return they facilitate essential physiological functions. Consequently, the gut microbiota is vital for human health and signifies a paradigm shift in modern medicine and our comprehension of various aspects of biology and human evolution (2–4).

In the small intestine, the proportion of microbiota in the duodenum is relatively low (10^4–10^5 cells/ml) due to the highly acidic chyme, the toxic effects of bile on many microorganisms, and

DOI: 10.1201/9781003398295-11

BOX 11.1 KEY CONCEPTS IN HOST-MICROBIOTA INTERACTIONS

Autochthonous Microbiota: Refers to the diverse microbial communities that stably colonize the epidermis, ducts, and the mucous membranes lining cavities that connect to the external environment.

Microbiome: The collection of microorganisms present in a specific environment, characterized by their biophysical and biochemical properties. This includes the microorganisms and their structural and genetic components, metabolites, and the surrounding environmental conditions.

Eubiosis: The state of health of the microbiota, characterized by a normal and balanced microbial community that meets the necessary conditions for the individual to benefit from its functions.

Dysbiosis: Alteration in microbial communities' composition, structure, and function, with a disruption or loss of the physiological symbiosis between the host and its microbiota.

Probiotics: Live microorganisms, typically bacteria or yeast, that, when administered in adequate amounts, provide health benefits to the host.

the presence of pancreatic fluids containing various enzymes that can digest microbial cells. In the jejunum, bacterial concentrations increase significantly (10^5–10^7 cells/ml). In the ileum, resident microorganisms' concentration and diversity rise rapidly, increasingly reflecting the composition found in the colon. Consequently, microbial density and diversity are immense in the large intestine, reaching levels of 10^{11}–10^{12} cells/ml of exudate. The acidity has nearly dissipated, the toxic components of bile have been reabsorbed through enterohepatic circulation, and, crucially, the slower flux of intestinal contents promotes the accumulation of microorganisms (Figure 11.1).

While the gut microbiota is mainly composed of bacteria, it also includes archaea (notably those responsible for methane production), yeasts of the genus *Candida*, and various protozoa, primarily

MICROBIOTA LOCATIONS

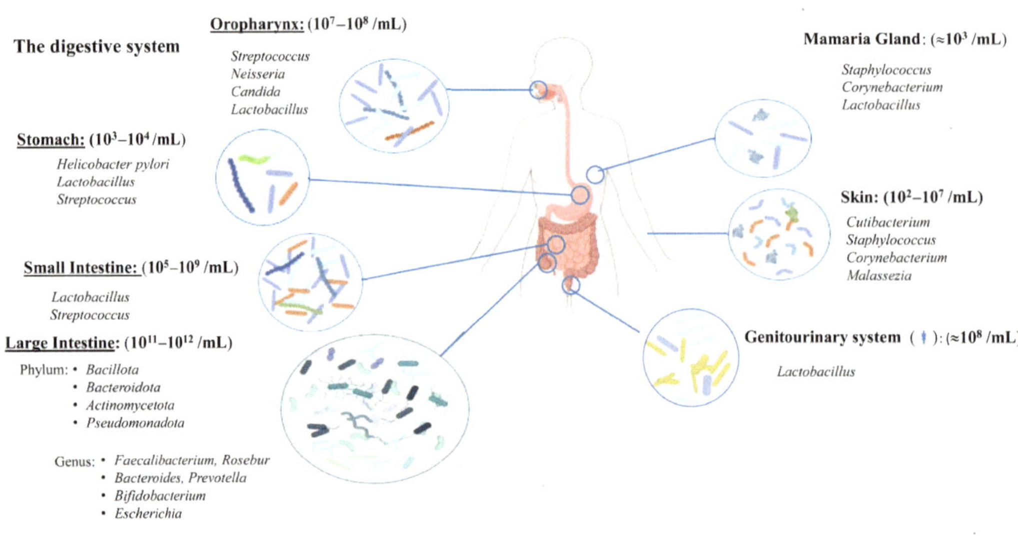

FIGURE 11.1 Locations, diversity, and concentration of the microbiota across the human body.

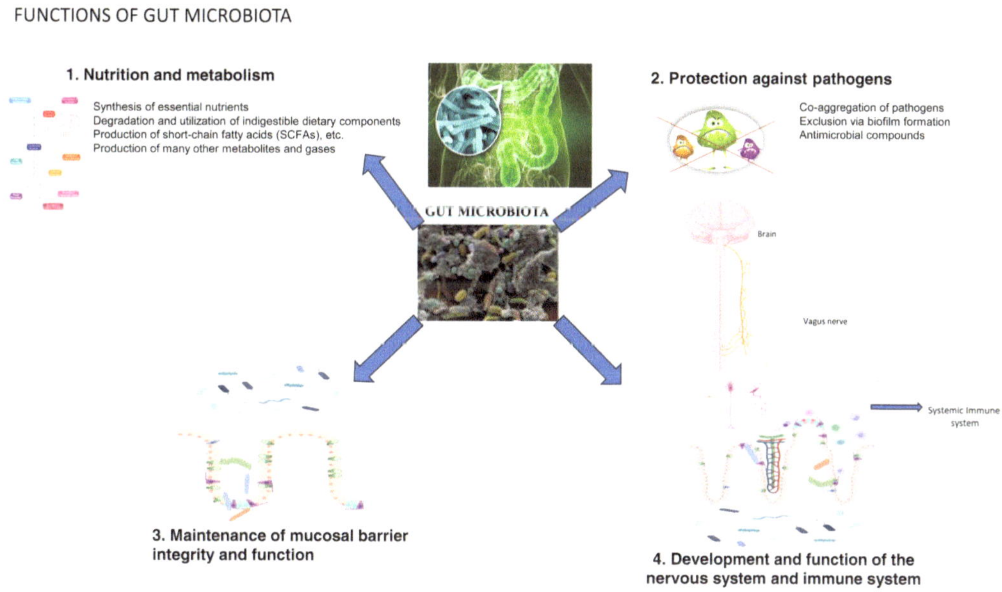

FIGURE 11.2 Different functions of the gut microbiota.

amoebae and coccidia. Among the bacterial phyla, *Firmicutes* (currently denominated *Bacillota*), which are predominantly represented by the order *Clostridiales* (such as *Faecalibacterium prausnitzii*), and *Bacteroidetes* (with *Bacteroides* and *Prevotella* as the principal representatives) are the most abundant, collectively constituting approximately 80% of the total microbial community (about 40% each). *Actinobacteria* (represented by *Bifidobacterium* spp.) may account for an additional 10%. Less abundant bacterial groups include *Proteobacteria* (primarily *Enterobacteriaceae*) and *Verrucomicrobia* (notably *Akkermansia muciniphila*), with concentrations typically below 3% in each case.

The functions of the gut microbiota can be summarized as follows (Figure 11.2):

1. **Nutrition and Metabolism**: The microbiota synthesizes essential nutrients, including vitamins (such as vitamin K), amino acids, and linoleic and α-linolenic fatty acids. Additionally, it facilitates the degradation and utilization of indigestible dietary components (e.g., cellulose and pectins), resulting in the production of short-chain fatty acids (SCFAs)—primarily acetate, propionate, and butyrate—as well as gases that serve important functions for the host.
2. **Protection Against Pathogens**: The microbiota plays a crucial role in defending against pathogens through several mechanisms, including exclusion via biofilm formation, co-aggregation that prevents pathogen adhesion to the mucosa, and destruction through the production of various antimicrobial compounds. Furthermore, it enhances the functionality of the immune system.
3. **Maintenance of Mucosal Integrity and Function**: The microbiota supports the integrity of the mucosal barrier by reinforcing Goblet cells (which secrete mucus that coats the mucosa), strengthening the tight junctions between enterocytes, and promoting the development of deeper crypts. This process facilitates the differentiation of base cells into Paneth cells, which produce antimicrobial peptides such as defensins and cathelicidins.
4. **Connection with Homeostatic Systems**: The gut microbiota is interconnected with the homeostatic systems, such as the immune, nervous, and endocrine systems, allowing their development and function, which will be elaborated upon in subsequent sections.

Gut Microbiota and Immunoneuroendocrine Interactions

The homeostatic systems—the nervous, endocrine, and immune systems—are essential for maintaining body homeostasis, which refers to the capacity for adaptation to constant changes. Consequently, they play a crucial role in health maintenance. These systems do not operate in isolation; they communicate continuously through mediators they each secrete, such as neurotransmitters, hormones, and cytokines, which act on the other systems' cells, forming a comprehensive neuroimmunoendocrine system. Thus, health maintenance relies not only on the proper functioning of each system and their effective bidirectional communication, established at both the systemic level and within individual tissues and organs. The interdisciplinary field studying these interconnections, known as psychoneuroimmunoendocrinology, has emerged relatively recently and is often called "the science of health" (5).

In addition, the maintenance of appropriate homeostasis depends not only on the genetic composition of each individual but also, more importantly, on environmental and lifestyle factors. The genome and the so-called "ambiome" influence individuals from the onset of existence. During fetal development, the interplay between genetic and environmental factors (shaped by the mother's health) conditions homeostatic systems' development and intercommunication. This environmental influence persists postnatally, particularly during childhood, and becomes increasingly evident in relation to lifestyle factors throughout the aging process (Figure 11.3) (6, 7).

Interaction of Gut Microbiota with the Immune System

The immune system located in the intestinal mucosa is arguably the most complex aspect of this physiological system, comprising numerous innate and adaptive immune cells with distinct characteristics. The epithelial barrier, which contains a variety of specialized cells—including lymphocytes integrated within it—is also considered a vital component of the intestinal defense system. This gut immune system is intricately connected to the organism's overall immunity. Intestinal

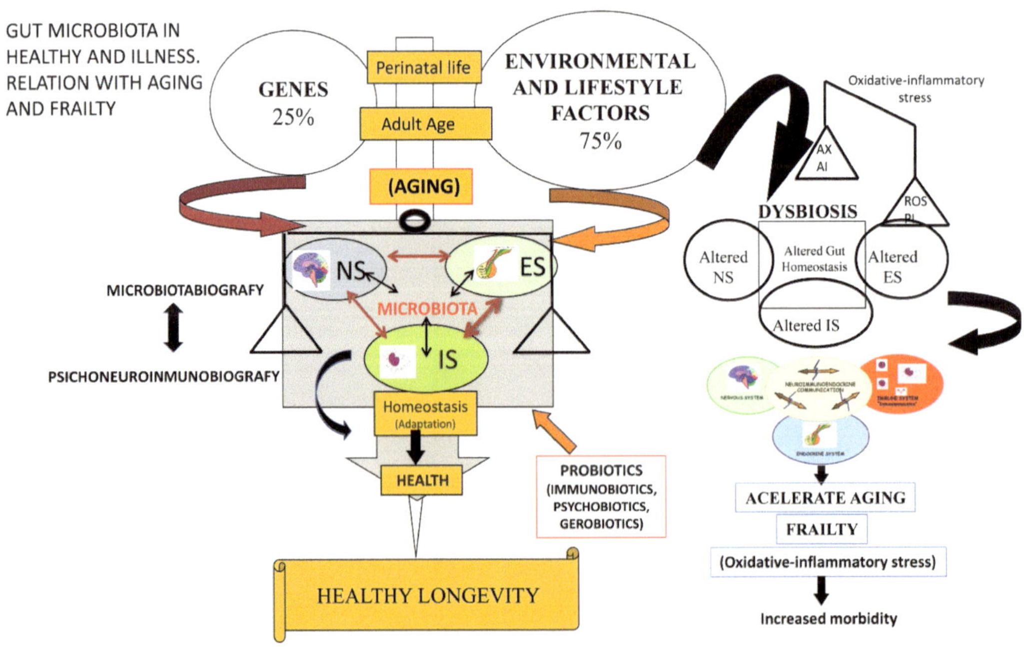

FIGURE 11.3 Gut microbiota in health and disease: its relation to aging and frailty. NS: Nervous system; ES: Endocrine system; Ax: Antioxidants; AI: Anti-inflammatory mediators; ROS: Reactive oxygen species; PI: Pro-inflammatory mediators.

microorganisms communicate with immune cells, primarily through indirect pathways involving the myriad metabolites they produce, and microvesicles. Mediators generated from the interaction between the microbiota and the epithelial barrier can also reach immune cells (7–9).

This interaction, beginning early in life, is crucial for the proper development and functionality of the immune system. Research has demonstrated that germ-free experimental animals exhibit impaired digestive function and compromised immune responses (10). The effective dialogue between the microbiota and the immune system enables the latter to learn to tolerate food antigens and autochthonous microorganisms while enhancing its capacity to defend against pathogens. This is achieved, in part, through an increased presence of cells that secrete anti-inflammatory cytokines, such as regulatory T cells (Tregs), which are abundant in the intestinal mucosa and produce IL-10, in contrast to cells that release pro-inflammatory mediators. Microbiota-derived metabolites, such as SCFAs, particularly butyrate, significantly regulate inflammation (11). Furthermore, the microbiota can epigenetically and metabolically reprogram immune cells, enabling them to acquire "memory" characteristics that enhance their subsequent responsiveness (12, 13).

Interaction of Gut Microbiota with the Nervous and Endocrine Systems

The intestine houses its nervous system, known as the enteric nervous system (ENS). It is characterized by a high density of neurons and intricate organization, earning it the designation of the "second brain." The ENS communicates with the central nervous system (CNS) through the autonomic nervous system, comprising both sympathetic and parasympathetic components; the latter, particularly via the vagus nerve, serves as the primary pathway for communication between the intestine and the brain. The intestinal microbiota influences the ENS by producing various mediators (metabolites and neurotransmitters) that affect the development and functioning of its neurons and glial cells (7, 14, 15).

Additionally, the intestine contains abundant enteroendocrine cells, which produce many hormones that act locally and can be disseminated throughout the body via the bloodstream. Moreover, intestinal microorganisms generate mediators characteristic of the endocrine system, which can interact with the intestinal mucosa cells, thereby modifying their functions (16).

Gut Microbiota and Extra-Intestinal Organ Axis

The intestinal microbiota interacts with the homeostatic systems in the gut, thereby influencing intestinal homeostasis. These interactions are established early in life, with specific "sensitive periods" or "critical windows" that are crucial for their consolidation (17). Furthermore, the microbiota's mediators and those generated through neuroimmunoendocrine communication involving intestinal microorganisms can transit from the gut to other organs and systems.

Numerous axes exist connecting the microbiota, intestine, and extra-intestinal organs, including the gut–brain axis, gut–muscle axis, gut–bone axis, gut–bone marrow axis, gut–liver axis, gut–lung axis, gut–adipose axis, gut–kidney axis, gut–heart axis, and gut–skin axis, among others (Figure 11.4). For instance, the gut–liver axis plays a crucial role in mediating the effects of the microbiota on various aspects of metabolism (18–20). The gut microbiota can also influence muscle mass and function through the "gut–muscle axis" (21, 22). The gut–bone axis is also significant in this context (23). Furthermore, these axes are interconnected; for example, the gut–brain–liver axis illustrates the complex interactions among these systems (24). Nevertheless, in this chapter only the microbiota–gut–brain axis will be briefly commented.

Microbiota–Gut–Brain Axis

As previously mentioned, mediators generated from the interaction between the microbiota and homeostatic systems in the gut can access the CNS via the nervous system, predominantly through the vagus nerve, but also through the bloodstream (encompassing intestinal circulation, systemic

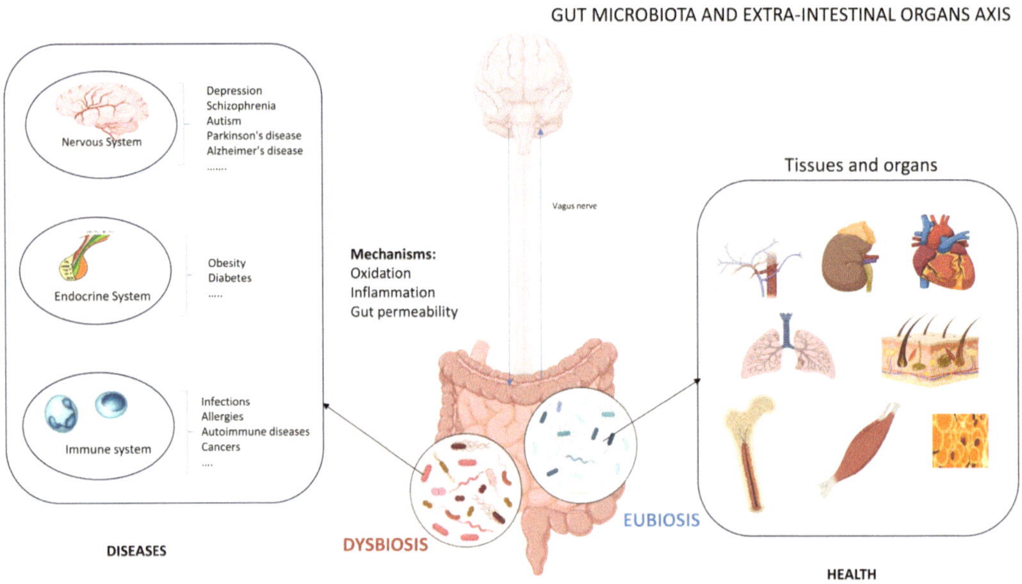

FIGURE 11.4 Gut microbiota and the extra-intestinal organs axis. Dysbiosis and diseases associated with subsequent alterations in homeostatic systems.

circulation, and the blood–brain barrier). Consequently, events occurring within the intestine can significantly influence brain development and function, ultimately reflecting in behavioral responses.

The intricate system of multidirectional interconnections between the microbiota and the brain is termed the "microbiota–gut–brain axis." This is an aspect of the "body–brain axis," which, given the bidirectional nature of these communications, modulates general homeostatic body responses, for example, that carried out by the immune system, regulating body inflammatory responses (25). In addition, mediators produced in the CNS can also reach the intestine through nerves or the bloodstream, thus modifying the neuroimmunoendocrine dialogue within the gut and the interactions between these systems and the microbiota. This interrelationship explains how inadequate responses to stress, emotional states, and other factors can impact intestinal function and the overall communication between microorganisms and homeostatic systems. Hence, this axis is also referred to as the "brain–gut–microbiota" axis (7, 26, 27).

Gut Microbiota in Healthy and Illness

The dialogues between the microbiota and homeostatic systems significantly influence an individual's health state, as well as their risk of illness and mortality throughout life. The critical role of the intestinal microbiota in maintaining health has led to extensive efforts in recent years to define its composition. This composition varies due to many factors, both individual (such as age and diet) and inter-individual (including genetic background and geographic location). Consequently, it is challenging to establish a universal definition of a "normal" healthy microbiota or the state of eubiosis, which is vital for health maintenance and disease prevention (28).

Each person possesses unique genetic traits and has experienced distinct environmental and lifestyle factors over their lifetime, leading to what can be termed their own "psychoneuroimmunobiography" and "microbiotabiography" (Figure 11.3). Thus, the dialogue between an individual's microbiota and homeostatic systems is unique to each person (7). Despite these individual differences, several characteristics indicate an optimal gut microbiota. These include functional competence (the ability to produce beneficial trophic and metabolic factors for the host), ecological stability (resilience and resistance to changes), and microbial composition stability, which depends on richness and diversity.

Dysbiosis, which extends beyond mere alterations in the composition of gut microorganisms, may be the basis of hundreds of diseases that affect the digestive system and all organs and systems (29). These pathologies often stem from the impaired functioning of homeostatic systems. The immune system is often impacted, leading to infections, allergies, autoimmune diseases, and cancers. In other instances, the endocrine system is affected, as seen in disorders like obesity and anorexia. Additionally, dysbiosis can influence the nervous system, contributing to conditions such as depression, schizophrenia, bipolar disorder, autism, Parkinson's disease, and Alzheimer's disease (30–32). All of these illnesses are associated with inflammation and given that oxidative stress and inflammation often occur concurrently (33), both oxidative and inflammatory stress may play significant roles in many dysbiosis-related diseases (7, 34) (Figure 11.4).

AGING AND FRAILTY

Concepts of Aging: Biological Age

Aging represents a biological process that begins in adulthood, characterized by a progressive and generalized deterioration of the body's functions, leading to reduced adaptability to changes and diminished capacity to restore homeostasis. Maintaining homeostasis is fundamental to health, so it is understandable that as we age, we experience poorer health and an increased risk of morbidity and mortality (5–7).

Considering that the functional state at the onset of adulthood significantly influences the aging process is crucial. Individuals who enter adulthood with signs of premature aging tend to have a shorter life expectancy than those with greater physiological potential. Since aging ultimately culminates in death, the aging process directly affects the longevity of each individual. Depending on the condition of one's homeostatic systems, the rate of aging may vary, impacting the potential for achieving extended, healthy longevity. While genetics plays a role, environmental factors and lifestyle choices are even more influential. Although there is no universal formula for healthy aging, certain practices—such as adequate nutrition, physical and mental activity, social connections, stress management, and a strong sense of purpose—can optimize homeostatic function, promote health, and contribute to healthier longevity (5–7, 35, 36).

This perspective underscores the heterogeneous nature of the aging process, where no two individuals age in precisely the same manner, even when they share the same chronological age and environment. Consequently, the concept of "biological age" has emerged as a more informative measure of individual aging and its rate. A higher biological age indicates accelerated aging and a shorter life expectancy, while a lower biological age is associated with greater longevity. However, determining biological age is challenging, as it relies on various markers that have not been universally validated in relation to life expectancy. Our group has proposed the "Immunity Clock," a mathematical model that calculates biological age based on immune function parameters analyzed in peripheral blood leukocytes, providing a reliable assessment (37). Understanding biological age is advantageous because, unlike chronological age, which inevitably increases, biological age can be influenced by lifestyle changes, potentially allowing individuals to "rejuvenate" it (6, 37, 38).

At this point, it is essential to highlight the immune system's significant role in aging. Several theories explain the aging process, with one of the most widely accepted being the oxidative damage theory. This theory posits that oxidative damage to cellular structures results from the oxidants (free radicals and reactive oxygen species) produced during the normal use of oxygen. It is not the mere presence of oxidants that causes cellular damage, as our bodies utilize them for various functions; rather, it is their excess in the context of diminished antioxidant defenses that leads to oxidative stress. Additionally, the phenomenon of "inflamm-aging" suggests that chronic inflammation associated with aging contributes to physiological deterioration. More recently, our group has proposed the oxidation-inflammation theory, which posits that oxidative and inflammatory stress interplay affects all body components, particularly the homeostatic, nervous, endocrine, and immune systems, ultimately compromising health and increasing morbidity and mortality (6). Moreover, the immune system's need to produce oxidants and inflammatory responses for pathogen defense can, if poorly regulated, exacerbate oxidative and inflammatory stress, thereby accelerating aging. Thus, the immune system plays a crucial role in "oxi-inflamm-aging," modulating the aging rate (6, 36).

Various lifestyle factors can influence this oxidation-inflammation balance, either accelerating or decelerating aging. Negative factors include emotional stress, loneliness, a sedentary lifestyle, and poor dietary habits. Conversely, effective stress management, diets rich in antioxidants, regular physical and mental activity, social engagement, and connection with the natural environment can help control oxidation and inflammation, improving homeostatic function and health, ultimately enhancing a healthier lifespan (6, 35, 36).

Concepts and Dimensions of Frailty

We know that the proper functioning of the organism and optimal regulation of homeostasis require close bidirectional relationships among the immune, nervous, and endocrine regulatory systems. This is particularly important during processes of homeorhesis (the maintenance of a dynamic equilibrium during development and long-term adaptation in response to internal and external changes) and allostasis (stability through change via immediate adaptation to stressful situations). Colloquially, one might say—albeit with the risk of some inaccuracies—that while homeostasis seeks to maintain a constant state under changing conditions, homeorhesis and allostasis focus on adaptation over time and in response to specific situations. These two concepts, which relate to how organisms manage stress and environmental changes, are crucial for studying and assessing the processes of aging (39).

In this context, "Frailty" is a complex multisystem geriatric syndrome characterized by increased vulnerability to stressors and a decreased functional reserve and resilience against internal and external changes, leading to a widespread loss of homeostasis (40–42). This heightened vulnerability is associated with an increased risk of morbidity that exceeds what is typically observed in the aging process, significantly affecting the quality of life in older adults and hindering the attainment of healthy aging. Consequently, frailty is often linked to older individuals who do not age well; its prevalence increases with advancing age, and is more common in women than in men, posing risks of functional decline, falls, disability, dependency, institutionalization, and even death, as indicated by the Word Health Organization (WHO, 2001) (43). This underscores the importance of early detection of the "frailty phenotype," differentiating between robust, pre-frail, and frail individuals (44). Preventive interventions are feasible in the pre-frail stage, but only palliative measures may be possible once clear frailty is established. Notably, frailty is a better indicator of health status and predictor of mortality than chronological age among older adults (45).

Initially, frailty was understood primarily in terms of physical dimensions; however, it has since been recognized as encompassing cognitive, emotional, sensory, social, and metabolic/nutritional dimensions. The Physical Frailty phenotype, developed by Fried and colleagues, remains the most widely used measure. The associated scale includes criteria such as reduced grip strength, slow walking speed, low endurance, limited physical activity, and unintentional weight loss. An individual is considered frail if they meet three or more of these criteria, while those meeting one or two criteria are classified as pre-frail, identifying those at heightened risk of progression to frailty (44). This physical frailty phenotype has been validated as a predictor of hospitalization and mortality risk (46).

Cognitive Frailty represents a clinical syndrome marked by a loss of adaptability within the framework of brain activity (47). Cognitive frailty is associated with cognitive impairment and is considered an early stage of neurodegeneration (48). It is also a significant predictor of falls, various types of injuries, and disability in older adults, as well as increases the risk of dementia, dependency in activities of daily living, and mortality (49).

Sensory frailty addresses the decline of sensory systems, which can lead to motor and cognitive deterioration, potentially preceding loss of mobility and independence by several years. Deterioration in sensory functions—such as vision, hearing, smell, and touch—constitutes risk factors for frailty and mortality (50, 51). However, the relationship between sensory and motor skills and frailty can vary depending on the assessment measures (52).

Psychological frailty encompasses cognitive, emotional, and affective aspects. Nevertheless, they currently tend to separate (53). Emotional or affective fragility indicates a diminished subjective response to emotional challenges. Research has primarily focused on the relationship between emotional fragility and depression, characterized by symptoms such as low energy, fatigue, sleep disturbances, and reduced interest (54), along with perceived stress, which correlates with greater fragility (55).

Recently, the roles of metabolic and nutritional factors in frailty have gained attention, leading to the introduction of Metabolic and Nutritional Frailty. Evidence suggests an interplay between weight, strength, energy loss, sarcopenia, and metabolic changes that support motor functions. Thus, conditions like metabolic syndrome, type 2 diabetes, and dysregulation of energy-producing pathways, among others, are associated with frailty (56). Oral Fragility, characterized by reduced ability to chew or swallow, poses risks for both physical and cognitive frailty and is linked to poorer physical, psychological, and social conditions (57).

Social Frailty is a prevalent and complex condition among older adults, characterized by the state of vulnerability due to the lack or absence of social resources that enable health. This emerging concept is associated with declines in physical function, cognitive impairment, and increased physical frailty, as well as it is predictive of mortality. Despite the high prevalence of social frailty, there is insufficient consensus on its definition and measurement (58).

The issue lies in the fact that, despite acknowledging multidimensional fragility, there is significant controversy regarding the definition's components and the assessment scales used. This leads to varying results depending on which definitions and scales are applied.

Differences between Aging and Frailty

The terms aging and frailty are sometimes considered synonymous or at least closely related; however, they do not necessarily overlap. Frailty can be viewed as an aspect of aging. Still, while all individuals experience aging, frailty affects only a subset of older adults who, at some point, experience an accelerated rate of aging (59). Therefore, we propose that biological age is more indicative of an individual's aging process and life expectancy than frailty itself (59).

GUT MICROBIOTA IN AGING

Age-Related Changes in Microbiota

The Beginning

The initial and massive colonization of the gut by microorganisms occurs during childbirth, primarily through the vaginal canal, which causes an avalanche of antigens that contribute to the maturation of the neonatal immune system. This is why cesarean delivery is associated with various metabolic and immunological challenges in children. Consequently, the mother's microbiota and the mode of delivery influence the initial microbial inoculum. Following birth, maternal lactation provides probiotics and prebiotics essential for establishing a healthy gut microbiota. Thus, the development of intestinal microbiota during the early stages of life is a highly dynamic process, influenced by perinatal factors such as gestational age, delivery method, lactation practices, and exposure to antibiotics. Furthermore, the maternal diet and other lifestyle factors play a crucial role in shaping the gut microbiota and developing of the offspring's homeostatic systems properly. In addition, in this last aspect, it is widely accepted that the maternal microbiome plays an indirect, but key role in the immune and nervous system fetal development, through the microbial metabolites produced by the mother, which may cross the blood–placenta barrier.

Under optimal conditions—specifically, full-term vaginal delivery and breastfeeding—the diversity and complexity of the microbiota initially remain low, dominated by bacteria from the phyla Pseudomonadota and Actinomycetota during the first months of life. Infants are rapidly colonized by enterobacteria and lactic acid bacteria, including various members of the genera *Lactobacillus* and *Streptococcus*, creating an environment conducive to the proliferation of anaerobic bacteria such as *Bifidobacterium*, which quickly becomes the dominant microorganism. With weaning and the introduction of complementary foods, the microbiota becomes more complex, with phyla such as *Bacillota* and *Bacteroidota* proliferating to become predominant. Thus, gut microbiota undergoes relevant developmental changes during the first year of life, being influenced by a great variety of external factors, including diet as one of the most important. By approximately three years of age, the microbiota exhibits significant similarities, though not yet complete, to that of adults. This developmental trajectory is influenced by dietary patterns, household composition (e.g., presence of siblings or pets), and the environmental context, whether rural or urban (60–62).

Given that early gut microbiota, particularly in humans, significantly affects health outcomes in adulthood, early life presents a critical opportunity to promote optimal seeding and nurturing of the gut microbial ecosystem, thereby ensuring lifelong health (63). In fact, the first five years of life has key implications for long-term health outcomes (64).

General Composition of the Microbiota during Adulthood

In adulthood, in the absence of disruptive factors, such as poor dietary habits or antimicrobial exposure, the microbiota of a healthy adult tends to remain stable. Diet is the primary determinant of microbiota composition and function. Thus, diets rich in fruits and vegetables are associated with higher levels of the genus *Prevotella* and increased production of SCFAs. Additionally, microorganisms such as *Bifidobacterium* are linked to healthy dietary patterns, such as the Mediterranean diet. Conversely, a "Western" diet is associated with a predominance of Bacteroides (65). It is also essential to recognize that various factors can alter microbiota composition during adulthood, including the use of antibiotics and other medications, and physiological changes related to pregnancy (66, 67).

Changes in the Microbiota Associated with Aging

Intriguingly, the stages of greatest instability and least diversity of the microbiota occur at both the beginning and end of life. However, the underlying reasons for these changes are fundamentally different: in childhood, they relate to the maturation of the physiological and homeostatic state of the individual, whereas in old age, they are associated with a decline in physiological and homeostatic functions.

Defining the changes in the microbiota during aging poses numerous challenges, as this period is not extensively studied in this context. Despite significant variability influenced by geographic location and health status—factors we believe correlate with differing biological ages—some common characteristics have been identified in the microbiotas of old adult individuals. Research has shown a decrease in bacterial diversity compared to young adults, alongside shifts in the predominant species. Notably, there is an increase in facultative anaerobic and opportunistic pathogenic bacteria (e.g., *Eggerthella*, *Bilophila*, *Fusobacteria*, *Streptococcus*, and members of the *Enterobacteriaceae* family). Conversely, there is a marked reduction in dominant commensal bacteria typically found in healthy adult microbiotas (e.g., *Prevotella*, *Faecalibacterium*, *Eubacterium rectale*, *Lachnospira*, *Coprococcus*, and *Bifidobacterium*). In elderly populations, these beneficial bacterial communities may be supplanted by other commensals, including genera such as *Akkermansia*, *Christensenellaceae*, *Butyricimonas*, *Odoribacter*, and *Butyricicoccus* (68–70).

Factors Influencing Age-Related Differences in Gut Microbiota

Several factors influence the establishment and development of the microbiota, complicating the attribution of changes solely to aging. It is essential to recognize that the microbiota is affected by numerous variables, which can act as confounding factors in research (Figure 11.5). Moreover, discrepancies may arise from the diverse techniques employed in microbiota assessment across different studies.

1. Geographical Location. Many studies have shown that certain bacteria decrease in elderly individuals but not uniformly across all European demographics (71). This occurs, for example, with *Faecalibacterium prausnitzii*, a beneficial bacterium, and with *Clostridium cluster IV* group, which decrease or increase, respectively, in elderly individuals, but with opposite results depending on the country studied. Another example is the *Bacteroides-Prevotella* group, which declines in elderly individuals of several countries but increases in those of others. Controversy exists regarding the representation of the *Lactobacillus* group in the elderly, with some studies indicating a decrease while others report an increase with aging.
2. Social Conditions. Social factors also contribute to differences in the microbiota of older adults. Variations have been observed between the microbiota of elderly individuals residing in the community and those in senior centers, where a higher proportion of *Bacteroidota* and a lower proportion of bacteria typically associated with a healthy microbiota have been identified, adversely affecting their quality of life (72, 73). Furthermore, a study examining inter-individual variation among elderly individuals of the same age group and geographical location highlighted the pronounced uniqueness of microbiota composition accompanying aging (74), complicating the identification of common characteristics that define an aging microbiota profile.
3. Health State. The taxonomic composition of the microbiota also differs between healthy and unhealthy elderly individuals. While it is generally acknowledged that bifidobacteria and other beneficial bacteria decrease with aging, recent studies suggest that these bacteria may increase in certain healthy elderly populations. Differences in microbiota composition have been identified between healthy elderly individuals and those exhibiting signs of frailty, which could aid in identifying microbiota profiles associated with healthy aging. Longitudinal studies would ideally address this issue; however, cross-sectional studies comparing healthy and unhealthy individuals have revealed consistent microbiota characteristics linked to poor health in elderly populations (75, 76). These aging microbiotas often show a loss of beneficial commensal bacteria and an increase in potentially pathogenic species, especially members of the *Enterobacteriaceae* family, *Ruminococcus torques*, and certain *Clostridium* species (69, 72, 73, 77). Additionally, reducing commensal bacteria among sick elderly individuals compared to their healthy counterparts may serve as biomarkers of healthy aging. Beneficial bacteria such as *Akkermansia*, *Odoribacter*, *Butyricimonas*, *Butyrivibrio*, and *Oscillospira*, along with members of the families *Christensenellaceae* and *Barnesiellaceae*, have been identified as potential indicators of health (69, 77). A study conducted in 2020, which analyzed over 2,500 metagenomes from individuals aged 20 to 89, identified groups of bacteria associated with healthy aging and others correlated with various pathologies (78). Many of the bacteria that decline with aging and are associated with health loss are SCFA producers, such as *Faecalibacterium*, *Roseburia*, *Coprococcus*, and *Eubacterium* spp., which positively contribute to host homeostasis, with butyrate playing a particularly significant role (79, 80). In healthy elderly adults, reductions in these bacteria can be compensated by increases in other commensal bacteria that produce butyrate, such as *Odoribacter*, *Butyricimonas*, *Butyrivibrio*, and *Oscillospira* (70).
4. Biological Age. Biological age—significantly influenced by environmental and lifestyle factors—may have a greater impact on microbiota changes than chronological age.

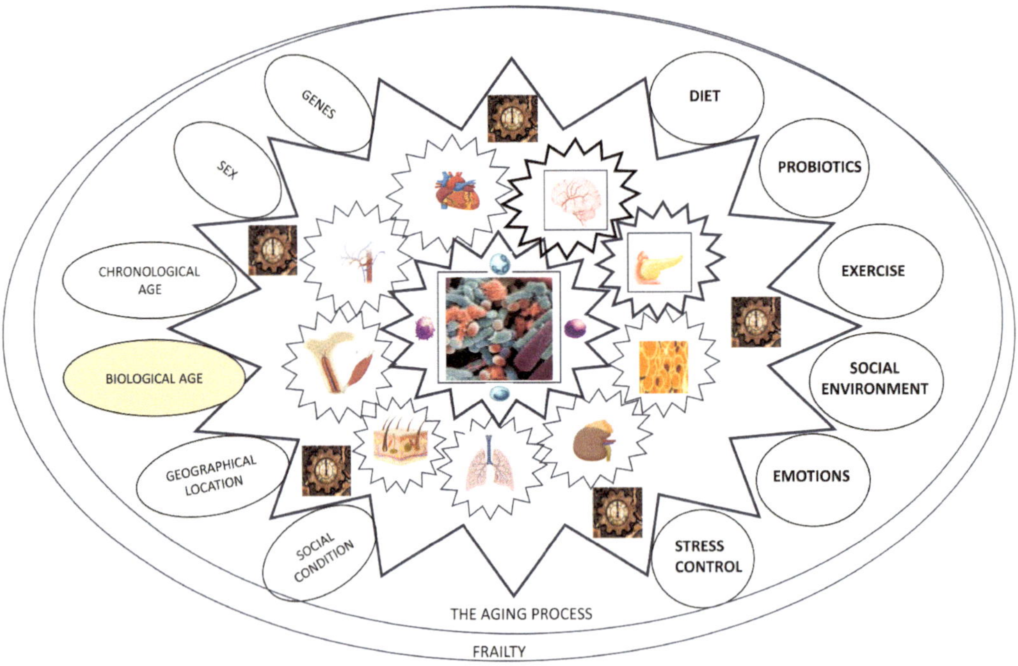

FIGURE 11.5 Characteristics and lifestyle factors influencing the gut microbiota in each individual, and its connection to various tissues and systems of the body, regulated by biological clocks. These factors collectively shape the aging process and the risk of frailty.

Preliminary unpublished findings from an ongoing study suggest that the intestinal microbiota of adults aged 58–60, analyzed over 14 years, show that those with a younger biological age, as determined by the Immunity Clock (37), do not exhibit the typical microbiota alterations associated with aging. Instead, their microbiota resembles that of earlier assessments, indicating a potential correlation with slower aging processes (Figure 11.5).

The Case of Centenarians

Centenarians and supercentenarians (individuals over 104 years of age) serve as models of healthy aging, characterized by a lower incidence of chronic diseases and reduced morbidity. Numerous studies involving this elderly demographic have identified significant differences in intestinal microbiota when compared to younger adults and elderly individuals (81, 82). A notable characteristic of centenarians' microbiota is the increased microbial diversity relative to younger elderly individuals. While a reduction in the core bacterial taxa that constitute the healthy adult microbiota—primarily from the families *Ruminococcaceae*, *Lachnospiraceae*, and *Bacteroidaceae*—has been observed, this decline appears to be offset by an increase in the abundance of less common commensal bacteria (75). Specifically, beneficial bacteria associated with host health, such as *Akkermansia*, *Bifidobacterium*, and members of the *Christensenellaceae* family, are more prevalent in centenarians (74). Similar findings have been reported across various centenarian populations in diverse geographical regions, including Russia, Italy, China, and India, suggesting that these microbiota characteristics may be universally associated with healthy aging (75, 83).

However, discerning whether the microbiota differences observed between centenarians and younger adults are causative factors of longevity or consequences of a healthy lifestyle remains challenging. To address this issue, a pilot study conducted in China in 2021 compared the microbiota of healthy centenarians with that of a small cohort of unhealthy centenarians.

Microbiota of the unhealthy group showed some similarity with the one present in other diseases, including an increase in pathogenic bacteria. Conversely, the microbiota of healthy centenarians demonstrated a heightened abundance of host-beneficial bacteria (77). While this last group of bacteria may play a pivotal role in promoting healthy aging, it remains unclear whether their presence is attributable to a healthy lifestyle or if they were inherently present in individuals at a younger age (75). Consequently, further studies involving healthy centenarians, employing rigorously defined criteria concerning the physiological states of the studied populations, are essential.

Role of Microbiota in Aging and Longevity

The significance of microbiota in the aging process and the associated diseases is now universally acknowledged. Recent literature on the "Hallmarks of Aging" has integrated microbiota as one of these hallmarks (84). Despite inconsistent results across human studies, substantial evidence supports the relationship between intestinal microbiota and healthy aging. Research utilizing animal models has demonstrated a connection between age-related dysbiosis in the gut microbiota and premature mortality (85).

Beyond the previously discussed alterations in intestinal microorganisms with aging, a novel area of inquiry is whether gut microbiota can influence aging and longevity. This aspect has received limited investigation. Recent findings indicate that intestinal microorganisms significantly contribute to achieving healthy aging and extreme longevity or, conversely, to the exacerbation of age-related pathologies (79, 86–88). The association between the aging process and gut microbiota is so well-established that a biological age calculated based on microbiota composition has been proposed, termed "gut microbiota Age" (gAge) (89). It is essential to consider taxonomic variations or the production of SCFAs, which are frequently analyzed, and other relevant factors. A crucial aspect is the interaction of the microbiome with various organs and systems, particularly homeostatic systems, which play a significant role in the aging rate and achieved longevity (6–9). Thus, alongside the nearly 400 theories of aging, we must now incorporate the microbiome theory (90), in which dysbiosis causes several problems, including increased inflammation and intestinal permeability, which allow lipopolysaccharides (LPS) to leak into systemic circulation producing the chronic systemic inflammation and accelerating biological aging. The microbiota can influence oxidative and inflammatory stress levels and the functionality of homeostatic systems, both at the intestinal level and through the microbiota–gut–organ axis, thereby impacting overall health. Certain microbial molecules, such as LPS, among others, can exacerbate oxidation-inflammation, while molecules with antioxidant and anti-inflammatory properties, such as butyrate, can mitigate it. Consequently, the composition of one's microbiota can either promote oxidative-inflammation or facilitate its regulation, thereby affecting the rate of aging and longevity.

One potential method to elucidate the role of intestinal microbiota in aging and longevity would involve fecal microbiota transfer from aged individuals to adults to observe the impact on their aging processes and life expectancy. While this is impractical in human subjects, it can be conducted in experimental animal models, such as mice, which have a relatively short mean lifespan (approximately two years compared to 74 years for humans). In a study, the fecal microbiota transplantation from wild type mice into progeroid mice increased the lifespan of these animals (91). We have carried out an experiment in which fecal microbiota transfer from aged mice to adult mice resulted in accelerated aging, as evidenced by behavioral changes (increased anxiety and reduced exploration), immunological alterations (enhanced immunosenescence), and oxidative stress (increased oxidative stress), compared to adult mice receiving microbiota from their peers. This was associated with a reduced life expectancy. In another recent study, we found that transferring fecal microbiota from long-lived animals to aged mice was particularly effective in enhancing behavior, immunity, redox status, and longevity (in publication).

GUT MICROBIOTA IN FRAILTY

When examining the role of gut microbiota in frailty, a syndrome characterized by an exacerbation of the processes seen in aging, the available studies are relatively limited, as this aspect has been explored more recently. Although various methods for determining frailty indices have been employed, and differences exist depending on the country of study, which complicates direct comparisons of results, consistent evidence supports the relationship between frailty and gut microbiota. This relationship is generally accepted as both causal and bidirectional (92–96). Furthermore, the potential role of gut microbiota in frailty syndrome has also been proposed (97).

Although it has been suggested that the differences in gut microbiota between frail and non-frail elderly individuals are not yet conclusive (98), it is currently understood that elderly people with frailty exhibit alterations in the composition, diversity, and metabolites produced by the gut microbiota when compared to their non-frail counterparts (98, 99). These differences generally align with the patterns observed between older adults and younger individuals, as outlined above. In general, frailty appears to be associated with a reduced abundance of beneficial species and an increased abundance of pathobionts. Specifically, this includes lower microbial diversity, a decreased proportion of several microorganisms—particularly those that produce SCFAs—along with an increase in pro-inflammatory microbes, leading to enhanced intestinal permeability (94, 99–102). Specifically, in frail older adults, compared to their non-frail counterparts, there is a lower abundance of *Prevotella, Faecalibacterium* (*Faecalibacterium prausnitzii*), *Roseburia* (*Roseburia inulinivorans*), *Coprococcus eutactus*, and *Blautia*. Additionally, certain beneficial bacteria, such as *Akkermansia mucinifera*, and those belonging to the *Bifidobacterium* and *Lactobacillus* genera, are also found in lower abundance. Conversely, *Bacteroides* (e.g., *Bacteroides fragilis* and *Bacteroides dorei*), *Clostridium* (*Clostridium hathewayi*), *Eubacterium* (e.g., *E. dolichum*, *E. cylindroides*, and *E. dolichum*), *Eggerthella lenta*, *Flavonifractor plautii*, and *Enterobacteriaceae*, which are associated with inflammation and the production of compounds such as trimethylamine (the precursor of trimethylamine N-oxide, TMAO), are found to be more abundant in frail older individuals (101). In a recent extensive study involving men and women aged 62 to 96 years, 18 microbial species and 17 metabolites were identified that changed with the severity of frailty, with these changes being more significant in women. The microbial composition was a predictor of mortality (103). Furthermore, several characteristics of frailty, such as walking speed and grip strength, which are included in the Frailty Index, show a clear relationship with the type and diversity of the gut microbiota and its metabolites (104, 105).

The previously mentioned gut aging clock, applied to an elderly cohort with varying levels of frailty, confirmed the association between gut microbiota and frailty, demonstrating that gut microbiota biological age (gAge) and its predictive residuals are closely related to the frailty stage (89). Thus, “microb-aging” in the gut is linked to frailty aging (106). Further corroboration of the role of gut microbiota in frailty was provided by the effects of fecal microbiota transplantation from young to old mice. This intervention improved frailty (grip strength, physical activity, running speed, and frailty score) and muscle mass. It enhanced intestinal barrier function and reduced systemic inflammation by increasing the abundance of SCFA-producing bacteria (107). Additionally, germ-free (GF) mice that received a microbiota transplant from frail patients exhibited a different microbiota composition compared to those that received transplants from healthy older individuals, including a decrease in *Akkermansia* and an increase in *Enterocloster*, *Parabacteroides*, and *Eisenbergiella*. These changes were accompanied by a decrease in SCFAs and an increase in inflammatory markers, such as C-reactive protein and TNF-α (108).

In the context of oxidation and inflammation’s role in aging and frailty, the gut microbiota is involved. Oxidative stress has been linked to frailty (109), and the gut microbiota regulates the body’s redox state (110), thus influencing frailty through this mechanism (111). Moreover, in elderly volunteers classified as non-frail or frail according to the Fried phenotype, concentrations of oxidative stress and oxidative damage were higher in frail individuals (112), demonstrating an association between oxidative imbalance and Frailty Syndrome (113). Inflammation is also associated with frailty syndrome (106), and alterations in the gut microbiota, as seen in frailty, promote intestinal inflammation and increased permeability. This allows pro-inflammatory metabolites to enter the body, creating a feedback loop that sustains a generalized inflammatory state (97, 100).

Furthermore, given that the gut microbiota is involved in multiple axes linking different organs, several of these—such as the gut–muscle axis and the gut–bone axis—may help explain the association between the microbiota and the pathophysiology of musculoskeletal disorders like sarcopenia (114) and osteoporosis (23). These two age-related diseases overlap in frailty and significantly impact physical health in the elderly (92, 96). Moreover, since these axes are interconnected, dysbiosis can affect organs such as the liver, potentially contributing to cognitive frailty (115). In summary, an unhealthy microbiota may accelerate age-related physiological changes and, consequently, exacerbate the characteristic complexity of frailty in older individuals.

MODULATION OF MICROBIOTA, AGING, AND FRAILTY

Based on the points outlined above, it is evident that any intervention aimed at improving gut microbiota will benefit aging and frailty. Conversely, lifestyle strategies that enhance aging outcomes will also impact the microbiota, which is highly responsive to lifestyle changes (116). To introduce the reader to potential interventions for improving health in aging and frailty through modulation of the microbiota, only a few lifestyle strategies will be briefly discussed. Specifically, those that we have previously published, which can modulate the aging process by affecting homeostatic systems, such as diet, physical exercise, social interaction, and the management of stress and emotions (36) (Figure 11.5).

Diet

Diet is a major determinant of gut microbiota composition throughout the entire lifespan. On one hand, malnutrition predisposes individuals to accelerated aging and, consequently, to frailty (117, 118). On the other hand, an appropriate diet promotes successful aging and prevents frailty (35, 36, 59). Moreover, several of the benefits of a healthy diet in aging may be mediated through the microbiome (119, 120). Thus, the Mediterranean diet (MD), a traditional dietary pattern widely regarded as a model for a healthy diet, has been shown in numerous studies to promote healthy aging. MD also exerts a positive modulatory effect on the intestinal microbiome, favoring bacterial taxa involved in synthesizing bioactive compounds, such as SCFAs, which counteract inflammation, anabolic resistance, and tissue degeneration. Intervention studies conducted in older populations have suggested that the individual response of older adults to MD, in terms of reduced frailty scores and improved cognitive function, is significantly mediated by the composition and functionality of the gut microbiota (120).

Probiotics

In this context, the use of probiotics represents a promising strategy for managing aging and frailty. The current definition, formulated in 2002 by the expert group of the Food and Agriculture Organization of the United Nations (FAO) and the WHO, describes probiotics as "live microorganisms that, when administered in sufficient amounts, confer a health benefit on the host" (121). Probiotics not only help by replenishing the altered intestinal microbiota and alleviating the symptoms of various diseases but also influence homeostatic systems, improving their function (122–124). Consequently, terms such as "Immunobiotics" and "Psychobiotics" have been coined to describe probiotics with specific positive effects on immunity and mental health (125–128).

Although studies in older populations are more limited, a body of work shows that probiotics can have promising effects on health maintenance, particularly as individuals age and become more vulnerable (124, 129, 130). In this context, the term "gerobiotics" has been proposed to define probiotic strains, as well as their derived postbiotics and para-probiotics, that can beneficially modulate aging processes and thus extend the host's healthspan (131). In frailty syndrome, emerging evidence suggests that the use of probiotics can improve the gut microbiota of older adults affected by this condition (132, 133).

The capacity of many probiotics to enhance antioxidant and anti-inflammatory defenses appears to underpin their potential therapeutic effects in the context of aging and frailty (134–137). Indeed, the hypothesis proposed by Elie Metchnikoff, the pioneer in phagocyte biology, cellular innate immunity, probiotics, and gerontology, regarding the role of intestinal microorganisms in promoting longevity, has since been supported by numerous researchers and experimentally validated in animal models (138). Recently, we have observed that ingesting a mixture of probiotic strains improved immune functions, redox balance, and inflammatory status, as well as rejuvenated the biological age of both men and women (38). Furthermore, in experimental animals, several probiotics that enhanced immunity, as well as decreased oxidative and inflammatory state, have been shown to positively influence lifespan extension (139–141).

Physical Exercise

The effects of physical exercise on various aspects of gut physiology are well-established. While these effects are dependent on exercise volume and intensity, they are generally regarded as beneficial to gastrointestinal health. Moderate exercise has been associated with improved immune function, whereas prolonged strenuous exercise has been linked to diminished immune cell function. Overall, physical exercise represents a key strategy for promoting healthy aging (35, 36) and is considered the most effective intervention for delaying frailty (142).

In this context, while it has been proposed that physical exercise, particularly endurance training, may influence intestinal transit, immune function, and modify the composition and metabolism of the gut microbiota, potentially increasing its diversity (143, 144) and even preventing frailty (145), it seems that diet is the most significant factor in modulating gut microbiota composition in athletes. Indeed, a robust relationship between exercise, diet, and gut microbiota has been observed (143, 146).

Moreover, there is evidence that probiotics, prebiotics, and synbiotics have positive health effects in physically active individuals (143, 146), supporting gastrointestinal health, preventing immunosuppressive effects, and reducing the incidence of respiratory tract infections associated with intense exercise (147). Conversely, dysbiosis may impair athletic performance by increasing fatigue, particularly in older individuals. However, the effects of probiotics on this aspect remain inconclusive. Nevertheless, within the framework of the microbiota–gut–brain axis, we have demonstrated that the administration of a synbiotic (rich in *Bifidobacterium lactis*, *Lactobacillus rhamnosus*, *Bifidobacterium longum*, and fructooligosaccharides) for one month was particularly beneficial in physically active individuals compared to the sedentary control group. It exerted an anti-inflammatory bioregulatory effect and improved stress and anxiety levels. Additionally, it enhanced sleep quality in physically active individuals (148). Furthermore, using the same synbiotic and protocol, we observed clear benefits in aged women with fibromyalgia, a condition that, due to musculoskeletal pain, severely limits daily physical activity. This condition is also associated with immunoneuroendocrine dysregulation. The administration of the synbiotic induced an anti-inflammatory effect and improved the interaction between the inflammatory and stress responses. Moreover, it enhanced physical activity levels, sleep quality, and various aspects of mental health in these patients (149).

Given the clear dysregulation of immune-neuroendocrine interactions during aging and frailty, and the downregulation of both inflammatory and stress responses, probiotics could potentially improve the bioregulation of these responses in older individuals, particularly when combined with physical exercise. However, studies on this topic in older populations are scarce, making it a critical avenue for future research in the coming years.

Social Interactions

Social vulnerability, characterized by a lack of adequate social connections, support, or interaction, becomes increasingly common with age and is associated with adverse outcomes such as physical decline, cognitive impairment, and increased mortality (150). Therefore, strategies aimed at mitigating these social issues appear to be beneficial for aging, frailty, and ultimately, longevity. In support of this, we have observed in mice that social interactions between chronologically older or prematurely aging animals and younger adult animals improve behavior, immune function, redox balance, and inflammatory status, while also reducing biological age and promoting healthier longevity (151–153).

Furthermore, maintaining frequent social interactions may help mitigate the negative effects of frailty on cognitive function in older adults (154). In this context, social interactions are closely linked to the gut microbiota, which may influence host social behavior (155, 156). As a result, factors that affect the microbiota can alter the neural mechanisms involved in behavioral responses (157). Additionally, social behavior and structure play a role in shaping the composition of the gut microbiota across individuals, and these reciprocal interactions can enhance our understanding of social dynamics (158). Moreover, the idea that the microbiota can influence mate choice and sexual attractiveness is now widely accepted (159, 160).

Control of Stress and Emotions

With the communication pathways of the gut–brain axis now well established, and recognizing that products released by the microbiota, such as SCFAs, can influence brain homeostasis by crossing the blood–brain barrier and reaching structures within the CNS, including the limbic system, it is reasonable to assume that this "language" between microbes and the brain influences both our thoughts and emotions. It is well established that a heightened perception of stress or negative emotions accelerates aging and the onset of related diseases, including frailty (36, 53–55, 161). Due to its bidirectional relationship with the brain, emotional and stressful situations directly affect the gut microbiota. It can modulate susceptibility and resilience to stress, influencing mood and mental health (162–164). Numerous examples demonstrate the emotional regulation of various gut microorganisms, some exerting positive effects, while others have negative impacts. However, most of these studies have been conducted in experimental animal models, highlighting the need for further human studies, particularly in older populations.

CONCLUSIONS AND FUTURE

Although numerous studies have been conducted on the microbiota and aging, much remains to be understood. The knowledge gap is even more pronounced when considering the relationship between microbiota and frailty. However, the importance of the gut microbiota in understanding health and aging is so significant that we can no longer regard these processes, along with associated conditions such as frailty and various age-related diseases, without accounting for the state of the microbiota. Therefore, understanding how to effectively modulate the microbiota is crucial for achieving healthy longevity. Thus, in each individual, genes, sex, chronological age, geographical location, social condition, and especially the biological age that all these factors and many others depending on personal attitude and lifestyle have conditioned are influencing the state of the intestinal microbiota, and its relationship with all tissues, organs, and body systems. In this context, the homeostatic systems, such as the nervous, endocrine, and immune systems, are crucial in how the aging process develops, and the immune system's role in this process can be highlighted. Furthermore, an important fact that has not been previously discussed is that all these interactions that will affect aging are carried out within the framework of biological rhythms that must function properly (165–167) (Figure 11.5).

In our opinion, two key issues face research and its translational application in the coming years. The first is that, although the majority of experimental studies have been conducted in animals, particularly germ-free models (which provide robust scientific data for investigating mechanisms of efficacy), and despite the well-recognized difficulty of conducting human studies—especially in aging and frailty—there is a pressing need to increase such research in human populations, rather than continuing to draw conclusions solely from the extensive body of review articles currently available in relation to original research. The second, arising from the first, is that conclusions must continue to be based on original studies, with an increasing emphasis on research in humans. To this end, conducting pilot clinical studies and clinical trials involving an adequate number of participants to evaluate the proposed interventions could provide greater insight into their benefits. This approach would further substantiate research on immune-neuroendocrine mechanisms observed in animal models while minimizing positive and negative biases often arising from overreliance on review and opinion-based articles.

REFERENCES

1. Margolles A, Suarez JE (2020) The human microbiota: General concepts, composition, distribution and functions. Approach Aging Control. 24: 8–16.
2. Sommer F, Bäckhed F (2013) The gut microbiota–masters of host development and physiology. Nat. Rev. Microbiol. 11: 227–238.
3. Kundu P, Blacher E, Elinav E, Pettersson S (2017) Our gut microbiome: The evolving inner self. Cell 171: 14811493.
4. McCallum G, Tropini C (2024) The gut microbiota and its biogeography. Nat Rev Microbiol. 22 (2): 105–118.
5. De la Fuente M (2018) Bio-psycho-social bridge: The psychoneuroimmune system in successful aging. In Fernández-Ballesteros R, Benetos A and Robie JM (eds), Cambridge Handbook of Successful Aging. New York: Cambridge University Press, pp. 265–280.
6. De la Fuente M, Miquel J (2009) An update of the oxidation-inflammation theory of aging: The involvement of the immune system in oxi-inflamm-aging. Curr Pharm Des. 15 (26): 3003–3026.
7. De la Fuente M (2021) The role of the microbiota-gut-brain axis in the health and illness condition: A focus on Alzheimer's disease. J Alzheimer's Dis. 81: 1354–1360.
8. Takeuchi T, Nakanishi Y, Ohno H (2024) Microbial metabolites and gut immunology. Annu Rev Immunol. 42 (1): 153–178.
9. Ullah H, Arbab S, Tian Y, Chen Y, Liu CQ, Li K (2024) Crosstalk between gut microbiota and host immune system and its response to traumatic injury. Frontier Immunol. 15: 1413485.
10. Aghighi F, Salami M (2024) What we need to know about the germ-free animal models. AIMS Microbiol. 10 (1): 107–147.
11. Mann ER, Lam YK, Uhlig HU (2024) Short-chain fatty acids: Linking diet, the microbiome and immunity. Nat Rev Immunol. 24 (8): 577–595.
12. Negi S, Das DK, Pahari S, Nadeem S, Agrewala JN (2019) Potential role of gut microbiota in induction and regulation of innate immune memory. Front Immunol. 10: 2441.
13. Takeuchi T, Nakanishi Y, Ohno H (2024) Microbial metabolites and gut immunology. Annu Rev Imunol. 42 (1): 153–178.
14. Gan Y, Chen Y, Zhong H, Liu Z, Geng J, Wang H, Wang W (2024) Gut microbes in central nervous system development and related disorders. Front Immunol. 14: 1288256.
15. Qu S, Yu Z, Zhou Y, Wang S, Jia M, Chen T, Zhang X (2024) Gut microbiota modulates neurotransmitter and gut-brain signaling. Microbiol Res. 287: 127858.
16. Lamaudiere MTF, Turner MC, Arasaradnam RP, Morozov IY (2024) Human gut microbiota and endocrinology: Paradigm shift from genome to its regulation. J Endocrinol. 260 (2): e230348.
17. Iza YAP, Brown CC (2024) Early life imprinting of intestinal immune tolerance and tissue homeostasis. Immunol Rev. 323 (1): 303–315.
18. Tilg H, Adolph TE, Trauner M (2022) Gut-liver axis: Pathophysiological concepts and clinical implications. Cell Metab. 34 (11): 1700–1718.
19. Hsu CL, Schnab B (2023) The gut–liver axis and gut microbiota in health and liver disease. Nat. Rev. Microbiol. 21: 719–733.
20. Sivaprasadan S, Anila KN, Nair K, Mallick S, Biswas L, Valsan A, Praseedom RK, Nair BKG, Sudhindran S (2024) Microbiota and gut-liver axis: An unbreakable bond? 81 (7): 193.
21. Zhao J, Huang Y, Yu X (2021) A narrative review of gut-muscle axis and sarcopenia: The potential role of gut microbiota. Int J Gen Med. 14: 1263–1273.
22. Li G, Jin B, Fan Z (2022) Mechanisms involved in gut microbiota regulation of skeletal muscle. Oxid Med Cell Longev. 18: 2151191.

23. Zhang YW, Wu Y, Liu XF, Chen X, Su JC (2024) Targeting the gut microbiota-related metabolites for osteoporosis: The inextricable connection of gut bone axis. Ageing Res Rev. 94: 102196.
24. Wang Z, Zeng M, Wang Z, Qin F, Chen J, He Z (2021) Dietary polyphenols to combat nonalcoholic fatty liver disease via the gut-brain-liver axis: A review of possible mechanisms. J Agric Food Chem. 69 (12): 3585–3600.
25. Jin H, Li M, Jeong E, Castro-Martinez F, Zuker ChS (2024) A body-brain circuit that regulates body inflammatory responses. Nature. 630 (8017): 695–703.
26. Cryan JF, O'Riordan KJ, Cowan CSM, Sandhu KV, Bastiaanssen TFS, Boehme M, et al. (2019) The microbiota-gut-brain axis. Physiol. Rev. 99: 1877–2013.
27. Shekarabi A, Qureishy I, Puglisi CH, Dalseth M, Vuong HE (2024) Host-microbe interactions: Communication in the microbiota-gut-brain axis. Curr Opin Microbiol. 80: 102494.
28. Gomaa EZ (2020) Human gut microbiota/microbiome in health and diseases: A review. Antonie Van Leeuwenhoek. 113: 2019–2040.
29. Rojo D, Mendez-Garcia C, Raczkowska BA, Bargiela R, Moya A, Ferrer M, Barbas C (2017) Exploring the human microbiome from multiple perspectives: Factors altering its composition and function. FEMS Microbiol Rev. 41: 453–478.
30. Levy M, Kolodziejczyk AA, Thaiss CA, Elinav E (2017) Dysbiosis and the immune system. Nat Rev Immunol. 17 (4): 219–232.
31. Clemente-Suarez VJ, Redondo-Florez L, Rubio-Zarapuz A, Martin-Rodriguez A, Tornero-Aguilera JF (2024) Microbiota implications in endocrine-relates diseases: From development to novel therapeutic approaches. Biomedicines. 12 (1): 221.
32. Borrego Ruiz A, Borrego JJ (2024) An update overview on the relationship between human gut microbiome dysbiosis and psychiatric and psychological disorders. Prog Neuropsychopharmacol Biol Psychiatry. 128: 110861.
33. De la Fuente M (2018) Oxidation and inflammation in the immune and nervous systems, a link between aging and anxiety. In Fulop T, Franceschi C, Hirokawa K, Pawelec G (eds), Handbook of Immunosenescence. Cham: Springer Nature, pp. 1–31.
34. Nohesara S, Abdolmaleky HM, Thiagalingam S, Zhou JR (2024) Gut microbiota defined epigenomes of Alzheimer's and Parkinson's diseases reveal novel targets for therapy. Epogenomics. 16 (1): 57–77.
35. De la Fuente M, Cruces J, Hernandez O, Ortega E (2011) Strategies to improve the functions and redox state of the immune system in aged subjects. Current Pharm. Des. 17 (36): 3966–3993.
36. Martinez de Toda I, Ceprian N, Diaz del Cerro E, De la Fuente M (2021) The role of immune cells in oxi-inflammaging. Cells. 10: 2974.
37. Martinez de Toda I, Vida C, Diaz-Del Cerro E, De la Fuente M (2021) The immunity clock. J Gerontol A Biol Sci Med Sci. 76 (11): 1939–1945.
38. Felix J, Baca A, Taboada L, Alvarez-Calatayud G, De la Fuente M (2024) Consumption of a probiotic blend with vitamin D improves immunity, redox, and inflammatory state, decreasing the rate of aging – A pilot study. Biomolecules. 14: 1360.
39. Cano-Martinez A, Rubio-Ruiz ME, Guarner-Lans V (2024) Homeostasis and evolution in relation to regeneration and repair. J Physiol. 602 (11): 2627–2648.
40. Campbell AJ, Buchner DM (1997) Unstable disability and the fluctuations of frailty. Age Aging. 26: 315–318.
41. Ghachem A, Fried LP, Legault V, Bandeen-Roche K, Presse N, Gaudreau P et al. (2021) Evidence from two cohorts for the frailty syndrome as an emergent state of parallel dysregulation in multiple physiological systems. Biogerontology 22(1): 63–79.
42. Taylor JA, Greenhaff PL, Bartlett DB, Jackson TA, Duggal NA, Lord JM (2023) Multisystem physiological perspective of human frailty and its modulation by physical activity. Physiol Rev. 103 (2): 1137–1191.
43. World Health Organization (2001) International Classification of Functioning, Disability and Health. World Health Organization, Geneva.
44. Fried LP, Tangen CM, Walston J, Newman AB, Hirsch C, Gottdiener J et al.; Cardiovascular Health Study Collaborative Research Group (2001) Frailty in older adults: Evidence for a phenotype. J Gerontol A Biol Sci Med Sci. 56: M146–156.
45. Clegg A, Young J, Iliffe S, Rikkert MO, Rockwood K (2013) Frailty in elderly people. Lancet. 381 (9868): 752–762.
46. Hoogendijl EO, Afilalo J, Ensrud KE, Kowal P, Onder G, Fried LP (2019) Frailty: Implications for clinical practice and public health. Lancet. 394 (10206): 1365–1375.
47. Arai H, Satake S, Kozaki K (2028) Cognitive frailty in geriatrics. Clin Geriatr Med. 34 (4): 667–675.

48. Cipolli GC, Ribeiro IC, Yasuda CL, Balthazar MLF, Fattori A, Yassuda MS (2024) Frailty and brain changes in older adults without cognitive impairment: A scoping review. Arch Gerontol Geriatr. 123: 105395.
49. Zhang Y, Xia H, Jiang X, Wang O, Hou L (2024) Prevalence and outcomes of cognitive frailty among community-dwelling older adults: A systematic review and meta-analysis. Res Gerontol Nurs. 17 (4): 2020–212.
50. Panza F, Lozupone M, Sardone R, Battista P, Piccininni M, Dibello V, et al. (2018) Sensorial frailty: Age-related hearing loss and the risk of cognitive impairment and dementia in later life. Ther Adv Chronic Dis.10: 1177–2040622318811000.
51. Tan BKJ, Man REK, Gan ATL, Fenwick EK, Varadaraj V, Swenor BK, et al. (2020) Is sensory loss an understudied risk factor for frailty? A ystematic review and meta-analysis. J Gerontol A Biol Sci Med Sci. 75(12): 2461–2470.
52. Beier F, Löffler M, Nees F, Hausner L, Frölich L, Flor H (2022) Sensory and motor correlates of frailty: Dissociation between frailty phenotype and frailty index. BMC Geriatr. 22 (1): 755.
53. Lameirinhas J, Gorostiaga A, Etxebarria I (2024) Definition and assessment of psychological frailty in older adults: A scoping review. Ageing Res Rev. 100: 102442.
54. Kim Y, Kim J, Kim M, Song K, Choi M (2023) Association of depressive mood and frailty with mortality and health care utilization: Korean national cohort study. J Am Med Dir Assoc. 24 (4): 504–510.
55. Lee SH, Shin J, Um S, Shin HR, Kim YS, Choi JK (2023) Perceived stress and frailty in older adults. Ann Geriatr Med Res. 27 (4): 310–314.
56. Mishra M, Wu J, Kane AE, Howlett SE (2024) The intersection of frailty and metabolism. Cell Metabolism. 36 (5): 893–911.
57. Zhao H, Wu B, Zhou Y, Yang Z, Zhao H, Tian Z, Jiang M, Huang D (2024) Oral frailty: A concept analysis. BMC Oral Health. 24 (1): 594.
58. Hanlon P, Wightman H, Politis M, Kirkpatrick S, Jones C, Andrew MK, et al. (2024) The relationship between frailty and social vulnerability: A systematic review. Lancet Healthy Longev. 5 (3): e214–e226.
59. Felix J, Martinez de Toda I, Diaz-Del Cerro E, Gonzalez-Sanchez M, De la Fuente M (2024) Frailty and biological age. Which best describes our aging and longevity? Mol Aspect Med. 98: 101291–101303.
60. Salazar N, Arboleya S, Valdes L, et al. (2014) The human intestinal microbiome at extreme ages of life. Dietary intervention as a way to counteract alterations, Front Gen. 5: 406.
61. Ch Tarracchini, Ch Milani, Lugli GA, Mancabelli L, Turroni F, van Sinderen D, Ventura M (2024) The infant gut microbiota as the cornerstone for future gastrointestinal health. Adv Appl Microbiol. 126: 93–119.
62. Samarra A, Flores E, Bernabeu M, Cabrera-Rubio R, Ch Bauerl, Selma-Royo M, Collado MC (2024) Shaping microbiota during the first 1000 days of life. Adv Exp Med Biol. 1449: 1–28.
63. Wang S, Cui J, Jiang S, Zheng Ch, Zhao J, Zhang H, Zhai Q (2024) Early life gut microbiota: Consequences for health and opportunities for prevention. Crit Rev Food Sci Nutr. 64 (17): 5793–5817.
64. Horwell E, Bearn P, Cutting SM (2024) A microbial symphony: A literature review of the factors that orchestrate the colonization dynamics of the human colonic microbiome during infancy and implications for future health. Microbiome Res Rep. 3: 50–70.
65. Ross FC, Patangia D, Grimaud G, Lavelle A, Dempsey EM, Ross RP, Stanton C (2024) The interplay between diet and the gut microbiome: Implications for health and disease. Nat Rev Microbiol. 22 (11): 671–686.
66. Vich Vila A, Collij V, Sanna S, et al. (2020) Impact of commonly used drugs on the composition and metabolic function of the gut microbiota. Nat Commun. 11:362.
67. Fuhler GM (2020) The immune system and microbiome in pregnancy. Best Pract Res Clin Gastroenterol. 44–45: 101671.
68. Salazar N, Valdés-Varela L, González S, et al. (2017) Nutrition and the gut microbiome in the elderly. Gut Microbes. 8: 82–97.
69. Ghosh TS, Das M, Jeffery IB, et al. (2020) Adjusting of age improves identification of gut microbiome alterations in multiple diseases. eLife. 9: e50240.
70. Ghosh TS, Shanahan F, O'Toole PW (2022) The gut microbiome as a modulator of healthy ageing. Nat Rev Gastroenterol Hepatol. 19: 565–584.
71. Yatsunenko T, Rey FE, Manary MJ, et al. (2012) Human gut microbiome viewed across age and geography. Nature. 486: 222–227.
72. Claesson MJ, Jeffery IB, Conde S, et al. (2012) Gut microbiota composition correlates with diet and health in the elderly. Nature. 488:178–184.
73. Jackson MA, Jeffery IB, Beaumont M, et al. (2016) Signatures of early frailty in the gut microbiota. Genome Med. 8: 8.

74. Wilmanski T, Diener C, Rapappport N, et al. (2021) Gut microbiome pattern reflects aging and predicts survival in humans. Nat Metab. 3: 274–286.
75. Biagi E, Nylund L, Candela M, et al. (2010) Through ageing, and beyond: Gut microbiota and inflammatory status in seniors and centenarians. PloS One. 5: e10667.
76. Shanahan F, Ghosh TS, O'Toole PW (2021) The healthy microbiome – What is the definition of a healthy gut microbiome? Gastroenterology. 160: 483–494.
77. Zhang S, Zeng B, Chen Y, et al. (2021) Gut microbiota in healthy and unhealthy long-living people, Gene. 779: 145510.
78. DeJong EN, Surette MG, Bowdish DME (2020) The gut microbiota and unhealthy aging: Disentangling cause from consequence. Cell Host Microbe. 28: 180–189.
79. Bradley E, Haran J (2024) The human gut microbiome and aging. Gut Microbes. 16 (1): 2359677.
80. Zhang M, Li RW, Yang H, Tan Z, Liu F (2024) Recent advances in developing butyrogenic functional foods to promote gut health. Crit Rev Food Sci Nutr. 64 (13): 4410–4431.
81. Kong F, Deng F, Li Y, et al. (2019) Identification of gut microbiome signatures associated with longevity. Gut Microbes 10: 210–215.
82. Wu L, Zeng T, Zinellu A, et al. (2019) A cross-sectional study of compositional and functional profiles of gut microbiota in Sardinian centenarians. Systems. 4: e00325-19.
83. Kong F, Hua Y, Zeng B, et al. (2016) Gut microbiota signatures of longevity. Curr Biol. 26: R832–R833.
84. Lopez-Otin C, Blasco MA, Patridge L, Serrano M, Kroemer G (2023) Hallmarks of aging: An expanding universe. Cell. 186 (2): 243–278.
85. Smith P, Willemsen D, Popkes M, et al. (2017) Regulation of life span by the gut microbiota in the short-lived African turquoise killifish. eLife. 6: e27014.
86. Araujo JR, Marques C, Rodrigues S, Calhau C, Faria A (2024) The metabolic and endocrine impact of diet-derived gut microbiota metabolites on ageing and longevity. Ageing Res Rev. 100: 102451.
87. Chen LA, Boyle K (2024) The role of the gut microbiome in health and disease in the elderly. Curr Gastroentero Rep. 26 (9): 217–230.
88. Lozada-Martinez ID, Lozada-Martinez LM, Anaya JM (2024) Gut microbiota in centenarians: A potential metabolic and aging regulator in the study of extreme longevity. Aging Med. 7 (3): 406–413.
89. Wang H, Chen Y, Feng L, Lu S, Zhu J, Zhao J, Zhang H, Chen W, Lu W (2024) A gut aging clock using microbiome multi-view profiles is associated with health and frail risk. Gut Microbes. 16 (1): 2297852.
90. Pelton R (2023) The microbiome theory of aging (MTA). Integr Med. 21 (6): 28–34.
91. Barcena C, Valdes.Mas R, Mayoral P, Garabaya C, Durand S, Rodriguez F, et al. (2019) Healthspan and lifespan extension by fecal microbiota transplantation into progeroid mice. Nat Med. 25 (8): 1234–1242.
92. D´Amico F, Barone M, Brigidi P, Turroni S (2023) Gut microbiota in relation to frailty and clinical outcomes. Curr Opin Clin Nutr Metab Care. 26 (3): 219–225.
93. Cui G, Li S, Ye H, Yang Y, Jia X, Lin M, et al. (2023) Gut microbiome and frailty: Insight from genetic correlation and mendelian randomizarion. Gut Microbes 15 (2): 2282795.
94. Strasser B, Ticinesi A (2023) Intestinal microbiome in normal ageing, frailty and cognitive decline. Curr Opin Clin Nutr Metab Care. 26 (1): 8–16.
95. Bo F, Teng H, Shi J, Luo Z, Xu Y, Pan R, et al. (2024) Exploring the causal relationship between gut microbiota and frailty: A two-sample mendelian randomization analysis. Front Med 11: 1354037.
96. Wang Z, Han S, Xiao Y, Zhang Y, Ge Y, Liu X, Gao J (2024) Genetically supports causality between gut microbiota and frailty: A two-sample mendelian randomization study. Front Microbiol. 15: 1324209.
97. Wang X, Wu M (2021) Research progress of gut microbiota and frailty syndrome. Open Med. 16 (1): 1525–1536.
98. Almeida HM, Sardeli AV, Conway J, Duggal NA, Cavaglieri CR (2022) Comparison between frail and non-frail older adults microbiota: A systematic review and meta-analysis. Ageing Res Rev. 82: 101773.
99. Xu YS, Liu XJ, Liu XX, Chen D, Eang MM, Jiang X, Xiong ZF (2021) The roles of the gut microbiota and chronic low-grade inflammation in older adults with frailty. Front Cell Infect Microbiol. 11: 675414.
100. Rashidah NH, Lim SM, Neoh Ch F, Majeed ABA, Tan MP, Khor HM, et al. (2022) Differential gut microbiota and intestinal permeability between and healthy older adults: A systematic review. Ageing Res Rev. 82: 101744.
101. Lim MY, Nam YD (2023) Gut microbiome in healthy aging versus those associated with frailty. Gut Microbes. 15 (2): 2278225.
102. Zhao L, Deng X, Ding N, Liu Y, Ma Y, Li J, et al. (2024) Diminished representation of vitamin-B12-producing bacteria in constipated elders with frailty. iScience. 27 (8): 110403.
103. Pu Y, Sun Z, Zhang H, Huang Q, Wang Z, Mei Z, et al (2024) Gut microbial features and circulating metabolomic signatures of frailty in older adults. Nat Aging 4 (9): 1249–1262.

104. Farsijani S, Cauley JA, Cawthon PM, Langsetmo L, Orwoll ES, Kado DM, et al. (2024) Associations between walking speed and gut microbiome composition in older men from the MrOS study. J Gerontol A Biol Sci Med Sci. 79 (4):glae030.
105. Guo Y, Wang Q, Lv Y, Xia F, Chen X, Mao Y, et al. (2024) Serum metabolome and gut microbiome alterations are associated with low handgrip strength in older adults. Aging. 16 (3): 2638–2656.
106. Le Cosquer G, Vergnolle N, Motta JP (2024) Gut microb-aging and its relevance to frailty aging. Microbes Infect. 26 (3): 105309.
107. Zhu M, Huang Y, Wang Z, Jin Z, Cao J, Zhong Q, Xiong Z (2024) Fecal microbiota transplantation attenuates frailty via gut-muscle axis in old mice. Aging Dis, 16 (2): 1180–1198.
108. Dong S, Zeng Q, He W, Cheng W, Zhang L, Zhong R, et al. (2024) Effect of *lactobacillus plantarum* BF1243 on a female frailty model induced by fecal microbiota transplantation in germ-free mice. Food Funct 15 (8): 3993–4009.
109. Nascimento CM, Ingles M, Salvador-Pascual A, Cominetti MR, Gomez-Cabrera MC, Viña J (2019) Sarcopenia, frailty and their prevention by exercise. Free Radic Biol Med. 132: 42–49.
110. Jones RM, Neish AS (2017) Redox signaling mediated by the gut microbiota. Free Radic Biol Med. 105: 41–47.
111. Chen SY, Wang TY, Zhao C, Wang HJ (2022) Oxidative stress bridges the gut microbiota and the occurrence of frailty syndrome. World J Gastroenterol. 28 (38): 5547–5556.
112. Wu IC, Shiesh SC, Kuo PH, Lin XZ (2009) High oxidative stress is correlated with frailty in elderly Chinese. J Am Geriatr Soc. 57 (9): 1666–71.
113. Serviddio G, Romano AD, Greco A, Rollo T, Bellanti F, Altomare E, Vendemiale G (2009) Frailty síndrome is associated with altered circulating redox balance and increased markers of oxidative stress. Int J Immunopathol Pharmacol. 22 (3): 819–27.
114. Mai X, Yang S, Chen Q, Chen K (2024) Gut microbial composition is altered in sarcopenia: A systematic review and meta-analysis of clinical studies. PloS One. 19 (8): e0308360.
115. Lauretani F, Longobucco Y, Maggio M (2021) Gut microbiota and motoric-cognitive frailty in hospitalized older persons. Curr Opin Clin Nutr Metab Care. 24 (3): 209–215.
116. Tamayo M, Olivares M, Ruas-Madiedo P, Margolles A, Espin JC, Medina I, et al. (2024) How diet and lifestyle can fine-tune gut microbiomes for healthy aging. Ann Rev Food Sci Technol. 15 (1): 283–305.
117. Dent E, Wright ORL, Woo J, Hoogendijk EO (2023) Malnutrition in older adults. Lancet. 401 (10380): 951–966.
118. Jensen GL, Cederholm T (2024) Exploring the intersection of frailty, sarcopenia, and cachexia with malnutrition. Nutr Clin Pract. 39 (6): 1286–1291.
119. Salazar N, Gonzalez S, Nogacka AM, Rios-Covian D, Arboleya S, Gueimonde M, Reyes-Gavilan CG (2020) Microbiome: Effects of ageing and diet. Curr Issues Mol Biol. 36: 33–62.
120. Ticinesei A, Nouvenne A, Serundolo N, Parise A, Mena P, Meschi T (2024) The interaction between Mediterranean diet and intestinal microbiome: Relevance for preventive strategies against frailty in older individuals. Aging Clin Exp Res. 36 (1): 58.
121. Hill C, Guarner F, Reid G, et al. (2014) Expert consensus document. The International Scientific Association for Probiotics and Prebiotics consensus statement on the scope and appropriate use of the term probiotic. Nat Rev Gastroenterol Hepatol. 11: 506–14.
122. Mateo D, Marques M, Domingo JL, Torrente M (2024) Influence of gut microbiota on the development of most prevalent neurodegenerative dementia and the potential effect of probiotics in elderly: A scoping review. Am J Med Genet B Neuropsychiatr Genet. 195 (2): e32959.
123. Recharla N, Choi J, Puligunndla P, Park SJ, Lee HJ (2023) Impact of probiotics on cognition and constipation in the elderly: A meta-analysis. Heliyon. 9 (7): e18306.
124. Hutchinson AN, Bergh C, Kruger K, Susserova M, Allen J, Ameen S, Tingo L (2021) The effect of probiotics on health outcomes in the elderly: A systematic review of randomized, placebo-controlled studies. Micrroganisms. 9 (6): 1344.
125. Clancy R (2003) Immunobiotics and the probiotic evolution. FEMS Immunol Med Microbiol. 38 (1): 9–12.
126. Szydlowska A, Sionek B (2022) Probiotics and postbiotics as the functional food components affecting the immune response. Microorganisms. 11 (1): 104.
127. Dinan TG, Staton C, Cryan JF (2013) Psychobiotics: A novel class of psychotropic. Biol Psychiatry. 74 (10): 720–726.
128. Binda S, Tremblay A, Iqbal UH, Kassem O, Le Barz M, Thomas V, et al. (2024) Psychobiotics and the microbiota-gut-brain axis: Where do we go from here? Microorganisms 12 (4): 634.

129. Ishaq M, Khan A, Bacha AS, Shah T, Hanif A, Ahmad AA, et al. (2021) Microbiota targeted interventions of probiotic *Lactobacillus* as an anti-ageing approach: A review. Antioxidants 10 (12): 1930.
130. Chenhuichen Ch, Cabello-Olmo M, Barajas M, Iaquierdo M, Ramirez.Velez R, Zambom-Ferraresi F, MArtinez-Velilla N (2022) Impact of probiotics and prebiotics in the modulation of the major events of the aging process: A systematic review of randomized controlled trials. Exp Gerontol. 164: 111809.
131. Tsai YCh, Cheng LH, Liu TW, Jeng OJ, Lee YK (2021) Gerobiotics: Probiotics targeting fundamental aging processes. Biosci Microbiota Food Health. 40 (1): 1–11.
132. Sanchez B, Martinez BE, Aguirre JF, Martinez R, Benitez AD, Valdes R, Soto AE (2022) Emerging evidence on the use of probiotics and prebiotics to improve the gut microbiota of older adults with frailty syndrome: A narrative review. J Nutr Health Aging. 26 (10): 926–935.
133. Giron M, Thomas M, Dardevet D, Chassard C, SavaryAuzeloux I (2022) Gut microbes and muscle function: Can probiotics make our muscles stronger? J Cachexia Sarcopenia Muscle. 13 (3): 1460–1476.
134. Kiousi DE, Kouroutzidou AZ, Neanidis K, Matthaios D, Pappa A, Galanis A (2022) Evaluating the role of probiotics in the prevention and management of age-related diseases. Int J Mol Sci. 23 (7): 3628.
135. Warman DJ, Jia H, Kato H (2022) The potential roles of probiotics, resistant starch, and resistant proteins in ameliorating inflammation during aging (inflammaging). Nutrients. 14 (4): 747.
136. Kumar H, Dhalaria R, Guleria S, Cimler R, Sharma R, Siddiqui SA, et al. (2023) Antioxidant potential of plants and probiotic spp. in alleviating oxidative stress induced by H_2O_2. Biomed Phatmacother. 165: 115022.
137. Colletti A, Pellizzato M, Cicero AF (2023) The possible role of probiotic supplementation in inflammation: A narrative review. Microorganisms. 11 (9): 2160.
138. Cavaillon JM, Legout S (2016) Centenary of the death of Elie Metchnikoff: A visionary and an outstanding team leader. Microbes Infect. 18 (10): 577–594.
139. Hunsche C, Cruces J, Garrido A, Herández O, De la Fuente M (2018) Dietary supplementation with fermented milk containing probioticas improves behavior and immune response of aged mice. J Probiotics & Health. 6 (1): 1–10.
140. Hunsche C, Cruces J, De la Fuente M (2019) Improvement of redox state and functions of immune cells as well as of behavior response in aged mice after two weeks supplementation of fermented milk with probiotics. Current Microbiol. 76 (11): 1278–1289.
141. Diaz-Del Cerro E, Lambea M, Felix J, Salazar N, Gueimonde M, De la Fuente M (2022) Daily ingestion of *Akkermansia muciniphila* for one month promotes healthy aging and increases lifespan in old female mice. Biogerontology. 23 (1): 35–52.
142. Millan-Domingo F, García-Dominguez E, Gambini J, Olaso-Gonzalez G, Viña J, Gomez-Cabrera MC (2024) Diet and exercise in frailty and sarcopenia. Molecular aspects. Mol Aspects Med. 100: 101322.
143. Costa A, Leite G, Resende A, Blachier F, Lancha Jr. L (2017) Exercise, nutrition and gut microbiota: Possible links and consequences. Int J Sports Exerc Med. 3: 69.
144. Clauss M, Gérard P, Mosca A, Leclerc M (2021) Interplay between exercise and gut microbiome in the context of human health and performance. Front Nutr. 8: 637010.
145. Shin HE, Kwak SE, Lee JH, Zhang D, Bae JH, Song W (2019) Exercise, the gut microbiome, and frailty. Ann Geriatr Med Res. 23 (3): 105–114.
146. Quero CD, Ortega E, Manonelles P (2020) Probiotics, prebiotics and synbiotics: Useful for athletes and active individuals? A systematic review. Benef Microbes. 11 (2): 135–149.
147. Martarelli D, Verdenelli M, Scuri S, Cocchioni M, Silvi S, Cecchini C, Pompei P (2011) Effect of a probiotic intake on oxidant and antioxidant parameters in plasma of athletes during intense exercise training. Curr Microbiol. 62: 1689–1696.
148. Quero CD, Manonelles P, Fernández M, Abellán-Aynés O, López-Plaza D, Andreu-Caravaca L, Hinchado MD, Gálvez I, Ortega E (2021) Differential health effects on inflammatory, immunological and stress parameters in professional soccer players and sedentary individuals after consuming a synbiotic. A triple-blinded, randomized, placebo-controlled pilot study. Nutrients. 13 (4): 1321.
149. Hinchado MD, Quero-Calero CD, Otero E, Gálvez I, Ortega E (2023) Synbiotic supplementation improves quality of life and inmunoneuroendocrine response in patients with fibromyalgia: Influence of codiagnosis with chronic fatigue syndrome. Nutrients. 15 (7): 1591.
150. Hanlon P, Wightman H, Politis M, Kirkpatrick S, Jones C, Andrew MK, et al. (2024) The relationship betwee frailty and social vulnerability: A systematic review. Lancet Healthy Longev. 5 (3): e214-1226.
151. Garrido A, Cruces J, Ceprian N, De la Fuente M (2018) Improvements in behavior and immune function and increases life span of old mice cohabitating with adult animals. J Gerontol A Biol Sci Med Sci. 73 (7): 873–881.

152. Garrido A, Martinez de, Toda I, Diaz-Del Cerro E, Felix J, Ceprian N, Gonzalez-Sanchez M, De la Fuente M (2022) Social environment as a modulator of immunosenescence. Expert Rev Mol Med. 24: e29.
153. Diaz-Del Cerro E, Felix J, De la Fuente M (2023) Skin-to-skin contact: Crucial for improving behavior, immunity, and redox state after short cohabitation of chronologically old mice and prematurely aging mice with adult mice. Int J Mol Sci. 24 (5): 4680.
154. Devita M, Marescalco E, Bordignon A, Trevisan S, Sergi G, Coin A (2024) Social interaction level modulates the impact of frailty on cognitive decline: A longitudinal study. Aging Ment Health. 28 (4): 652–657.
155. Sarkar A, Harty S, Johnson KVA, Moeller AH, Carmody RN et al. (2020) The role of the microbiome in the neurobiology of social behaviour. Biol Rev Camb Philos Soc. 95(5): 1131–1166.
156. Abavisani M, Faraji N, Ebadpour N, Kesharwani P, Sahebkar A (2024) Beyond digestion: Exploring how the gut microbiota modulates human social behaviors. Neuroscience. 565: 52–62.
157. Lozupene M, D´Urso F, Piccininni C, La Montagna M, Sardone R, Resta E, et al. (2020) The relationship between epigenetics and microbiota in neuropsychiatric diseases. Epigenomics 12 (17): 1559–1568.
158. Münger E, Montiel-Castro AJ, Langhans W, Pacheco-Lopez G (2018) Reciprocal interactions between gut microbiota and host social behavior. Front Integr Neurosci. 12: 21.
159. Heys C, Lize A, Lewis Z, Price TAR (2020) Drosophila sexual attractiveness in older males is mediated by their microbiota. Microorganisms. 8 (2): 168.
160. Yi X, Cha M (2022) Gut dysbiosis has the potential to reduce the sexual attractiveness of mouse female. Front Microbiol. 13: 916766.
161. Martínez de Toda I, Miguélez L, Siboni L, Vida C, De La Fuente M (2019) High perceived stress in women is linked to oxidation, inflammation and immunosenescence. Biogerontology. 20 (6): 823–835.
162. Warren A, Nyavor Y, Beguelin A, Frame LA (2024) Dangers of the chronic stress response in the context of the microbiota-gut-immune-brain axis and mental health: A narrative review. Front Immunol. 15: 1365871.
163. Wang J, Zhou T, Liu F, Huang Y, Xiao Z, Qian Y, Zhou W (2023) Influence of gut microbiota on resilience and its possible mechanisms. Int J Biol Sci. 19 (8): 2588–2598.
164. Toader C, Dobrin N, Costea D, Glavan LA, Covache-Busuioc RA, Dumitrascu DI, et al. (2024) Mind, mood and microbiota-gut-brain axis in psychiatric disorders. Int J Mol Sci. 25 (6): 3340.
165. Mate I, Madrid JA, De la Fuente M (2014) Chronobiology of the neuroimmunoendocrine system and aging. Curr Pharm Des. 20 (29): 4642–55.
166. Oliveira Melo NC, Cuevas-Sierra A, Souto VF, Martinez A (2024) Biological rhythms, chrononutrition, and gut microbiota: Epigenomics insights for precision nutrition and metabolic health. Biomolecules. 14: 559.
167. Ferreira LL, Rosatto N, Marzullo P, Bellan M (2024) Circadian variations in the elderly: A scoping review. Chronobiol Int. 41 (3): 311–328.

12 Frailty in Experimental Animals
Models and Scores

LaDora V. Thompson

INTRODUCTION

The global population tripled over the past seven decades, growing from 2.9 billion in 1950 to 8.2 billion in 2023 (Figure 12.1). Over the same period, the average life expectancy—an important indicator of population health—increased by 26 years, from 47 to 73 years (Figure 12.2). This significant rise in longevity led to a shift in the demographic structure, with a marked increase in the proportion of people aged 65 and older. As a result, more and more countries now have greater than 10% of their populations over the age of 65, which is expected to hit 16% in 2025 and eventually 24% by 2100 (Figure 12.3). The increase in life expectancy is also reflected in the projection that the number of people aged 80 and older will triple from 2020 to 2050, reaching 426 million. While

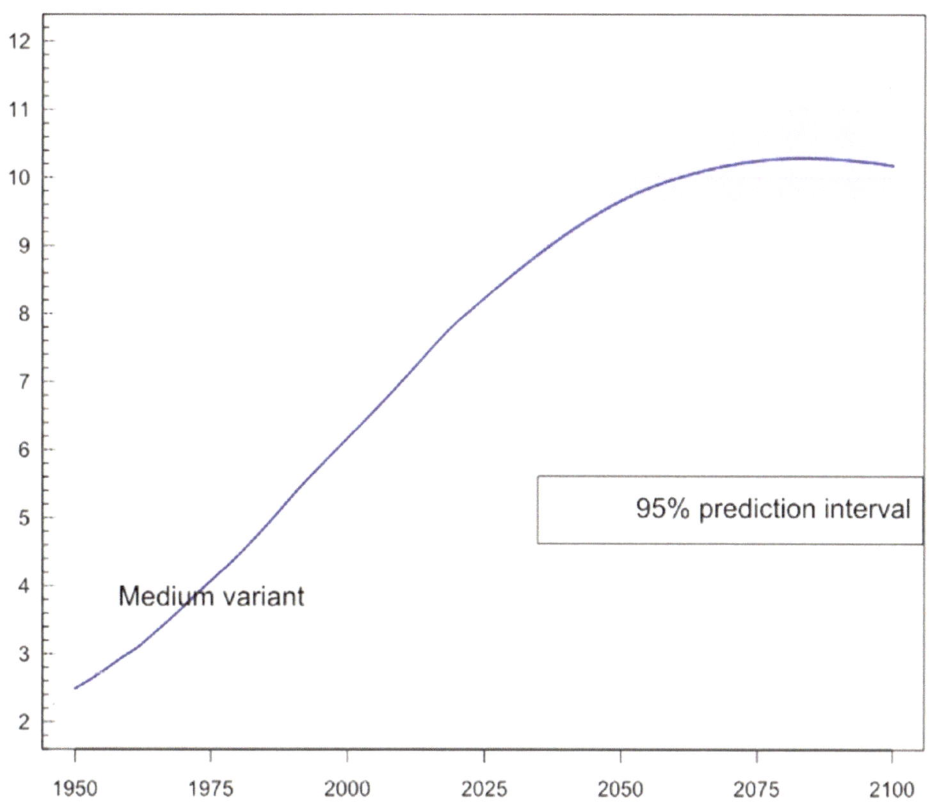

FIGURE 12.1 Population, 1950–2100. The world population has increased rapidly in recent decades. The world's population is more than three times larger than it was in 1950. The world population is expected to increase by nearly 2.5 billion persons in the next 25 years, from the current 8.2 billion to 9.7 billion in 2050, and could peak at nearly 10.4 billion in the mid-2080s. (Data source: UN, World Population Prospects (2024). OurWorldinData.org/population-growth.)

DOI: 10.1201/9781003398295-12

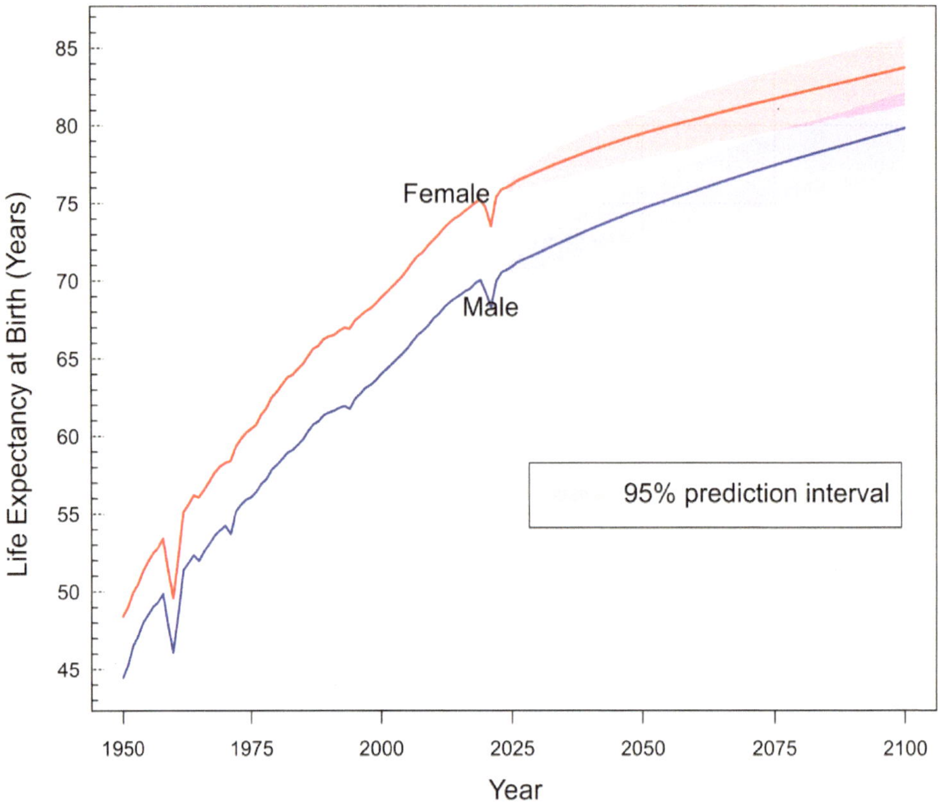

FIGURE 12.2 Life expectancy, 1950–2100. Across the world, people are living longer. In 1950, the average life expectancy of a newborn was 48.4 and 44.5 years for females and males, respectively. By 2023, life expectancy has almost doubled, with an increase to 75.9 and 70.5 years for females and males, respectively. This extraordinary rise is the result of a wide range of advances in health—nutrition, clean water, sanitation, neonatal healthcare, antibiotics, vaccines, and other technologies and public health efforts—and improvements in living standards, economic growth, and poverty reduction. (Data source: UN, World Population Prospects (2024). OurWorldinData.org/population-growth.)

the extension of life is a remarkable achievement, it also brings challenges: aging populations often face rising levels of chronic illness. Age-related chronic illnesses are often compounded by frailty, a multisystem decline that increases vulnerability (Perazza et al., 2022). Despite its significant impact on quality of life, frailty is frequently under-recognized, yet it contributes to disability, falls, hospitalizations, long-term care needs, and increased mortality.

Scientists are seeking to uncover the mechanisms driving frailty and its progression and to develop interventions that can slow, prevent, or even reverse it. Although progress is being made, research into the molecular and cellular mechanisms of frailty is limited by the lack of preclinical studies to assess frailty status and to assess the stages or the progression of frailty. The development of reliable assessment tools for determining frailty in experimental animal models enables scientists to investigate this important clinical condition in preclinical studies. Indeed, experimental animal models offer a controlled setting, which helps minimize the influence of confounding variables or external factors such as lifestyle and socioeconomic status that exist in human studies. Thus, frailty assessment tools in experimental animal models not only support investigations into the biological mechanisms underlying frailty (e.g., inflammation, altered intercellular communication, stem cell exhaustion, mitochondrial dysfunction, deregulated nutrient sensing, cellular senescence, telomere and genomic instability, epigenetic alterations, loss of proteostasis, compromised autophagy, microbiome disturbance) but also facilitate the use of frailty as a primary endpoint in testing therapeutics.

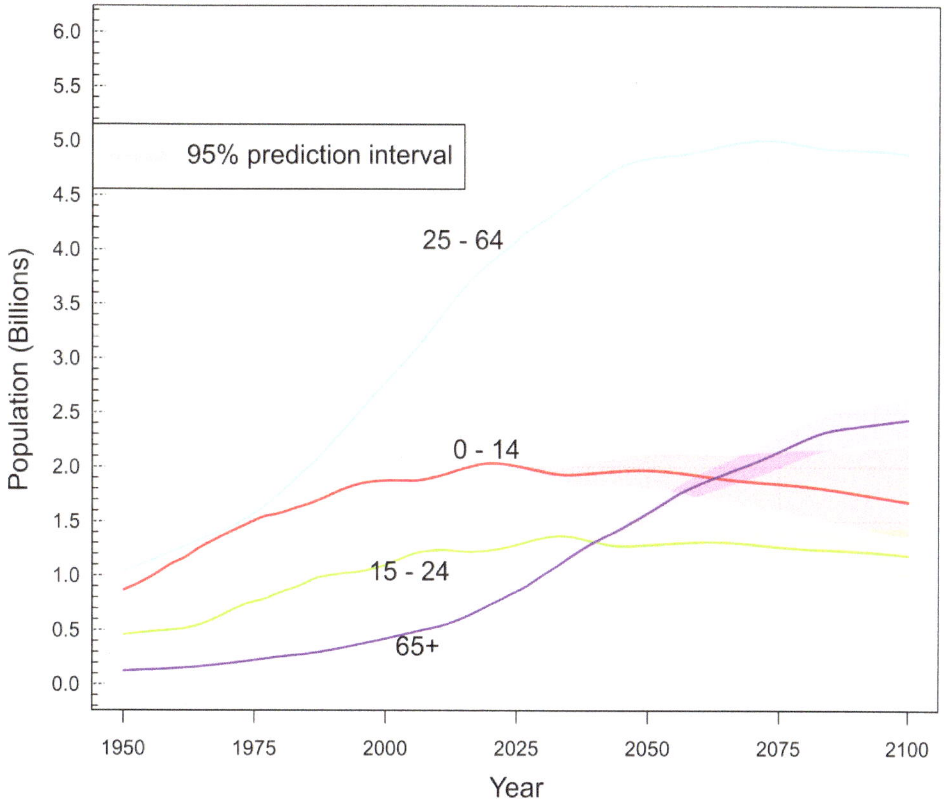

FIGURE 12.3 Population by age group 1950 to 2100. Since 1950, the world population has seen a significant increase across all age groups, with the most notable growth occurring in the "65 and older" demographic group (65+), meaning the population of older adults has grown at a much faster rate than the younger age groups, 0–14 group and 15–24 group. Forecasts for 2100 imply that over 80% of the population will exhibit an "aging" or "advanced aging" composition. (Data source: UN, World Population Prospects (2024). OurWorldinData.org/population-growth.)

To date, over 50 frailty assessment tools are utilized in clinical practice (Buta et al., 2016). Among these, two widely recognized tools for assessing the multidimensional nature and complexity of frailty in humans are the Frailty Index and the Physical Frailty Phenotype, both of which are validated across numerous clinical populations and diverse clinical settings (Kwak & Thompson, 2021). These tools differ in their conceptualization of frailty and the methods or approaches employed to assess it. The Physical Frailty Phenotype views frailty as a biological syndrome and categorizes individuals as frail, pre-frail, or non-frail based on their status in the lowest quintile for at least three out of five physical function criteria to be categorized as frail: unintentional weight loss, exhaustion, low physical activity, slowed gait speed, and decreased grip strength (Fried et al., 2001). In contrast, the Frailty Index conceptualizes frailty as a multidimensional risk state, measured quantitatively as the proportion of deficits (e.g., symptoms, signs, functional impairments, disease states, and laboratory markers) across multiple domains, relative to the total number of deficits assessed (Frailty Index score ranges from 0 to 1.0.) A higher Frailty Index score indicates greater frailty (Rockwood & Mitnitski, 2007). The Frailty Index emphasizes the number of health deficits and a deficit must meet the following criteria to be included in the index: (1) increase with age; (2) be health-related; (3) be present in at least 1% of the study population; (4) not be present in 80% of the study population before age 80; and (5) not be missing in more than 5% of the study population (Rockwood et al., 2017). The research field has yet to reach a consensus on the specific Frailty Index values that should be used to categorize individuals as robust, pre-frail, or frail. For example, some classifications designate individuals with a Frailty

Index of ≤0.20 as robust, those with a Frailty Index between 0.21 and 0.35 as pre-frail, and those with a Frailty Index > 0.35 as frail (Kulminski et al., 2008; Pereira et al., 2019). There is a limit to the number of deficits that an individual can accumulate (Frailty Index ≅ 0.70), above which an individual struggles to survive (Rutenberg et al., 2018).

Both frailty assessment tools provide valuable insights, viewing frailty as a correlate of aging, and each uses multiple parameters or criteria to define it. They both acknowledge that aging rates vary from person to person and that individuals of the same age do not share the same mortality risks. While these two well-established clinical frailty assessment tools are similar in some ways, they differ significantly in their conceptualization of frailty (physical vs. multidimensional) and their assessment methods (categorical vs. quantitative). Thus, these two frailty assessment tools are considered complementary.

During the last decade, the Frailty Index and the Physical Frailty Phenotype assessment tools were reverse-translated and developed for experimental animal models. Recognizing how frailty assessment tools in experimental animal models are developed and optimized, this chapter offers how these two well-established tools were applied to experimental animal models. The overall goal of this chapter is to highlight key publications, with a special emphasis on frailty criteria or parameters and the determination of frailty cut-off values. This chapter does not aim to provide a comprehensive review of the growing preclinical experimental frailty assessment tools and their published studies but aims to demonstrate how the field values using reverse-translation to explore the multifactorial and multidimensional nature of frailty.

VALUE OF PRECLINICAL STUDIES

Preclinical studies form the foundation of translational research and offer unique advantages in studying frailty that are difficult to achieve in human studies (Bartolomucci et al., 2024). Experimental animals such as the murine models offer convenience, affordability, short lifespans, and genetic plasticity. The importance of investigating frailty in experimental animal models (Table 12.1) includes: (1) *Understanding the Biological Mechanisms of Frailty*: Frailty is a complex, multifactorial syndrome that involves various physiological systems, including the musculoskeletal, endocrine, immune, and nervous systems. Animal models, especially rodents, allow scientists to dissect these biological pathways and their interconnectedness in detail. With the ability to manipulate genetic and environmental factors, animal studies help identify the molecular, cellular, and physiological mechanisms that drive frailty, which is critical for developing targeted therapies. (2) *Mimicking Human Frailty*: Animal models, particularly mice and rats, can be induced to exhibit traits like human frailty, such as muscle weakness, impaired mobility, cognitive decline, and increased vulnerability to stressors. This makes animals an invaluable tool for studying how frailty develops and progresses over time, and for testing potential interventions before they are used in humans. (3)

TABLE 12.1
Importance of Investigating Frailty in Experimental Animal Models

- Understanding the biological mechanisms of frailty
- Mimicking human frailty
- Studying genetic and environmental factors
- Longitudinal sampling
- Testing interventions
- Identifying early indicators and biomarkers
- Understanding aging and frailty
- Ethical and practical considerations
- Translating findings to human populations

Studying Genetic and Environmental Factors: Frailty is influenced by both genetic and environmental factors. Animal models offer the opportunity to study how specific genes and environmental exposures contribute to frailty. Scientists can use genetically modified animals or controlled environmental conditions to examine the role of external factors (e.g., diet, exercise, or toxins) in the onset and progression of frailty. This helps identify potential biomarkers and pathways that may be targeted for intervention. (4) *Longitudinal Sampling*: With longitudinal sampling, there is increased statistical power because of within-subject mixed-model analysis. Phenotyping and assessments (cellular and molecular) collected longitudinally across the lifespan within the same animal can be used to train predictive algorithms of healthspan and longevity. Further, it is possible to investigate how outcomes change over time and focus on these outcome trajectories, rather than values at single points in time. (5) *Testing Interventions*: One of the most important aspects of studying frailty in animal models is the ability to test interventions in a controlled setting. Animal studies allow for the investigation of pharmaceutical treatments, nutritional interventions, physical activity programs, and other therapeutic strategies to assess their effectiveness in preventing or reversing frailty. Such preclinical studies are critical for determining whether a therapy is safe and effective before moving to human clinical trials. (6) *Identifying Early Indicators and Biomarkers*: Frailty often develops subtly over time, making it difficult to detect in humans at early stages. Animal models allow researchers to study frailty from its onset and identify early biological markers or functional changes that could serve as indicators of frailty in humans. This is essential for developing early detection strategies and preventive measures. (7) *Understanding Aging and Frailty*: Frailty is closely related to aging, and animal models, particularly those of accelerated aging or age-related diseases, provide insights into how frailty develops as part of the aging process. By studying frailty in animal models, scientists can better understand the underlying aging mechanisms (e.g., inflammation, oxidative stress, and mitochondrial dysfunction) that contribute to frailty, helping to identify potential strategies for slowing or preventing its onset in humans. (8) *Ethical and Practical Considerations*: Conducting experiments on frailty in humans poses significant ethical and logistical challenges. While it is difficult to manipulate or control many of the factors that contribute to frailty in human studies, animal models provide a more ethical and feasible approach to studying frailty. Scientists can manipulate specific factors in a way that would not be possible in human studies, such as inducing frailty or testing novel treatments in a controlled and repeatable manner. (9) *Translating Findings to Human Populations*: One of the ultimate goals of studying frailty in animal models is to translate the findings to human populations. By establishing reliable frailty models and assessing biomarkers, therapies, and environmental factors in animals, scientists can generate data that help inform clinical practices for frailty management and intervention in humans. Animal studies are a crucial step in bridging the gap between basic scientific research and clinical application. In summary, experimental animal models provide an invaluable platform for exploring the biological mechanisms of frailty, testing potential therapies, and examining the role of genetic and environmental factors. Ultimately, findings from animal studies help inform clinical approaches to frailty in humans, with the potential to improve health outcomes for the aging population.

SELECTING AN ASSESSMENT TOOL TO INVESTIGATE FRAILTY IN PRECLINICAL STUDIES

When selecting a frailty assessment tool for an experimental study, it is important first to identify an approach that aligns with the study's objectives—whether it is to primarily assess the physical characteristics of the experimental animals or to focus on specific deficits in particular physiological systems. Utilizing more than one frailty assessment tool in a research study has the potential to offer numerous advantages and help mitigate the potential limitations of relying on a single method. The advantages in using multiple frailty assessment tools (Table 12.2) include: (1) *Comprehensive Assessment*: Different frailty tools determine distinct aspects of frailty, such as physical, cognitive, and social domains. By using multiple assessment tools, scientists can gain a more comprehensive and holistic understanding of

TABLE 12.2
Advantages in Using Multiple Frailty Assessment Tools

- Comprehensive assessment
- Validity and reliability
- Addressing tool-specific limitations
- Capturing different dimensions of frailty
- Cross-validation across different frailty models
- Improved predictive power

frailty, capturing a broader range of relevant factors. (2) *Validity and Reliability*: No single frailty assessment tool is universally applicable or perfect in all contexts. Different assessment tools vary in terms of validity, reliability, and sensitivity across diverse populations. Using multiple frailty assessment tools helps improve the overall validity of the study and ensures more reliable results. (3) *Addressing Tool-Specific Limitations*: Every frailty assessment tool has its strengths and weaknesses. Some tools are highly sensitive but may have lower specificity (leading to more false positives), while other assessment tools are more specific but less sensitive (leading to missed cases of frailty). By employing multiple frailty assessment tools, scientists can balance these strengths and weaknesses, leading to more robust and dependable findings. (4) *Capturing Different Dimensions of Frailty*: Frailty is a multidimensional phenomenon that extends beyond physical symptoms. It includes psychological well-being, nutritional status, and social engagement, among other factors. Using multiple frailty assessment tools allows researchers to capture a fuller and more accurate picture of frailty, addressing its various components. (5) *Cross-Validation across Different Frailty Models*: Frailty models may prioritize different aspects, such as biological markers, functional outcomes, or self-reported measures. By using multiple frailty assessment tools, scientists can cross-validate their findings and compare how different frailty models apply to the specific study population, ensuring a more complete assessment. (6) *Improved Predictive Power*: Frailty is a key predictor of adverse outcomes, such as mortality and functional decline. Using more than one frailty assessment tool enhances the predictive power of the assessments, providing stronger evidence to predict future outcomes and improve the study's conclusions. In summary, using multiple frailty assessment tools in a research study strengthens the findings by providing a more thorough, valid, and reliable measure of frailty. This approach helps address the complexity and multifaceted nature of frailty, improving the accuracy of results and enhancing the predictive power of the study.

REVERSE-TRANSLATION OF HUMAN FRAILTY ASSESSMENT TOOLS TO PRECLINICAL STUDIES

The concept of "reverse-translation" in the context of medical research refers to the process of taking clinical observations, insights, or results, such as biomarkers, behaviors, or disease mechanisms identified in human studies, and applying them to laboratory models (e.g., animals or cell cultures) to gain deeper insights or test hypotheses. This is the opposite of forward translation, which involves taking discoveries from basic science (preclinical studies) and translating them into human applications (e.g., therapies). Although reverse-translation is well-established in the fields of cancer, neurodegenerative diseases, and cardiovascular diseases, the reverse-translation of human frailty assessment tools to preclinical studies is an evolving area of research.

DETERMINING CUT-OFF VALUES

As mentioned earlier, the Frailty Index and the Physical Frailty Phenotype assessment tools were reverse-translated and adapted for use in experimental animal models (Table 12.3). A key challenge in this reverse-translation process was establishing appropriate cut-off values for these two widely

TABLE 12.3
Main Characteristics of the Physical Frailty Phenotype and the Clinical Frailty Index

Mouse Physical Frailty Phenotype—Stresses Physical Decline	
Criteria	**Assessments**
Pre-defined physical function domains	
Strength	Grip test/wire hang time
Walking speed	Rota-rod, walking speed, or treadmill running speed
Exhaustion / Endurance	Treadmill running time, inclined screen test, and tight rope
Physical activity	Voluntary wheel running distance, open field test
Unintentional weight loss	Weight loss, gain
Considerations when selecting domains:	
Assess different aspects of physical function.	
Metabolic demands of physiological systems.	
Physiological functions decline differently.	
Manifestation of frailty is different between mice.	
Categorical variable	The degree of frailty is expressed as the number of the five criteria that are present. Three of these five criteria define frail, two criteria define pre-frail, and 0–1 criterion defines non-frail.
Cut-off value for positive marker for frailty	Selection of a statistical parameter and selection of cohort group at which this parameter is applied.
Considerations when selecting statistical parameter:	
A stringent cut-off statistical parameter may lead to underestimation of frailty.	
A liberal cut-off statistical parameter may lead to an overestimation of frailty.	
Considerations when selecting cohort group to which the chosen statistical method is applied:	
Overall research question and study design.	
If using age: select an age at survival rate is 100%, first observation of physical function decline, and consistent with humans.	
Use healthy, non-frail mice of same strain and sex, use all mice in study, or employ a reference group.	

Mouse Clinical Frailty Index—Stresses State of Poor Health	
Criteria	**Assessments**
At least 30 unspecified set of measures required	• Physical/musculoskeletal includes kyphosis, distended abdomen, gait disorders, tremor, tumors, tail stiffing, forelimb grip strength, and body condition score • Digestive/urogenital includes diarrhea, malocclusions, rectal prolapse, and uterine/vaginal/penile prolapse • Nasal/ocular includes nasal discharge, corneal opacity, cataracts, eye discharge/swelling, vision loss, microphthalmia, and menace reflex • Integument includes coat condition, dermatitis, loss of fur color, loss of whiskers, and alopecia • Vestibulocochlear/auditory includes vestibular disturbance, and hearing loss • Respiratory includes breathing rate/depth • Discomfort includes mouse grimace scale and piloerection • Other includes temperature and body weight
Continuous variable	The degree of frailty is expressed as the fraction of deficits that are present to the number that were considered. If at least 30 age-related and adverse-outcome-associated items are considered, the degree of frailty does not depend on which items are counted.
Cut-off value for frail status	0.18

recognized human frailty assessment tools. Determining cut-off values for the Physical Frailty Phenotype and Frailty Index assessment tools in experimental animal models involves setting specific thresholds to distinguish frail animals from those that are non-frail or less frail.

For the Physical Frailty Phenotype assessment tool, several approaches are used to define these cut-off values, but most begin with identifying the relevant frailty domains of physical function. In mice, the physical frailty phenotype assessment typically includes the following domains: strength, walking speed, endurance (exhaustion), activity level, and body weight. Additional measures, such as balance and coordination, may also be considered. When selecting these physical function domains, it is important to: (1) include a range of tests that assess different aspects of physical function; (2) consider the metabolic demands placed on the animal, ensuring its physiological systems are appropriately challenged; (3) recognize that the decline in different physical functions occurs at different rates and ages; and (4) acknowledge that frailty may manifest differently in each individual mouse (Baumann et al., 2020). Each selected domain is assessed separately using specific performance tests, and cut-off values are established for each to identify a "positive marker" of frailty.

To objectively determine the most appropriate cut-off value for each frailty domain, two key steps must be considered: (1) selecting a cut-off statistical parameter and (2) identifying the specific age or cohort population at which this parameter is applied (Table 12.3). Statistical methods can be used to define the cut-off parameter, such as percentile-based measures (e.g., the lowest 20%), Z-scores, or standard deviations (SD) (e.g., 1, 1.5, or 2 SD below the mean if the data is normally distributed), or operating characteristic curves. The choice of how stringent or lenient the cut-off value is can significantly influence various frailty characteristics, such as the prevalence or incidence of frailty. In fact, using more rigorous cut-offs (e.g., 1 SD from the mean) may lead to an underestimation of frailty, while more liberal cut-offs (e.g., 2 SD from the mean) may result in overestimating frailty.

The next step in determining the appropriate cut-off values is selecting the cohort or population group to which the chosen statistical method (e.g., percentile or SD) will be applied. Defining this cohort group can be challenging, and the approach depends on the overall research question and study design (cross-sectional studies vs. longitudinal/longevity studies). One approach is to apply statistical methods or determine cut-off values using data from healthy, non-frail mice of the same strain and sex, without considering age. Alternatively, this can be done while considering age. Another option is to use age-matched controls and base the cut-off values on poor performance within that group. Another approach is to apply statistical methods using all the mice within the study, without considering age. As the study of frailty in experimental animal models is still an emerging field, best practices for selecting the cohort group for setting cut-off values have yet to be established.

The following guidelines can be used when selecting a cohort based on age: (1) The chosen age should correspond to a time point when the survival rate of the mice is 100% or nearly 100%. (2) It should be the age when physical function deficits are first observed. (3) It should correspond to an age range that is roughly equivalent to 60–75 human years. These criteria ensure that the frail mice are still alive, that physical function deficits are reversible, and that the selected age aligns with those used in human frailty studies. These guidelines were employed in three landmark longitudinal (longevity) studies where frailty was assessed across the lifespan in both male and female C57BL/6 mice. Based on the data obtained from lifespan cohort studies, cut-off values were determined using the statistical method of the lowest 20% from 23-month-old mice (male) and 20-month-old mice (female) with the exception of body weight (highest 20%) (Baumann et al., 2018, 2019; Kwak et al., 2020). Other landmark studies have also applied these age ranges and percentiles (Gomez-Cabrera et al., 2017; Kane et al., 2017) (Table 12.4).

In studies using cross-sectional research designs, which assess frailty status in different experimental groups at a single point in time (rather than tracking the same group of animals over a longer period), selecting a statistical cut-off value for each functional domain (e.g., 20th percentile or 1 SD) is similar to the approach used in longitudinal studies. The age or cohort at which this cut-off value is applied can be determined using several approaches. One option is to use data from healthy, non-frail mice of the same strain and sex, without factoring in age. Another approach is to include all

TABLE 12.4
Landmark Reverse-Translation Studies: Physical Frailty Phenotype

Animal Model	Physical Frailty Phenotype			
	Criteria and Tests	**Cut-Off Value for Positive Marker for Frailty**	**Designation of Frailty Status**	**Major Findings**
Male C57BL/6 mice 27–28 month old (50% survival rate) (Liu et al. 2014)	*Strength*: Four paw inverted hang test (seconds to fall) *Endurance*: Derived from four paw inverted hang test and rotarod test (seconds) *Walking speed*: Rotarod test (maximum speed) *Activity*: Voluntary wheel running (daily running distance)	1.5 SD below cohort mean (lowest 7th%)	Frail: three or more positive markers Mild-Frail (Pre-Frail): two positive markers Non-Frail: zero or one positive marker	Predictive of age associated physical decline. Prevalence of frailty is 9% within this old cohort.
Male C57BL/6 mice 24-month-old (75% survival rate) (Seldeen et al. 2018)	*Strength*: Grip meter test with fore and hind paws (grams) *Endurance*: Treadmill time to fatigue test (seconds) *Walking speed*: Gait speed analysis to darkened safe house box (seconds) *Activity*: Activity monitor (quadrant crossings) *Body weight*	1.5 SD below the group mean (approximately lowest 7%) >5% body weight change within 1-week period	Frail: three or more positive markers Pre-Frail: one or two positive markers Non-Frail: no positive marker	At 24 months 4.1% of the mice are frail and 33.3% of the mice are pre-frail. High-intensity interval training improves frailty status in old mice.
Male C57BL/6 mice Lifespan/longevity study with frailty criteria assessment every 3 months starting at 14 months of age (100% survival rate) until natural death. (Baumann et al. 2018)	*Strength*: Grip meter test with fore and hind paws (grams) *Endurance*: Treadmill time to fatigue test (seconds) *Walking speed*: Rotarod test (seconds) *Activity*: Voluntary wheel running (daily running distance) *Body weight*	0.8 SD below the reference group mean (performance at 23-months of age for the aging cohort) (approximately lowest 20%) Top 20% for body weight	Frail: three or more positive markers Pre-Frail: two positive markers Non-Frail: zero or one positive marker	Onset of frailty is at 17 months of age. Prevalence of frailty increases across the lifespan within the aging cohort. Frailty status predicts mortality risk.
Female C57BL/6 mice Lifespan/longevity study with frailty criteria assessment every 3 months starting at 14 months (100% survival rate) of age until natural death. (Kwak et al. 2020)	*Strength*: Grip meter test with fore and hind paws (grams) *Endurance*: Treadmill time to fatigue test (seconds) *Walking speed*: Rotarod test (seconds) *Activity*: Voluntary wheel running (daily running distance) *Body weight*	0.8 SD below the reference group (performance at 20 months of age for the aging cohort) (approximately lowest 20%) Top 20% for body weight	Frail: three or more positive markers Pre-Frail: two positive markers Non-Frail: one or no positive marker	Onset of frailty is at 17 months of age. Prevalence of frailty increases across the lifespan within the aging cohort. Frailty status predicts mortality risk.

(Continued)

TABLE 12.4 *(Continued)*
Landmark Reverse-Translation Studies: Physical Frailty Phenotype

Physical Frailty Phenotype				
Animal Model	**Criteria and Tests**	**Cut-Off Value for Positive Marker for Frailty**	**Designation of Frailty Status**	**Major Findings**
Male C57BL/6 mice and $Nrf2^{-/-}$ mice 20-24 months (90-75% survival rate for C57BL/6 mice) (Huang et al. 2019)	*Strength*: Four paw inverted hanging test (seconds to fall) *Endurance*: Treadmill running (distance) *Walking speed*: Rotarod test (maximum speed) *Activity*: Open-field assessment (distance moved in 5 min) *Body weight*	0.8 SD below the cohort mean (approximately lowest 20%)	Frail: three or more positive markers Mild-Frailty: two positive markers Non-Frail: one or no positive marker	$Nrf2^{-/-}$ mice higher frailty scores compared with old WT mice. 12% of the WT mice are frail. 43.5% of the $Nrf2^{-/-}$ are frail.
Male C57BL/6 mice 24 month old (75% survival rate) (Kane et al. 2017a)	*Strength*: Front paw wire hang test (seconds to fall) *Endurance*: Derived from wire hang and rotarod measures (seconds) *Walking speed*: Rotarod test (maximum speed) *Activity*: Open-field assessment (distance moved in 5 min) *Body weight*	0.8 SD below the cohort mean (approximately lowest 20%)	Frail: three or more positive markers Pre-Frail: two positive markers	50% of the mice are frail or pre-frail within this cohort.
Male C57BL/6 mice 17, 20, 23, 26, and 28 months (100% to 50% survival rate) *Valencia Score* (Gomez-Cabrera et al. 2017)	*Strength*: Grip meter test with fore paws (grams) *Endurance and slowness*: Incremental treadmill test (running time, maximum running speed) *Activity*: Motor coordination test with a Tight-rope test (percentage of mice passing the test) *Body weight*	20th percentile for each criterion at every specific age >5% body weight change from the mean of 17-month-old mice	A frailty score is determined for each experimental group by total number of tests failed by the animals at each age group, divided by the total number of tests performed by these animals, expressed in percentage.	Valencia Score increases with age at the group level. Sedentary mice had higher Valencia Scores than those with exercise.
Female outbred ICR/CD1 mice 40-, 56-, and 80-week-olds *Valencia Score* (Martinez de Toda et al. 2018)	*Strength*: Grip meter test with fore paws, (grams/body weight) *Endurance and slowness*: Incremental treadmill test (running time, maximum running speed) *Activity*: Motor coordination test with a Tight-rope test (percent of mice passing the test) *Body weight*	20th percentile for each criterion at every specific age >5% body weight change from the previous month	A frailty score is determined for each experimental group by total number of tests failed by the animals at each age group, divided by the total number of tests performed by these animals, expressed in percentage. Individual mouse frailty score: Frail: failure of three or more components Pre-Frail: failure of one or two criteria Robust- no failure	Valencia Score increases with age in outbred ICR/CD1 mice. Prevalence of frailty at the individual animal level increased with age.

mice within the cohort study, while considering strain and sex. Alternatively, age-matched controls could be used, with the cut-off values set based on poor performance within that specific group. However, using age-matched controls has some drawbacks. For example, in a cross-sectional study with an older cohort, a disabled mouse might be incorrectly classified as robust, while a healthy mouse might be classified as frail within a younger cohort. To avoid these issues, it is recommended to use a reference group in cross-sectional or short-term intervention studies. The reference group should be chosen based on an age when frailty is more likely to occur, ensuring 100% survival and a suitable window for therapeutic interventions. Collectively, for any physical domain to be a reliable marker of frailty, it is essential to have a well-established baseline. In this context, reference groups play a crucial role in providing a standard for assessing experimental outcomes.

In summary, determining reliable and accurate cut-off values for frailty criteria in experimental animals, such as mice, is a complex process, influenced by a variety of factors that affect frailty expression. In experimental animal models, the employment of reverse-translation of the two well-established human frailty assessment tools provides a starting point to investigate this complex process, which is discussed in the following section.

PHYSICAL FRAILTY PHENOTYPE

Mouse Physical Frailty Phenotype

The mouse Physical Frailty Phenotype follows the general criteria outlined by the clinical Physical Frailty Phenotype and focuses on physical function (Tables 12.3 and 12.4). Defining frailty with the mouse Physical Frailty Phenotype begins with identifying physical performance tests in mice that measure strength, endurance, walking speed, and physical activity levels as well as body weight changes. The next step in defining frailty requires determining the cut-off values or points. The original Physical Frailty Phenotype investigated mice at the 50% survival rate age (27–28-month old C57BL/6 mice) (Liu et al., 2014). Following the human Physical Frailty Phenotype, mice with three or more positive frailty markers are identified as frail, mice with two positive markers are identified as pre-frail, and mice with one or no positive frailty markers are considered non-frail. In this initial Physical Frailty Phenotype, a cut-off percentile corresponding to 1.5 SD below the cohort mean (i.e., the lowest seventh percentile of the group) was selected, resulting with a prevalence of frailty of 9% in mice at this age, which is comparable to the prevalence in humans at similar survival ages. In a subsequent study (Seldeen et al., 2018), which examined 24-month-old mice with a 75% survival rate, the researchers used cut-off percentiles set at 1.5 SD below the cohort mean. However, they classified frail, pre-frail, and non-frail mice in a different manner. Specifically, mice with three or more positive frailty markers were categorized as frail, those with one or two positive markers as pre-frail, and mice with no positive frailty markers as non-frail. Using this classification, 33% of the 24-month-old mice were identified as pre-frail, while only 4% were categorized as frail (Seldeen et al., 2018). Indeed, using a different number of positive markers to determine frailty, pre-frail, or non-frail status can influence the frailty classification within a cohort, using a cut-off statistical parameter of 1.5 SD.

Many additional characteristics of frailty such as onset of frailty, prevalence of frailty at specific ages, and mortality risk of frail mice are possible in studies using a longitudinal or longevity research design (Baumann et al., 2018, 2019, 2020; Kwak et al., 2020) (Table 12.4). Using the mouse Physical Frailty Phenotype assessment tool and a cut-off percentile corresponding to 0.8 SD (more liberal cut-off percentile) below the mean from data collected from 23 months old (males) and 20 months old (females), frailty is first observed at 17 months of age in both male and female mice and the prevalence of frailty increases from 3% to over 60% across the lifespan or life course (evaluating the same cohort of mice every 3 months starting at 14 months of age until death, with a maximum age of 32 months) in C57BL/6 mice. Importantly, these two longevity studies clearly showed that non-frail mice live longer than frail mice in both sexes, with non-frail male and female mice living 20% (6.7 months) and 7% (2.1 months) longer than frail mice, respectively.

Because the mouse Physical Frailty Phenotype is evaluating physical functions of strength, walking speed, endurance/exhaustion, and activity, there are several performance tests available to evaluate these specific functional characteristics in mice. For instance, the mouse Physical Frailty Phenotype employed physical performance tests of grip strength test, four paw inverted hanging test, rotarod analysis, open-field analysis, and treadmill endurance test (Table 12.4) (Huang et al., 2019). Frailty scores are calculated by the number of criteria (0–5) below the cut-off point, where the cut-off point is defined as 0.8 SD below the total cohort mean. In this Physical Frailty Phenotype assessment, frailty is defined if three or more of the criteria measures are below the cut-off values, whereas mild frailty is defined if two factors are below the cut-off points. If there is one criterion below the cut-off values or none, the mouse is identified as non-frail (Table 12.4) (Huang et al., 2019). Using these performance tests, cut-off values, and frailty status classifications, 12% of male mice (C57BL/6) aged 20–24 months were identified as frail. The results highlight how varying performance tests, statistical cut-off parameters, and the number of positive markers used to determine frailty, pre-frail, or non-frail status can impact the prevalence of frailty in mice with a 75% survival rate. Notably, the mouse Physical Frailty Phenotype proves to be a valuable tool for identifying frail mice.

The mouse Physical Frailty Phenotype is used to determine the prevalence of frailty in aged male Fischer 344 rats (17 months old, approximately 80% survival rate) (Miller et al., 2017) using the forelimb wire suspension test, rotarod test, open-field test, and the inclined screen test. In this study evaluating rats, the cut-off point is 1.5 SD below the mean, with the lowest quintile (20%) on three or four of the tests being considered frail. Rats are identified as mildly frail if they rank in the lowest 20% of performance on two of the four tests and identified as non-frail if they rank in the lowest 20% of performance on one or none of the tests. The prevalence of mild frailty and frailty is 17.3% and 2.3%, respectively. Further, the rats identified as mildly frail are 3.8 times and the rats identified as frail are 27.5 times more likely to die during a 100-day follow-up period than the non-frail rats.

The mouse Physical Frailty Phenotype assessed frailty status in geriatric Labrador and Golden Retriever dogs, using the lowest or slowest 25% cut-off values for each criterion. At baseline, a dog is considered frail if it presents with three or more of the five criteria, pre-frail if it presents with one or two, and non-frail if it presents with none. In geriatric dogs at the 5-year follow-up, this assessment tool predicted all-cause mortality (Hua et al., 2016). The mouse Physical Frailty Phenotype was also adapted for nonhuman primates in a calorie restriction study with cut-off values set at 5% weight loss and the lowest 20–30% for the other criteria (Yamada et al., 2018). The identification of frail occurs when three of the five individual frailty criteria are present. The nonhuman primates were followed for an 8-year period in this intervention study. Total prevalence of frailty is higher in the ad libitum animals, and the intervention of calorie restriction reduces the prevalence of frailty and increases healthy lifespan in nonhuman primates.

Valencia Score

The Valencia Score is derived from the physical frailty phenotype (Fried et al., 2001) and includes five components: unintentional weight loss, weakness (assessed by forelimb grip strength using a horizontal bar), poor endurance and slowness (evaluated through running time and maximal running speed on a graded treadmill test), and low activity level (measured by motor coordination using the tightrope test) (Gomez-Cabrera et al., 2017; Martinez de Toda et al., 2018). According to this scoring system, a mouse is considered frail if it fails three or more of these five components. If it fails one or two components, it is classified as pre-frail, and if it passes all components, it is classified as robust. The cut-off values for each criterion are determined for each age group studied, with the 20th percentile used as the threshold for identifying a positive frailty marker. For example, at ages 17, 20, 23, 26, and 28 months, mice that fall below the 20th percentile for grip strength (weakness), running time (poor endurance), running speed (slowness), and motor coordination (low activity) are considered frail. For unintentional weight loss, a mouse is flagged as frail if it loses more than 5% of its body weight at 17 months or experiences a weight loss greater than 5% compared to the previous month. To calculate the frailty score for each age group, the total number of tests failed by the mice is divided by the total number of tests conducted, yielding a percentage. As mice age, the percentage of frail mice is significantly higher in sedentary groups compared to those with access to wheel running. The Valencia Score is a useful tool for assessing frailty in intervention studies and when working with outbred mouse populations (e.g., ICR/CD1 mice), allowing for a comprehensive measure of overall frailty. Similar to the mouse Physical Frailty Phenotype, which was modified for rats, the Valencia Score was also modified for Wistar rats (Todorovic et al., 2018). The modified Valencia Score included 11 items associated with physical health, mental function, and weight loss. In Wistar rats, there is an increase in frailty scores with aging.

In summary, the reverse-translation of the physical frailty phenotype conceptualizes frailty as a biological syndrome, incorporating assessments of various physical function domains to quantify the three stages of frailty in experimental animal models. The process of selecting cut-off values, statistical parameters, and cohort groups to identify positive markers for frailty, as well as determining the number of positive markers needed to classify frailty stages, will impact key characteristics of frailty, such as its prevalence. The field would greatly benefit from the identification of best practices and a consensus among scientists.

Mouse Clinical Frailty Index

The mouse Clinical Frailty Index, developed in consultation with a veterinarian, follows the cumulative deficit model (human clinical Frailty Index) and evaluates various physiological systems (Whitehead et al., 2014) (Table 12.5). These include the integumentary, musculoskeletal, ocular, nasal, digestive, vestibulocochlear/auditory, urogenital, and respiratory systems, as well as indicators of discomfort, body weight, and body surface temperature. The mouse Clinical Frailty Index consists of 31 health-related visible signs: body weight, temperature, alopecia, fur color changes, dermatitis, whisker loss, coat condition, respiratory rate and depth, the mouse grimace scale, piloerection, tumors, abdominal distension, kyphosis, gait abnormalities, tremors, vestibular disturbances, tail rigidity, cataracts, corneal opacity, eye discharge or swelling, microphthalmia, nasal discharge, malocclusions, rectal prolapse, vaginal/uterine/penile prolapse, diarrhea, body condition score, forelimb grip strength, menace reflex, vision loss, and hearing loss.

Each health-related visible sign or criterion is scored as 0, 0.5, or 1, reflecting the severity of the deficit: 0 indicates no deficit, 0.5 indicates a mild deficit, and 1 indicates a severe deficit. To determine body weight and surface temperature changes, young adult mice are used as references. The Clinical Frailty Index shows high reproducibility between trials. The Clinical Frailty Index score, which is a continuous scale, increases progressively with age, and the increase is most pronounced in the oldest age group of mice (reported between 5 months and 28 months, 100% survival rate to 25% survival rate). The average frailty index is approximately 0.020 in young adult mice, nearly 0.112 in older adult mice, and a frailty index of 0.329 is reported in aged mice (e.g., scores between 0.3 and 0.4 are reported in 28-month-old female mice) (Whitehead et al., 2014). In fact, it is noted that the highest Frailty Index score in mice of 0.55 is close to the frailty limit in humans (Rockwood et al., 2017). Although the Frailty Index score is not commonly used to delineate the stages of frailty, cut-off values are applied to the frailty index scores to distinguish frail and non-frail mice with <0.18 = non-frail and ≥0.18 = frail (Kane et al., 2017).

Due to the evaluation of visible signs, which provides easy and timely administration, the utilization of the mouse Clinical Frailty Index is widespread among different rodent models, such as rats (e.g., Scheuren et al., 2020; Yorke et al., 2017). In aged Fischer 344 rats, the survival rate in frail rats with a Frailty Index > 0.21 (e.g., 21-month-old rats) is significantly lower than that in rats with a Frailty Index < 0.21 (37.5% vs. 82.4%, respectively) (Yorke et al., 2017).

The mouse Clinical Frailty Index described above was originally developed using 31 invasive and non-invasive health-related parameters, which included information about activity levels, body

TABLE 12.5
Landmark Reverse-Translation Studies: Frailty Index

Frailty Index				
Animal Model	**Criteria and Tests**	**Cut-Off Value for Positive Marker for Frailty**	**Designation of Frailty Status**	**Major Findings**
Male and Female C57BL/6 mice 12- and 30-month-old (100% and 25% survival rate for males; 100% and 10% survival rate for females.) (Parks et al. 2012)	31 health-related deficits based on: activity level (open field activity), body composition for bone mass density and percent fat (DEXA scan), hemodynamic (blood pressure, heart rate), metabolism (blood chemistries)	Deficits identified by number of SD from mean values derived from the middle-aged mice with: >1 SD scored as 0.25 >4 SD scored as 1.0	A frailty score is calculated as the sum of the deficits divided by the maximum number of deficits measured to give a score between 0 and 1 for each mouse.	Frailty Index scores increase with age, regardless of sex
Mostly female C57BL/6 mice 5, 19, and 28 months (100%, 100%, and 25% survival rates) *Clinical Frailty Index* (Whitehead et al. 2014)	31 health-related deficits derived from a visual inspection of domains including: integument, ocular and nasal systems, musculoskeletal system, digestive system, urogenital system, respiratory system, vestibulocochlear/auditory systems, and signs of discomfort Body mass and body surface temperature	Deficits identified with: a score of 0 (no deficit) a score of 0.5 (mild deficit) a score of 1 (severe deficit) Body mass and temperature scored in 0.25 units, based on how many SDs the item is from the mean reference values in young animals	A frailty score is calculated as the sum of the deficits divided by the maximum number of deficits measured to give a score between 0 and 1 for each mouse	Clinical Frailty Index score increases with age
Male and Female NIH Swiss mice 52, 78, and 104 weeks of age Cross-sectional experimental design 26-week-old mice – reference group *Physiological Frailty Index* (Antoch et al. 2017)	29-item physiological parameters including: physical health (e.g., grip strength), cardiovascular system (e.g., blood pressure parameters, heart rate, body temperature, tail blood flow and volume), blood cell composition (e.g., counts of whole blood cell, neutrophil, lymphocyte, monocyte, eosinophil, and red blood cell counts; hemoglobin, hematocrit), metabolic (e.g., glucose, triglycerides), and immune (e.g., circulating pro-inflammatory cytokines)	Value for each parameter is compared with corresponding value from the reference group and assigned a score Values differ by: <1SD scored as 0 (no deficit) 1 SD scored as 0.25 (minimal deficit) 2 SD scored as 0.5 3 SD scored as 0.75 >3 SD scored as 1 (extreme deficit)	A frailty score is calculated as the ratio of the number of deficits/total number of parameters measured	Physiological Frailty Index score predicts health and lifespan Physiological Frailty Index scores are higher in older mice and in obese mice Therapeutic interventions impact Physiological Frailty Index scores
Male C57BL/6 21 months of age until natural death (initial 80% survival rate) Frailty index measured every 6 weeks until the end of their lives *Mouse FI based on machine learning analysis* (Schultz et al. 2020)	29 health-related deficits derived from a visual inspection of domains including: integument, ocular and nasal systems, physical/musculoskeletal system, digestive/urogenital system, respiratory system, vestibulocochlear/auditory systems, signs of discomfort Body weight (malocclusions and body temperature not assessed)	Deficits identified with: a score of 0 (no deficit) a score of 0.5 (mild deficit) a score of 1 (severe deficit)	A frailty score is calculated as the sum of the deficits divided by the maximum number of deficits measured to give a score between 0 and 1 for each mouse	Frailty Index scores are correlated and predictive of age and survival Development of two clocks based on components of the Frailty Index: FRIGHT age (models apparent chronological age) and AFRAID clock (predicts life expectancy)

composition, and various hemodynamic and metabolic measures, to generate a frailty score for each mouse (Parks et al., 2012). Briefly, activity levels were measured with an open-field behavior test with the analysis of the following parameters: total distance moved in 10 min (cm), Maximum distance moved between bouts of inactivity (cm), Total duration of movement (seconds), Percentage of total time spent moving, Meander (degrees/cm), which measures the change in direction per unit of distance moved, ranging from 0° to 180°, Average velocity of movement over 10 min (cm/s), and Rearing frequency (number of occurrences/10 min).

A major strength of this original study was the use of reference values. Reference values for each of the parameters were established using three male and three female adult mice. Each individual mouse's parameter values were then compared to the sex-matched mean reference values for adult animals. A graded scale was then utilized to assess frailty: values differing by 1 SD from the mean were assigned a frailty score of 0.25; values differing by 2 SD received a score of 0.5; those differing by 3 SD were given a score of 0.75; and values deviating more than 3 SD from the mean were assigned the maximum frailty score of 1. Parameters with differences of less than 1 SD from the reference values received a score of 0. Lastly, the scores for each parameter were summed and divided by the total number of parameters to calculate the frailty index for each mouse, with a score of 0 indicating no deficits and a score of 1 representing maximal deficits. Adult female mice (e.g., 12 months of age, 100% survival rate) had significantly lower frailty indices than aged female mice (e.g., 30 months of age, 10% survival rate), with mean values of 0.08 and 0.43, respectively. Because this approach was labor-intensive, time-consuming, invasive procedures, and required specialized equipment, it was modified to the mouse Clinical Frailty Index discussed above (31 health-related visible signs). The mouse Clinical Frailty Index has been modified by other groups, such as a 27-item clinical measures, with the deficits scored such that a score of 0 indicates no deficit, a score of 0.5 indicates a mild deficit, and a score of 1 indicates a severe deficit (Sukoff Rizzo et al., 2018).

Further research is needed because in human studies there is evidence of ceiling effects with the Frailty Index. Briefly, when frailty indices are constructed with fewer than 30 variables, a ceiling effect occurs where people can have all possible deficits and graded levels of frailty cannot be distinguished. In contrast, if a frailty index is made up of only low-prevalence items, the mean value of the frailty index is low (Whitehead et al., 2014). Indeed, a non-invasive performance-based eight-item Frailty Index in mice showed increased variance, increased test-to-test variability, and an inability to distinguish gradations of frailty in very old mice (Parks et al., 2012) (Table 12.6).

In conclusion, the reverse-translation of the deficit accumulation model of frailty, which incorporates a wide range of non-invasive clinical parameters, is easy to implement and can effectively quantify frailty in experimental animal models. Similar to the Mouse Physical Frailty Phenotype, the field would greatly benefit from establishing best practices and reaching consensus among scientists on the process to define cut-off points for the different stages of frailty.

OTHER FRAILTY ASSESSMENT TOOLS

Physiological Frailty Index (PFI)

The Physiological Frailty Index (PFI) is a quantitative, non invasive tool used to assess the health status of physiological systems, reflecting biological age (Antoch et al., 2017). Briefly, the PFI evaluates 29 parameters across several physiological domains: physical fitness (e.g., body weight, grip strength), cardiovascular health (e.g., blood pressure, heart rate, tail blood flow and volume), blood cell composition (e.g., white and red blood cell counts and differentials), and plasma levels of CXCL1/KC, triglycerides, and glucose.

To calculate the PFI for each mouse (e.g., an old mouse), the value for each parameter is compared to the corresponding value in a reference group, and a score is assigned based on the difference. If a parameter's value differs by less than SD from the reference, it receives a score of 0 (no deficit). A difference of one SD results in a score of 0.25 (minimal deficit), two SDs corresponds

TABLE 12.6
Frailty Assessment Tools Identifies Different Mice as Frail

Comparing the Two Frailty Assessments Tools in Mice with a Disease

Animal Model	Criteria and Tests	Cut-Off Value for Positive Marker for Frailty	Designation of Frailty Status	Major Findings
Male and Female 5xFAD mice 3- and 11-month-old mice (Todorovic et al. 2020)	Physical Frailty Phenotype- 8 item: *Motor coordination*: Tight-rope test *Strength/endurance*: Wire hang test *Walking speed*: Total distance travelled (cm)/ambulatory time (s) *Activity*: Open-field assessment (total distance traveled; total duration of movement-ambulatory time; percent of total time spent in moving, average velocity of movement, rearing/vertical frequency). *Body weight*	Physical Frailty Phenotype-20% for each item 5% change in body weight	Physical Frailty Phenotype: A frailty score is calculated for each experimental group (based on age and/or sex) as total number of tests failed by the experimental group divided by the total number of tests performed by the same group of animals and expressed as a percentage	Physical Frailty Phenotype: The percent of old mice that failed all tests is higher (46%) than the percent of younger mice (12%).
	Frailty Index: 24 health-related deficits derived from a visual inspection of domains including: Integument system, ocular system, physical/musculoskeletal system, digestive/urogenital system, respiratory system, vestibulocochlear/auditory systems, and neurological system	Frailty Index: Deficits rated with: a score of 0 (no deficit) a score of 0.5 (mild deficit) a score of 1 (severe deficit) Deficit in body weight is scored based on deviation from reference values determined in 3-month-old animals. Values differ by: <1SD scored as 0 ± 1 SD scored as 0.25 ± 2 SD scored as 0.5 ± 3 SD scored as 0.75 >3 SD scored as 1	Frailty Index: All the values are totaled and the total is divided by the number of parameters measured, score between 0 and 1 for each animal.	Frailty Index: Frailty Index scores increase in both sexes with aging. Older male mice have 20 times higher Frailty Index scores than the young mice, while the increase in frailty in 11-month-old females in comparison with 3-month-old is less. ** The two frailty assessment tools did not detect the same mice as frail.

Male C57BL/6 mice 24-month-old (75% survival rate) (Kane et al. 2017)	Physical Frailty Phenotype-*Strength*: Front paw wire hang test (seconds to fall) *Endurance*: Derived from wire hang and rotarod measures (seconds) *Walking speed*: Rotarod test (maximum speed) *Activity*: Open-field assessment (distance moved in 5 min) *Body weight*	Physical Frailty Phenotype-0.8 SD below the cohort mean (approximately lowest 20%)	Physical Frailty Phenotype-Frail: 3 or more positive markers Pre-Frail: 2 positive markers	Physical Frailty Phenotype-50% of the mice are frail or pre-frail within this cohort.
	Frailty Index: 31 health-related deficits derived from a visual inspection of domains including: integument, ocular and nasal systems, musculoskeletal system, digestive system, urogenital system, respiratory system, vestibulocochlear/auditory systems, and signs of discomfort. Body mass and body surface temperature.	Frailty Index: Deficits identified with: a score of 0 (no deficit) a score of 0.5 (mild deficit) a score of 1 (severe deficit) Body mass and temperature scored in 0.25 units, based on how many SDs the item is from the mean reference values in young animals.	Frailty Index: A frailty score is calculated as the sum of the deficits divided by the maximum number of deficits measured to give a score between 0 and 1 for each mouse. <0.18 = Non-Frail ≥ 0.18 = Frail	Frailty Index: 44.4% of mice are frail within this cohort. ** The two frailty assessment tools did not detect the same mice as frail.

to a score of 0.5, and a three-SD difference receives a score of 0.75. If the value deviates by more than three SDs, the score is 1 (extreme deficit). The total number of deficits observed in a mouse is then normalized by the number of parameters assessed, yielding a ratio that reflects the overall frailty.

Both male and female mice show an increase in PFI with age (reference group was 26 weeks of age), with female mice exhibiting a more rapid accumulation of health deficits. At each corresponding age (26 to 104 weeks), the average PFI score is higher in females than in males. Further, mice on a high-fat diet show an accelerated increase in PFI, while those treated with rapamycin (an mTOR inhibitor) experience a slower rise in frailty. These findings suggest that the PFI is a reliable, non-invasive, and quantitative tool for evaluating the effects of anti-aging treatments in pre-clinical studies focused on longevity and health.

Fragility Index

The Fragility Index is a frailty assessment tool developed from existing frailty indices, with additional deficits identified through the expertise of veterinarians (Luciano et al., 2024). The Fragility Index quantifies the accumulation of harmful health outcomes, incorporating a wide range of variables that are easy to implement and can be repeated at high frequency, including: integument (pallor and/or cyanosis), digestive/urogenital health (urinary expression and dehydration), behavior (activity levels, responsiveness to external stimuli, and paralysis), ocular/nasal health (peri-retro-orbital swelling), physical/musculoskeletal health (palpable thoracic mass, kyphosis and hunched posture), responsiveness to analgesic (corneal opacity, cataracts, and micropthalmia are collapsed into changes to eye globe), vestibulocochlear health and discomfort (piloerection subdivided into head piloerection and other forms.) Notably, the Fragility Index does not include assessments for alopecia, whisker loss, fur color change, forelimb grip strength, hearing loss, vision loss, menace reflex, or the mouse grimace scale—common components of other frailty indices.

Specifically, the Fragility Index is based on 32 deficits, including two quantitative measures (bodyweight and rectal temperature) and 30 ordinal assessments. Most of the ordinal assessments are rated on a 3-level scale (absent, mild, severe), with some exceptions scored on a 2-level scale (absent or severe). Most deficits are evaluated on a 2-level [0, 1] or 3-level [0, 0.5, 1] scale, where 0 indicates no deficit, 0.5 represents a mild deficit, and 1 indicates a severe deficit. Dermatitis is an exception, with four levels, but is simplified to three during data processing. Each deficit is assigned an ordinal value, and the final Fragility Index score is computed by calculating the mean of these values, yielding a score between 0 and 1. For C57BL/6 mice, Fragility Index scores range from 0.02 to 0.48. Several indicators within the Fragility Index are never observed in the mice, as none show mild or severe nasal discharge. However, the cumulative incidence of other indicators is high, meaning that nearly all mice eventually display clinical signs of severe frailty in the health domain. These signs include piloerection, kyphosis, and poor grooming (coat condition).

When comparing male and female mice, a few notable differences emerge. Males are more likely to show deficits in activity, head piloerection, and hunched posture, while peri-retro-orbital swelling, although rare, is slightly more common in females. As expected, Fragility Index scores generally increase with age. The following factors are among the top ten most strongly associated with life expectancy: chronological age, body weight, head piloerection, body condition score, Fragility Index score, coat condition, and the presence of palpable tumors. The Fragility Index defines the healthspan endpoint as the time when four or more severe health deficits are observed. Healthspan, in this context, is defined as the length of time an individual lives with fewer than four severe health deficits. Notably, sex differences in healthspan are pronounced in C57BL/6J mice, with females typically exhibiting a longer healthspan than males.

In summary, the PFI and the Fragility Index are variations of the mouse Clinical Frailty Index, designed to quantify the cumulative decline across multiple physiological systems. Both indices offer a valuable measure of the animal's frailty and overall health status.

COMBINING THE CONCEPTUAL FRAMEWORKS OF PHYSICAL FRAILTY PHENOTYPE AND FRAILTY INDEX

VITALITY SCORE

The Vitality Score is a comprehensive health assessment for mice, combining a physical function score and a frailty index score (Marcozzi et al., 2023). The physical function score is derived from five composite scores based on specific tests in different domains: body size (body weight, body length), strength (forelimb grip strength determined by grip strength meter with a plastic grid, grip strength meter with an iron bar, home cage lift test, gripping weights lift test), endurance (treadmill distance to exhaustion, mean latency time on rotarod test, gripping weights lift test normalized to body weight), speed (peak speed traveled for at least 3 s in the 5 min locomotor activity test, maximum speed on rotarod test, mean stride length using footprint method and video tracking), and activity (total distance traveled during 5 min open field test, time not in resting state during 5 min open field test.) An overall score representative of physical decline is computed as the mean of the composite scores of the five criteria. Mice scoring in the bottom 20% of the cohort for the composite score computed for each composite are assigned one frailty point. If a mouse meets the frailty threshold in three or more of these criteria, it is classified as frail (Baumann et al., 2018; Kwak et al., 2020; Kane et al., 2017; Liu et al., 2014). In the Vitality Score frailty assessment tool, the scores for each physical function are converted to a scale from 0 (lowest functionality) to 1 (highest functionality). Frailty is then defined by a physical function score below 0.55, with a score between 0.55 and 0.63 indicating pre-frailty. In the Vitality Score, the mouse frailty index score (Whitehead et al., 2014) is used to compute the Clinical Health Score (calculated as 1 – mouse frailty index score), with values ranging from 0 (poor health) to 1 (optimal health). The Vitality Score is subsequently calculated by averaging the Clinical Health Score and physical function scores, resulting in a health assessment score ranging from 0 (poor health) to 1 (optimal health).

Vitality Score declines gradually with age until 28 months, after which the decline accelerates. Both the physical function score and frailty index, as well as the Vitality Score, are strong predictors of mortality. A Vitality Score cut-off of 0.68 is suggested to classify frailty. The Vitality Score is also associated with biomarkers linked to the buildup of senescent cells and the epigenetic clock.

VITALITY PHENOTYPE

The Vitality Phenotype is a comprehensive frailty assessment tool that integrates both the Frailty Index and the Physical Frailty Phenotype. Specifically, the health questionnaire used to generate the Frailty Index serves as a "health status marker" and is recognized as a criterion (accumulation of health deficits) within the Vitality Phenotype (Perazza et al., 2024). Therefore, the Vitality Phenotype encompasses six criteria: endurance, grip strength, walking speed, physical activity, weight gain/loss, and accumulation of health deficits. The lowest 20th percentile for each frailty criterion corresponds to the cut-off value of that specific criterion. Animals whose performance would be equal to or lower than the cut-off value for a given criterion would score positive for that criterion. Four or more positive frailty markers identified an animal as frail, two or three positive markers indicated pre-frail, and one or zero markers indicated robust. The Vitality Phenotype increased the Pre-frail classification by nearly 10% as compared to the Physical Frailty Phenotype due to a reduction of nearly 2% of mice classified under the Frail category and 8% from Robust. Indeed, because the Vitality Phenotype considers the heterogeneity of the frailty syndrome and the complementary nature of the Frailty Index and Physical Frailty Phenotype assessment tools, it has the potential to be employed to identify the risk of frailty development.

Collectively, the Vitality Score and the Vitality Phenotype assessment tools are comprehensive measures of an animal's physical and behavioral health. These two tools provide valuable insights into the animal's general well-being and are important tools in aging, disease, and interventional research. Indeed, the reverse-translation of the Physical Frailty Phenotype and Frailty Index assessment tools and the development of new assessment tools, which are based on these well-established assessments, have significant strengths in advancing the understanding of frailty and its biological underpinnings.

FRAILTY ASSESSMENT TOOLS INCLUDING MULTIPLE PHYSIOLOGICAL DOMAINS

Physical-Cognitive Frailty Score

Cognition plays a critical role in identifying and understanding frailty, as it is deeply interconnected with physical and functional health. Cognitive decline is increasingly recognized as a key factor contributing to the development and progression of frailty. The Physical-Cognitive Frailty Score combines cognitive and physical measures to assess overall frailty (Mladenovic & Pracer, 2024). Cognitive function is evaluated using three behavioral tests: the Open Field Test, which assesses spontaneous exploratory behavior as an indicator of preserved cognition; the Novel Object Test, which measures both short-term and long-term memory; and the Y-Maze Test, which specifically evaluates short-term memory. The Open Field Test provides five parameters: total distance traveled, total duration of movement (ambulatory time), percentage of time spent moving, average velocity, and rearing frequency. The Novel Object Test provides two parameters: exploration ratio at 1 h and 24 h. The Y-Maze Test offers five parameters: total number of entries, time spent in the novel arm, number of entries in the novel arm, percentage of novel arm entries, and spontaneous alterations from the second trial.

The Physical-Cognitive Frailty Score is calculated at the group level, not for individual animals. First, animals from all experimental groups are pooled together, and a cut-off value of 20% is established for each frailty assessment parameter. Animals whose scores fall below this cut-off value are considered to have "failed" the test, while those with scores equal to or above the cut-off value are considered to have "passed" for each parameter. The next step is to calculate the percentage of success for each parameter. For each parameter, the percentage of success is calculated as:

$$\text{Percentage of Success} = \text{Number of Passed Animals}/(\text{Number of Passed} + \text{Failed Animals}) \times 100$$

Subsequently, the number of failed animals for each parameter within each experimental group is determined. The frailty score for each group is then calculated by dividing the number of animals that failed all tests by the total number of animals that passed and expressing this as a percentage.

This process is repeated for each parameter. After determining the number of animals that failed or passed each parameter in each group, the total number of failed and passed animals across all tests for each group is calculated. Finally, the group frailty score is determined by dividing the total number of failed tests across all parameters by the total number of tests performed and multiplying by 100 to express the result as a percentage. The group Physical-Cognitive Frailty Score increases with age in mice. Although the Physical-Cognitive Frailty Score is based on the open field test, novel object recognition, and the Y-maze test, it is not necessary to use these specific tests to evaluate cognition and physical movement. However, the selection of the tests is based on strain-specific and behavioral changes in various rodent strains.

Mouse Social Frailty Index

Frailty is not only a physical condition but also a multifactorial syndrome influenced by social, psychological, and environmental factors. Social isolation, reduced social interactions, and a lack of social support are key contributors to frailty and can significantly affect an individual's physical, mental, and emotional well-being.

The mouse Social Frailty Index is a non-invasive tool designed to assess social communication, interaction, and motivation in mice, based on the deficit accumulation model (Collinge et al., 2025). It is optimized for use in longevity studies and comprises seven behavioral assays:

1. Olfactory Test (Behavioral Assay No. 1): This assay evaluates the mouse's ability to distinguish between social and nonsocial odors. Social odor discrimination is quantified by calculating the percentage of time spent sniffing urine compared to all odors, using the formula:

$$\text{Urine Preference} = (\text{Time spent sniffing urine}/\text{Total time spent sniffing all odors}) \times 100$$

2. Urine Marking and Countermarking (Behavioral Assays No. 2 and No. 3): These assays measure the mouse's spontaneous marking behavior in response to mixed-sex urine. Urine marking is a form of social communication that releases pheromones for territorial defense and social signaling. The number of urine marks deposited is quantified.
3. Juvenile Social Interaction (Behavioral Assays No. 4 and No. 5): These assays assess social interactions between the test mouse and a same-sex juvenile in either the home cage or a novel cage. Behaviors like sniffing (pursuit, olfactory investigation), aggression (attack bites, offensive postures), and allo-grooming are scored. Social interaction is expressed as a percentage:

$$\text{Social Interaction} = (\text{Total social intervals}/\text{Total intervals observed}) \times 100$$

4. Social/Novel Object Preference (Behavioral Assay No. 6): In this test, mice choose between a social stimulus (juvenile mouse) and a novel object in a three-chamber social approach task. The percentage of time spent interacting with the social stimulus versus the object is calculated:

$$\text{Social Approach} = (\text{Intervals with mouse} - \text{Intervals with object}/\text{Total intervals}) \times 100$$

5. Nest Building (Behavioral Assay No. 7): This assay assesses the mouse's motivation and ability to build a nest, an important behavior for shelter, social interactions, and territory defense. Nest quality is scored on a 5-point scale.

Each assay produces a quantitative score reflecting the mouse's social behavior in that specific domain (e.g., a percentage of urine preference from the olfactory test). The mouse Social Frailty Index score ranges from 0 to 1, where 0 indicates no deficits in social functioning and 1 represents maximal deficits. Intermediate scores reflect varying levels of social frailty. To calculate the mouse Social Frailty Index, each behavioral assay score is converted into an "index score" based on its deviation from the population or cohort mean. The index score system is as follows: scores within ±1 SD of the mean receive an index score of 0. Scores within 1–2 SD receive 0.25. Scores within 2–3 SD receive 0.5. Scores within 3–4 SD receive 0.75. Scores greater than 4 SD receive 1. For nest building, the following index scores are applied: nest quality score of 5 = 0. Nest quality score of 4 = 0.25. Nest quality score of 3 = 0.5. Nest quality score of 2 = 0.75. Nest quality score of 1 = 1.

The final mouse Social Frailty Index is calculated by averaging the index scores across all seven assays. The resulting value ranges from 0.00 to 0.57 in C57BL/6 mice. The mouse Social Frailty Index detects a progressive, age-dependent increase in social frailty over the lifespan of group-housed mice. Female mice exhibit a delayed onset and lower level of social frailty compared to males.

TECHNOLOGY TO ADVANCE FRAILTY ASSESSMENT IN PRECLINICAL STUDIES

Current approaches to mouse frailty assessment often require extensive handling and/or exposure to non-standard environments. Both handling and exposure to novel environments are known to induce stress in laboratory mice, with even routine husbandry procedures leading to increased activity and elevated corticosterone levels. Moreover, several parameters within frailty indices, such as posture, response to stimuli, exploratory behavior, and blood chemistry, are known to be influenced by handling-related stress. Most frailty indices also rely on subjective scoring by a trained researcher, presenting two main challenges for practical use. First, the wide range of traits to be assessed demands significant time, labor-intensive, from the researcher. Second, while frailty index rater agreement is typically high under ideal conditions, human factors—such as communication gaps or varying levels of training—can introduce biases that lower inter-rater reproducibility. These obstacles restrict its scalability limit the size of studies using the frailty index metric. While inter-scorer reliability is generally high, achieving consistent agreement across scorers often necessitates substantial discussion and refinement—an effort that may not be feasible in multi-site or long-term studies.

Digital Frailty Index

The Digital Frailty Index improves throughput and reduces handling-related stress, which can otherwise confound frailty assessments. The Digital Frailty Index also addresses the human factors by eliminating potential inconsistencies across raters.

The Digital Frailty Index uses computational analysis of continuous home-cage video recordings to monitor mice in their natural environment (Ruby et al., 2023). This approach enables ongoing assessment across four functional areas: physical fitness, environmental engagement, circadian rhythm, and body composition. The Digital Frailty Index evaluates eight frailty parameters, with two measurements taken per area: average daily distance run on the wheel, gait speed on the wheel, gait speed on the cage floor, circadian distribution of wheel-running activity, circadian distribution of movement on the cage floor, rate of change in body weight, coat quality, and average displacement of nesting material. Each of these eight parameters is scored on a scale from 0 to 1, where 0 indicates normal behavior and 1 indicates frail behavior (observation of an exceptional value that reflects an abnormal, pathological state). Intermediate scores are linearly assigned based on their proximity to these two scores. The overall Digital Frailty Index score is calculated as the average of the eight individual parameter scores. The Digital Frailty Index correlates with both chronological age and the manual frailty index (31(29)-item mouse clinical frailty index), making it a reliable measure of frailty in mice. Generally, the gait/wheel and circadian components are highly correlated with age; while components exhibit strong correlations with age, parameters related to nesting material displacement, coat quality, and changes in body weight show weaker correlations.

Although there are many advantages with continuous home-cage video recordings, the Digital Frailty Index requires the mice to be singly housed and may have negative psychological and physiological consequences because mice are social animals.

Visual Frailty Index

Advances in machine learning techniques have significantly expanded the types of metrics that can be extracted from the open-field assay, including image segmentation, tracking in complex environments, action detection, and pose-based measurements of gait and whole-body coordination (Hession et al., 2022; Sabnis et al., 2024). By analyzing video data from an open-field assay of C57BL/6J mice, morphometric, gait, and behavioral features that correlate with manual frailty index scores and age are extracted, ultimately creating a Visual Frailty Index.

These extracted features are used to train a regression model that accurately predicts the normalized frailty index score with a mean absolute error of 0.04 ± 0.002 (using statistical machine learning methods). When unnormalized, the prediction error is 1.08 ± 0.05, which corresponds to approximately one frailty index item being mis-scored by one point, or two frailty index items mis-scored by 0.5 points. The Visual Frailty Index thus provides an effective means of predicting frailty. In addition to predicting frailty with machine learning techniques, the relationship between frailty and lifespan showed that frailty predicts mortality in aged male mice (Schultz et al., 2020).

Thus, this successful application of machine learning, the Visual Frailty Index, offers improved reproducibility and scalability, enabling large-scale mechanistic and interventional studies on aging in mice.

FRAILTY ASSESSMENT IN SPECIAL POPULATIONS

Recent studies on frailty in animal models of age-related diseases have been increasing. One of the pioneering studies examined frailty in a transgenic mice model carrying five mutations associated with Familial Alzheimer's Disease (5xFAD), a widely used model of Alzheimer's disease, by employing both the Physical Frailty Phenotype and Clinical Frailty Index assessment tools (Todorovic et al., 2020). In this study, the Frailty Phenotype includes five parameters: body weight, physical activity (measured via the open field test), walking speed (calculated as total distance traveled (cm) divided by ambulatory time (s) in the open field test), grip strength (assessed using the cage-top test, where the mouse's ability to hang upside down is scored as positive or negative), and motor coordination (evaluated through the tightrope test). The physical activity component involves five parameters assessed in the open field test: total distance traveled, total duration of movement (ambulatory time), the percentage of time spent moving, average velocity, and rearing frequency. A 20% cut-off value is applied for all parameters, except for body weight, for which a 5% cut-off is used. The frailty score for each experimental group is calculated by dividing the total number of tests failed by the total number of tests performed, yielding a percentage. In this study, 12% of young male mice were classified as frail compared to 46% of older male mice. In females, the frailty percentage was 20% in 3-month-old mice and 36% in 11-month-old mice. When males and females were grouped together, 13% of young mice were classified as frail, compared to 36% in the older cohort. These data clearly indicate a significantly lower number of frail mice in the younger group, as well as a lower frailty score in the young compared to the older mice.

The Clinical Frailty Index employed in the study includes 24 parameters, with each deficit rated on a scale: 0 (no deficit), 0.5 (mild deficit), and 1 (severe deficit). The body weight deficit is scored based on deviations from reference values measured in young mice. If values differ from reference values by less than 1 SD, they receive a score of 0. Deviations of ±1 SD are scored 0.25, ±2 SD score 0.5, ±3 SD score 0.75, and values differing by more than 3 SD from the mean receive the maximum score of 1. The individual scores for each parameter are summed and divided by the total number of parameters measured, yielding a frailty index score between 0 and 1 for each animal. A significant increase in frailty was observed in both sexes as the mice aged. Older male mice showed almost a 20-fold increase in frailty index compared to their younger counterparts ($p<0.0001$), while the increase in frailty among 11-month-old females, though somewhat less pronounced, was still significant ($p<0.0001$). When males and females were analyzed together, the effect of aging on the frailty index was similarly detectable and significant ($p<0.0001$).

A side-by-side comparison of the two frailty assessment tools reveals no correlation between the Physical Frailty Phenotype and the Clinical Frailty Index. This indicates that the two tools identify different subsets of animals as frail. These findings suggest that the two frailty tools have different sensitivities to age and gender, implying that both approaches should be considered in frailty research. Indeed, the Vitality Phenotype discussed above could be employed, which combines both phenotypic and clinical parameters (Perazza et al., 2024).

Another example of assessing frailty in a disease model includes the auto-immune encephalomyelitis mouse model (EAE mouse model), with a 34-item assessment of health deficits, where the frailty scores increased with disease severity (Ribeiro et al., 2022). As expected, due to the shortened lifespans, transgenic strains accumulate age-related deficits at increasing rates and reach higher frailty scores at earlier ages compared to their wild type counterparts.

Assessing frailty in experimental animals with disease offers invaluable insights into the complex interactions between aging, disease, and overall health. It helps to identify early disease markers, evaluate potential treatments, understand disease mechanisms, and develop more personalized therapeutic approaches. By integrating frailty assessment into disease models, researchers can improve our understanding of the aging process and design more effective interventions to improve the health and quality of life for aging populations with chronic diseases.

CONCLUSION

Developing reliable and valid frailty assessment tools that capture various domains of frailty in experimental animal models holds great potential for identifying early biomarker predictors of different trajectories of functional impairments (e.g., physical, overall health, cognitive, social). These tools can also be particularly sensitive to age acceleration or the effects of therapeutics. While progress has been made in reverse-translating human frailty assessment tools to animal models, there is still a need for consensus on how to assess and categorize frailty. Additionally, the current approach of using a single "cut-off statistical parameter" for each criterion in the Mouse Physical Frailty Phenotype (e.g., the lowest 20th percentile) and the use of "three scores" (e.g., 0, 0.5, 1.0) in the Clinical Frailty Index may not be ideal. This is because the underlying physiological systems that these criteria and items reflect can vary significantly. The next step in developing frailty assessment tools for experimental animals will be to determine specific cut-off values tailored to each criterion.

ACKNOWLEDGMENT

This work is supported, in part, by the Travis Roy Endowed Professorship (to L.V. Thompson), the National Institute on Aging (K07 AG-072124 and R56 AG-067724 to L.V. Thompson), and Hevolution (HF-AGE-003 to L.V. Thompson).

REFERENCES

Antoch, M. P., Wrobel, M., Kuropatwinski, K. K., Gitlin, I., Leonova, K. I., Toshkov, I., … Gudkov, A. V. (2017). Physiological Frailty Index (PFI): Quantitative in-life estimate of individual biological age in mice. *Aging (Albany NY)*, *9*(3), 615–626. https://doi.org/10.18632/aging.101206

Bartolomucci, A., Kane, A. E., Gaydosh, L., Razzoli, M., McCoy, B. M., Ehninger, D., … Snyder-Mackler, N. (2024). Animal models relevant for geroscience: Current trends and future perspectives in biomarkers, and measures of biological aging. *J Gerontol A Biol Sci Med Sci*, *79*(9). https://doi.org/10.1093/gerona/glae135

Baumann, C., Kwak, D., & Thompson, L. (2018). Assessing onset, prevalence and survival in mice using a frailty phenotype. *Aging-Us*, *10*(12), 4042–4053. https://doi.org/10.18632/aging.101692

Baumann, C., Kwak, D., & Thompson, L. (2019). Sex-specific components of frailty in C57BL/6 mice. *Aging-Us*, *11*(14), 5206–5214. https://doi.org/10.18632/aging.102114

Baumann, C., Kwak, D., & Thompson, L. (2020). Phenotypic frailty assessment in mice: Development, discoveries, and experimental considerations. *Physiology*, *35*(6), 405–414. https://doi.org/10.1152/physiol.00016.2020

Buta, B. J., Walston, J. D., Godino, J. G., Park, M., Kalyani, R. R., Xue, Q. L., … Varadhan, R. (2016). Frailty assessment instruments: Systematic characterization of the uses and contexts of highly-cited instruments. *Ageing Res Rev*, *26*, 53–61. https://doi.org/10.1016/j.arr.2015.12.003

Collinge, C. W., Razzoli, M., Mansk, R., McGonigle, S., Lamming, D. W., Pacak, C. A., … Bartolomucci, A. (2025). The mouse Social Frailty Index (mSFI): A novel behavioral assessment for impaired social functioning in aging mice. *Geroscience*, *47*(1), 85–107. https://doi.org/10.1007/s11357-024-01263-4

Fried, L. P., Tangen, C. M., Walston, J., Newman, A. B., Hirsch, C., Gottdiener, J., … McBurnie, M. A. (2001). Frailty in older adults: Evidence for a phenotype. *J Gerontol A Biol Sci Med Sci*, *56*(3), M146–156. https://doi.org/10.1093/gerona/56.3.m146

Gomez-Cabrera, M. C., Garcia-Valles, R., Rodriguez-Mañas, L., Garcia-Garcia, F. J., Olaso-Gonzalez, G., Salvador-Pascual, A., … Viña, J. (2017). A new frailty score for experimental animals based on the clinical phenotype: Inactivity as a model of frailty. *J Gerontol A Biol Sci Med Sci, 72*(7), 885–891. https://doi.org/10.1093/gerona/glw337

Hession, L. E., Sabnis, G. S., Churchill, G. A., & Kumar, V. (2022). A machine-vision-based frailty index for mice. *Nat Aging, 2*(8), 756–766. https://doi.org/10.1038/s43587-022-00266-0

Hua, J., Hoummady, S., Muller, C., Pouchelon, J. L., Blondot, M., Gilbert, C., & Desquilbet, L. (2016). Assessment of frailty in aged dogs. *Am J Vet Res, 77*(12), 1357–1365. https://doi.org/10.2460/ajvr.77.12.1357

Huang, D. D., Fan, S. D., Chen, X. Y., Yan, X. L., Zhang, X. Z., Ma, B. W., … Yu, Z. (2019). Nrf2 deficiency exacerbates frailty and sarcopenia by impairing skeletal muscle mitochondrial biogenesis and dynamics in an age-dependent manner. *Exp Gerontol, 119*, 61–73. https://doi.org/10.1016/j.exger.2019.01.022

Kane, A. E., Huizer-Pajkos, A., Mach, J., Mitchell, S. J., de Cabo, R., Le Couteur, D. G., … Hilmer, S. N. (2017). A comparison of two mouse frailty assessment tools. *J Gerontol A Biol Sci Med Sci, 72*(7), 904–909. https://doi.org/10.1093/gerona/glx009

Kulminski, A. M., Ukraintseva, S. V., Culminskaya, I. V., Arbeev, K. G., Land, K. C., Akushevich, L., & Yashin, A. I. (2008). Cumulative deficits and physiological indices as predictors of mortality and long life. *J Gerontol A Biol Sci Med Sci, 63*(10), 1053–1059. https://doi.org/10.1093/gerona/63.10.1053

Kwak, D., Baumann, C. W., & Thompson, L. V. (2020). Identifying characteristics of frailty in female mice using a phenotype assessment tool. *J Gerontol A Biol Sci Med Sci, 75*(4), 640–646. https://doi.org/10.1093/gerona/glz092

Kwak, D., & Thompson, L. V. (2021). Frailty: Past, present, and future? *Sports Med Health Sci, 3*(1), 1–10. https://doi.org/10.1016/j.smhs.2020.11.005

Liu, H., Graber, T. G., Ferguson-Stegall, L., & Thompson, L. V. (2014). Clinically relevant frailty index for mice. *J Gerontol A Biol Sci Med Sci, 69*(12), 1485–1491. https://doi.org/10.1093/gerona/glt188

Luciano, A., Robinson, L., Garland, G., Lyons, B., Korstanje, R., Di Francesco, A., & Churchill, G. A. (2024). Longitudinal fragility phenotyping contributes to the prediction of lifespan and age-associated morbidity in C57BL/6 and Diversity Outbred mice. *Geroscience, 46*(5), 4937–4954. https://doi.org/10.1007/s11357-024-01226-9

Marcozzi, S., Bigossi, G., Giuliani, M. E., Giacconi, R., Cardelli, M., Piacenza, F., … Malavolta, M. (2023). Comprehensive longitudinal non-invasive quantification of healthspan and frailty in a large cohort (n=546) of geriatric C57BL/6 J mice. *Geroscience, 45*(4), 2195–2211. https://doi.org/10.1007/s11357-023-00737-1

Martinez de Toda, I., Garrido, A., Vida, C., Gomez-Cabrera, M. C., Viña, J., & De la Fuente, M. (2018). Frailty quantified by the "Valencia Score" as a potential predictor of lifespan in mice. *J Gerontol A Biol Sci Med Sci, 73*(10), 1323–1329. https://doi.org/10.1093/gerona/gly064

Miller, M. G., Thangthaeng, N., & Shukitt-Hale, B. (2017). A clinically relevant Frailty Index for aging rats. *J Gerontol A Biol Sci Med Sci, 72*(7), 892–896. https://doi.org/10.1093/gerona/glw338

Mladenovic, A., & Pracer, S. (2024). Evaluating frailty scores across experimental groups in rodent models: Bridging physical and cognitive domains. *FEBS Open Bio*. https://doi.org/10.1002/2211-5463.13955

Parks, R. J., Fares, E., Macdonald, J. K., Ernst, M. C., Sinal, C. J., Rockwood, K., & Howlett, S. E. (2012). A procedure for creating a Frailty Index based on deficit accumulation in aging mice. *J Gerontol A Biol Sci Med Sci, 67*(3), 217–227. https://doi.org/10.1093/gerona/glr193

Perazza, L. R., Bougher, C. M., & Thompson, L. V. (2024). Evaluating frailty assessment tools and biological frailty markers in C57BL/6 female mice. *Geroscience*. https://doi.org/10.1007/s11357-024-01430-7

Perazza, L. R., Brown-Borg, H. M., & Thompson, L. V. (2022). Physiological systems in promoting frailty. *Compr Physiol, 12*(3), 3575–3620. https://doi.org/10.1002/cphy.c210034

Pereira, A. A., Borim, F. S. A., Aprahamian, I., & Neri, A. L. (2019). Comparison of two models of frailty for the prediction of mortality in Brazilian community-dwelling older adults: The FIBRA study. *J Nutr Health Aging, 23*(10), 1004–1010. https://doi.org/10.1007/s12603-019-1264-0

Ribeiro, A. R., Barros, C., Barateiro, A., Howlett, S. E., & Fernandes, A. (2022). Improved assessment of overall health in variably aged murine models of multiple sclerosis with a novel frailty index tool. *J Gerontol A Biol Sci Med Sci, 77*(1), 1–9. https://doi.org/10.1093/gerona/glab185

Rockwood, K., Blodgett, J. M., Theou, O., Sun, M. H., Feridooni, H. A., Mitnitski, A., … Howlett, S. E. (2017). A Frailty Index based on deficit accumulation quantifies mortality risk in humans and in mice. *Sci Rep, 7*, 43068. https://doi.org/10.1038/srep43068

Rockwood, K., & Mitnitski, A. (2007). Frailty in relation to the accumulation of deficits. *J Gerontol A Biol Sci Med Sci, 62*(7), 722–727. https://doi.org/10.1093/gerona/62.7.722

Ruby, J., Di Francesco, G., Ylagan, A., Luo, P., Keyser, A., Williams, R., … Karlsson, O. (2023). An automated, home-cage, video monitoring-based mouse frailty index detects age-associated morbidity in C57BL/6 and diversity outbred mice. *J Gerontol A Biol Sci Med Sci*, *78*(5), 762–770. https://doi.org/10.1093/gerona/glad035

Rutenberg, A. D., Mitnitski, A. B., Farrell, S. G., & Rockwood, K. (2018). Unifying aging and frailty through complex dynamical networks. *Exp Gerontol*, *107*, 126–129. https://doi.org/10.1016/j.exger.2017.08.027

Sabnis, G. S., Churchill, G. A., & Kumar, V. (2024). Machine vision based frailty assessment for genetically diverse mice. *bioRxiv*. https://doi.org/10.1101/2024.10.13.617922

Scheuren, A. C., D'Hulst, G., Kuhn, G. A., Masschelein, E., Wehrle, E., De Bock, K., & Müller, R. (2020). Hallmarks of frailty and osteosarcopenia in prematurely aged PolgA. *J Cachexia Sarcopenia Muscle*, *11*(4), 1121–1140. https://doi.org/10.1002/jcsm.12588

Schultz, M. B., Kane, A. E., Mitchell, S. J., MacArthur, M. R., Warner, E., Vogel, D. S., … Sinclair, D. A. (2020). Age and life expectancy clocks based on machine learning analysis of mouse frailty. *Nat Commun*, *11*(1), 4618. https://doi.org/10.1038/s41467-020-18446-0

Seldeen, K. L., Lasky, G., Leiker, M. M., Pang, M., Personius, K. E., & Troen, B. R. (2018). High intensity interval training improves physical performance and frailty in aged mice. *J Gerontol A Biol Sci Med Sci*, *73*(4), 429–437. https://doi.org/10.1093/gerona/glx120

Sukoff Rizzo, S. J., Anderson, L. C., Green, T. L., McGarr, T., Wells, G., & Winter, S. S. (2018). Assessing healthspan and lifespan measures in aging mice: Optimization of testing protocols, replicability, and rater reliability. *Curr Protoc Mouse Biol*, *8*(2), e45. https://doi.org/10.1002/cpmo.45

Todorovic, S., Loncarevic-Vasiljkovic, N., Jovic, M., Sokanovic, S., Kanazir, S., & Mladenovic Djordjevic, A. (2020). Frailty index and phenotype frailty score: Sex- and age-related differences in 5XFAD transgenic mouse model of Alzheimer's disease. *Mech Ageing Dev*, *185*, 111195. https://doi.org/10.1016/j.mad.2019.111195

Todorovic, S. T., Smiljanic, K. R., Ruzdijic, S. D., Djordjevic, A. N. M., & Kanazir, S. D. (2018). Effects of different dietary protocols on general activity and frailty of male wistar rats during aging. *J Gerontol A Biol Sci Med Sci*, *73*(8), 1036–1044. https://doi.org/10.1093/gerona/gly015

Whitehead, J. C., Hildebrand, B. A., Sun, M., Rockwood, M. R., Rose, R. A., Rockwood, K., & Howlett, S. E. (2014). A clinical Frailty Index in aging mice: Comparisons with Frailty Index data in humans. *J Gerontol A Biol Sci Med Sci*, *69*(6), 621–632. https://doi.org/10.1093/gerona/glt136

Yamada, Y., Kemnitz, J. W., Weindruch, R., Anderson, R. M., Schoeller, D. A., & Colman, R. J. (2018). Caloric restriction and healthy life span: Frail phenotype of nonhuman Primates in the Wisconsin National Primate Research Center Caloric Restriction Study. *J Gerontol A Biol Sci Med Sci*, *73*(3), 273–278. https://doi.org/10.1093/gerona/glx059

Yorke, A., Kane, A. E., Hancock Friesen, C. L., Howlett, S. E., & O'Blenes, S. (2017). Development of a rat clinical Frailty Index. *J Gerontol A Biol Sci Med Sci*, *72*(7), 897–903. https://doi.org/10.1093/gerona/glw339

Index

R

S

T